METHODS IN MOLECULAR BIOLOGY™

For further volumes:
http://www.springer.com/series/7651

Guanylate Cyclase and Cyclic GMP

Methods and Protocols

Edited by

Thomas Krieg

Department of Medicine, Addenbrooke's Hospital
University of Cambridge, Cambridge, UK

Robert Lukowski

Department of Pharmacology, Toxicology and Clinical Pharmacy
Universität Tübingen, Tübingen, Germany

Editors
Thomas Krieg
Department of Medicine
Addenbrooke's Hospital
University of Cambridge
Cambridge, UK

Robert Lukowski
Department of Pharmacology
Toxicology and Clinical Pharmacy
Universität Tübingen
Tübingen, Germany

ISSN 1064-3745 ISSN 1940-6029 (electronic)
ISBN 978-1-62703-458-6 ISBN 978-1-62703-459-3 (eBook)
DOI 10.1007/978-1-62703-459-3
Springer New York Heidelberg Dordrecht London

Library of Congress Control Number: 2013937718

Printed on acid-free paper

Humana Press is a brand of Springer
Springer is part of Springer Science+Business Media (www.springer.com)

Preface

Since the Nobel Prize in Physiology or Medicine was awarded to Robert Furchgott, Louis Ignarro, and Ferid Murad in 1998 for the discovery of nitric oxide (NO) as an important signalling molecule, the downstream pathway of NO has attracted considerable interest in various physiological and pathophysiological conditions. To date, soluble guanylate cyclase(s), cyclic guanosine 3'-5'-monophosphate (cGMP), protein kinase G (PKG) (also known as cGMP-dependent protein kinase), and cGMP-activated or -inactivated phosphodiesterases (PDEs) are by far the best-characterized elements of the downstream signalling. In the past years each of these structures has been of intense scientific interest not only as important signalling molecules but also as highly promising drug targets.

It is immensely challenging to measure NO and the spatiotemporal profile of its downstream effectors and targets in vitro or in vivo to unravel their roles in physiological conditions as well as various diseases. Recently, many groundbreaking steps have been made towards a better understanding of the NO/cGMP/PKG pathway, its components, substrates, and its localization within a given cell. These advances were possible only due to the development of sophisticated new techniques in the field.

This book seeks to provide an overview of novel techniques to identify various elements of the NO/cGMP/PKG pathway and further characterize their function, signalling, localization, and importance on a cellular level and in whole animal models providing a higher patho-/physiological integration and relevance.

The first two chapters briefly review the current state of research and methodology in the field and might serve as a reminder for the expert or an introduction for anybody new in this fast-evolving, exciting area.

The following 14 chapters offer detailed step-by-step instructions of each method, including a full list of materials and reagents, as well as useful tips to avoid common pitfalls. We hope that readers will find *Guanylate Cyclase and Cyclic GMP: Methods and Protocols* a comprehensive overview of current methods and a useful guide towards the possibility to apply these techniques to their own research. In addition, some of the chapters use disease models in order to demonstrate the applicability of the method in a currently ongoing research area.

Finally, we would like to thank all the authors for their excellent contributions to this volume and all the time and effort that went into it. We are particularly grateful for the guidance and support from the senior editor of the *Methods in Molecular Biology* series, John Walker.

Cambridge, UK — *Thomas Krieg*
Tübingen, Germany — *Robert Lukowski*

Contents

Contributors

Joseph A. Beavo • *Department of Pharmacology, University of Washington, Seattle, USA*
Noomen Bettaga • *Physiologisches Institut I, Universität Würzburg, Würzburg, Germany*
Martin Biel • *Center for Integrated Protein Science Munich (CIPSM), Ludwig-Maximilians-Universität, Munich, Germany; Department of Pharmacy, Center for Drug Research, Ludwig-Maximilians-Universität, Munich, Germany*
Andrea L. Britain • *Department of Pharmacology, University of South Alabama, Mobile, USA*
Joseph Robert Burgoyne • *Cardiovascular Division, The Rayne Institute, St. Thomas' Hospital, King's College London, London, UK*
Xi-Qin Ding • *Department of Cell Biology, University of Oklahoma Health Sciences Center, Oklahoma City, USA*
Wolfgang R. Dostmann • *Department of Pharmacology, University of Vermont, Burlington, USA*
Philip Eaton • *Cardiovascular Division, The Rayne Institute, St. Thomas' Hospital, King's College London, London, UK*
Robert Feil • *Interfakultäres Institut für Biochemie, Universität Tübingen, Tübingen, Germany*
Natalie Fomin • *Interfakultäres Institut für Biochemie, Universität Tübingen, Tübingen, Germany; Graduate School of Cellular and Molecular Neuroscience, Universität Tübingen, Tübingen, Germany*
Andreas Friebe • *Physiologisches Institut I, Universität Würzburg, Würzburg, Germany*
Hyun-Soon Geisler • *Department of Otolaryngology, Hearing Research Centre Tübingen (THRC), Molecular Physiology of Hearing, Universität Tübingen, Tübingen, Germany*
Konrad R. Götz • *Emmy Noether Group of the DFG, Department of Cardiology and Pneumology, European Heart Research Institute Göttingen, Universität Göttingen, Göttingen, Germany*
Dieter Groneberg • *Physiologisches Institut I, Universität Würzburg, Würzburg, Germany*
Bodo Haas • *Institute of Pharmacology and Toxicology, Universität Bonn, Bonn, Germany; Institute for Drugs and Medical Devices, Bonn, Germany*
Kara F. Held • *Department of Pharmacology, Yale University, New Haven, USA*
Thomas R. Hinds • *Department of Pharmacology, University of Washington, Seattle, USA*
Franz Hofmann • *FOR 923, Institut für Pharmakologie und Toxikologie, der Technischen Universität München, Munich, Germany*
Ronald Jäger • *Physiologisches Institut I, Universität Würzburg, Würzburg, Germany*
Katja Jennissen • *Institute of Pharmacology and Toxicology, Universität Bonn, Bonn, Germany*
Marlies Knipper • *Department of Otolaryngology, Hearing Research Centre Tübingen (THRC), Molecular Physiology of Hearing, Universität Tübingen, Tübingen, Germany*

Doris Koesling • *Pharmakologie und Toxikologie, Medizinische Fakultät, Ruhr-Universität Bochum, Bochum, Germany*
Christian Krawutschke • *Pharmakologie und Toxikologie, Medizinische Fakultät, Ruhr-Universität Bochum, Bochum, Germany*
Michaela Kümmel • *Physiologisches Institut I, Universität Würzburg, Würzburg, Germany*
Silas J. Leavesley • *Department of Chemical and Biomolecular Engineering, University of South Alabama, Mobile, USA*
Barbara Lies • *Physiologisches Institut I, Universität Würzburg, Würzburg, Germany*
Ruirui Lu • *Pharmazentrum Frankfurt/ZAFES, Institut für Klinische Pharmakologie, Klinikum der Johann Wolfgang Goethe-Universität, Frankfurt am Main, Germany; Institut für Pharmakologie und Toxikologie, Universität Witten/Herdecke, Witten, Germany*
Stylianos Michalakis • *Center for Integrated Protein Science Munich (CIPSM), Ludwig-Maximilians-Universität, Munich, Germany; Department of Pharmacy, Center for Drug Research, Ludwig-Maximilians-Universität, Munich, Germany*
Michaela M. Mitschke • *Institute of Pharmacology and Toxicology, Universität, Bonn, Bonn, Germany*
Evanthia Mergia • *Pharmakologie und Toxikologie, Medizinische Fakultät, Ruhr-Universität Bochum, Bochum, Germany*
Viacheslav O. Nikolaev • *Emmy Noether Group of the DFG, Department of Cardiology and Pneumology, European Heart Research Institute Göttingen, Universität Göttingen, Göttingen, Germany*
Alexander Pfeifer • *Institute of Pharmacology and Toxicology, Universität, Bonn, Bonn, Germany*
Fritz G. Rathjen • *Max-Delbrück-Centrum für Molekulare Medizin, Berlin, Germany*
Thomas C. Rich • *Department of Pharmacology, University of South Alabama, Mobile, USA*
Michael Russwurm • *Pharmakologie und Toxikologie, Medizinische Fakultät, Ruhr-Universität Bochum, Bochum, Germany*
Corina Russwurm • *Pharmakologie und Toxikologie, Medizinische Fakultät, Ruhr-Universität Bochum, Bochum, Germany*
Sergei D. Rybalkin • *Department of Pharmacology, University of Washington, Seattle, USA*
Katharina Salb • *Pharmakologie und Toxikologie, Institut für Pharmazie, Universität Regensburg, Regensburg, Germany*
Peter M. Schmidt • *CSL Limited, BIO21 Institute, Parkville, VIC, Australia*
Hannes Schmidt • *Max-Delbrück-Centrum für Molekulare Medizin, Berlin, Germany*
Achim Schmidtko • *Pharmazentrum Frankfurt/ZAFES, Institut für Klinische Pharmakologie, Klinikum der Johann Wolfgang Goethe-Universität, Frankfurt am Main, Germany; Institut für Pharmakologie und Toxikologie, Universität Witten/Herdecke, Witten, Germany*
Jens Schlossmann • *Pharmakologie und Toxikologie, Institut für Pharmazie, Universität Regensburg, Regensburg, Germany*
Franziska Siegel • *Institute of Pharmacology and Toxicology, Universität Bonn, Bonn, Germany; Pharma-Center, Universität, Bonn, Bonn, Germany*
Wibke Singer • *Department of Otolaryngology, Hearing Research Centre Tübingen (THRC), Molecular Physiology of Hearing, Universität Tübingen, Tübingen, Germany*

JOHANNES-PETER STASCH • *Cardiovascular Research, Bayer Pharma AG, Wuppertal, Germany*
TIFFANY STEDMAN • *Department of Chemical and Biomolecular Engineering, University of South Alabama, Mobile, USA*
GOHAR TER-AVETISYAN • *Max-Delbrück-Centrum für Molekulare Medizin, Berlin, Germany*
MARTIN THUNEMANN • *Interfakultäres Institut für Biochemie, Universität Tübingen, Tübingen, Germany*
JÖRG W. WEGENER • *FOR 923, Institut für Pharmakologie und Toxikologie, der Technischen Universität München, Munich, Germany*
JIANHUA XU • *Department of Cell Biology, University of Oklahoma Health Sciences Center, Oklahoma City, USA*

Chapter 1

NO/cGMP: The Past, the Present, and the Future

Michael Russwurm, Corina Russwurm, Doris Koesling, and Evanthia Mergia

Abstract

The NO/cGMP signalling cascade participates in the regulation of physiological parameters such as smooth muscle relaxation, inhibition of platelet aggregation, and neuronal transmission. cGMP is formed in response to nitric oxide (NO) by NO-sensitive guanylyl cyclases that exist in two isoforms (NO-GC1 and NO-GC2). Much has been learned about the regulation of NO-GC; however the precise role of cGMP in complex physiological and especially in pathophysiological settings and its alteration by biological factors needs to be established. Despite reports on a variety of cGMP-independent NO effects, KO mice with a complete lack of NO-GC provide evidence that the vasorelaxing and platelet-inhibiting effects of NO are solely mediated by NO-GC. Isoform-specific KOs demonstrate that low cGMP increases are sufficient to induce smooth muscle relaxation and that either NO-GC isoform is sufficient in most instances outside the central nervous system. In the neuronal system, however, the NO-GC isoforms obviously serve distinct functions as both isoforms are required for long-term potentiation and NO-GC1 was shown to enhance glutamate release in excitatory neurons in the hippocampal CA1 region by gating HCN channels. Future studies have to clarify the role of NO-GC2, to show whether HCN channels are general targets of cGMP in the nervous system and whether the NO/cGMP signalling cascade participates in synaptic transmission in other brain regions.

Key words Nitric oxide, Guanylyl cyclase, Cyclic guanosine monophosphate, Phosphodiesterases, Vasorelaxation, Synaptic plasticity, Hyperpolarization-activated cyclic nucleotide-gated channels

1 The Past

Cyclic GMP was first detected in the urine of rats a few years after the discovery of cAMP and in the following was detected throughout the body [1]. In contrast to cAMP whose generation was restricted to plasma membranes, cGMP-forming activity in membranous and cytosolic cell fractions indicated the existence of two cGMP-forming enzymes [2]. Although considerable efforts to characterize and isolate the two forms of the enzyme had been made, the identification of the physiological activators of the guanylyl cyclases (GCs), i.e., nitric oxide (NO) for the soluble enzyme [3] and atrial natriuretic peptide for the membrane forms [4], turned

Thomas Krieg and Robert Lukowski (eds.), *Guanylate Cyclase and Cyclic GMP: Methods and Protocols*, Methods in Molecular Biology, vol. 1020, DOI 10.1007/978-1-62703-459-3_1,

out to be major clues for the better understanding of the enzymes. The membrane-bound GCs and their respective activators are not discussed here but in a recent review by Michaela Kuhn [5].

NO-releasing compounds had been shown to activate cGMP formation in cytosolic fractions already in the 1970s [6]; the impact of this stimulatory NO effect remained unclear until NO was identified as the endothelium-derived relaxation factor to occur endogenously in the body [3]. Subsequently, NO as a signalling molecule received much attention and Furchgott, Ignarro, and Murad who made major contributions in the NO research were awarded with the Nobel price. NO was found to be produced enzymatically by the NO synthases (NOS) with the neuronal (nNOS or NOSI) and endothelial NOS (eNOS or NOSIII) being responsible for the production of NO as a signalling molecule. In contrast, the high amounts of NO produced by the inducible NOS (iNOS or NOSII) exert direct cGMP-independent toxic effects. The NO-sensitive guanylyl cyclase (NO-GC) was identified as the most important effector molecule for NO and with the heme moiety the enzyme is very well equipped to sense and to translate low amounts of NO into cGMP increases. Downstream in the NO/cGMP signalling pathway, the cGMP-dependent protein kinase (cGKI) which is reviewed in Chapter 2 of this book [7] mediates most of the known cGMP effects in smooth muscle relaxation, platelet aggregation, and synaptic plasticity but other cGMP targets (cGMP-regulated PDEs, cGMP-gated channels) may also execute cGMP functions. Next, we will introduce two isoforms of NO-GC on the protein level and explain the activation of the enzyme.

1.1 Two Isoforms of NO-GC, NO-GC1 and NO-GC2

NO-GC has been first purified from bovine lung, a tissue with a very high GC content [8]. The enzyme was shown to consist of two different subunits α_1 and β_1 with molecular weights of 73 kDa and 70 kDa, respectively [9] and to contain a prosthetic heme group which acts as the receptor for NO and is required to mediate the NO stimulation [10]. With peptide sequences derived from the purified protein, the primary structures of the β_1 subunit and α_1 subunit were determined; homology screening yielded the α_2 subunit. In "ex vivo" precipitation experiments from placenta, the β_1 subunit was identified as the dimerizing partner of the α_2 subunit [11]. Thus, two isoforms of the NO receptor GC exist, the heterodimers $\alpha_1\beta_1$ and $\alpha_2\beta_1$; these will be termed NO-GC1 and NO-GC2 in the following. An additionally identified subunit of NO-GC, β_2, does not yield a catalytically active GC upon coexpression with the other subunits and does not appear to be expressed. Therefore, the β_2 may represent a pseudogene.

Extensive biochemical and kinetic analysis of both NO-GC1 and NO-GC2 did not reveal any significant differences in regard to enzyme activity or NO sensitivity; thus, the enzymatic regulation of the isoforms appears to be similar [11]. Structurally, NO-GC2 was shown to be able to interact with adapter domains (PDZ domains) of synaptic proteins such as PSD95 via its α_2 C-terminal peptide, and a special intracellular localization may represent the special feature of this isoform [12]. Studies on the expression revealed a ubiquitous distribution of NO-GC1 and a more limited occurrence of NO-GC2 [13]. The highest expression of NO-GC2 is found in brain where both isoforms are present in similar amounts.

1.2 Mechanism of Activation of NO-GC

GC contains a prosthetic heme group which is required for NO stimulation. The heme has an absorption maximum as 431 nm indicating a five-coordinated heme with the amino acid histidine as the axial ligand [14]. After different intermediate steps proposed earlier, NO binding to the sixth coordination position of the heme iron finally results in a five-coordinated heme in which the bond to the proximal histidine is broken. The activated enzyme exhibits a 300-fold increased catalytic rate [11]. This NO-bound five-coordinated heme shows an absorption maximum at 399 nm. Because of the pronounced change in absorbance, binding of the ligand, NO, to its receptor, GC, can be monitored spectrophotometrically; NO binding and activation have been assumed to always coincide.

However, recent results indicate the presence of two spectrophotometrically indistinguishable NO-bound states of the GC which differ in catalytic activity [15]. NO binding in the presence of the substrate GTP or the reaction products (cGMP, pyrophosphate) yields the highly activated GC species, whereas NO binding in the absence of the substrate as usually performed in spectral studies leads to the formation of the non-activated NO-bound GC state that requires additional free NO in the presence of the substrate/reaction products to adopt the active conformation. Under physiological conditions, GTP should always be present in the GC environment. Therefore, the non-activated NO-bound state that was studied for a long time probably represents an artifact due to the absence of the substrate in spectral studies. As the catalytic rate is not the only difference between the two NO-bound states, GTP should be included in further spectral studies of the enzyme to ensure that the naturally occurring NO-bound activated state is studied. Recently, binding of a second NO molecule to a cysteine residue has been proposed to be required for the activation of the enzyme; however, the in vivo relevance of a second non-heme NO target awaits further confirmation [16].

In addition to NO, *NO-GC activators* have been identified which due to the beneficial cGMP effects are predicted to have pharmacotherapeutical relevance [17, 18]. Compounds, like the prototype YC-1 [3-(5′-hydroxymethyl-2′-furyl)-1-benzylindazole], sensitize NO-GC towards NO by slowing down the fast NO dissociation rate and are called *NO-sensitizers* [19]. Another group of substances is supposed to act by substituting for the activated heme conformation. Chapter 13 of this book describes an experimental approach to characterize these NO- and heme-independent activators [20]. Compounds of both groups are being tested in clinical trials.

2 The Present

2.1 Phosphodiesterases as Critical Determinants of NO-Induced cGMP Signals

cGMP signals in cells are determined by not only guanylyl cyclases but also cGMP-degrading phosphodiesterases (PDEs). The maximal activity of the cGMP-degrading PDEs by far exceeds guanylyl cyclase activity in most tissues. However, because PDEs suppress the concentration of their substrates cGMP or cAMP to levels below their kM, they typically do not operate at maximal speed. The potentially high cGMP-degrading activity together with the low substrate concentration turns PDEs into excellent drug targets as PDE inhibitors do not have to compete with the substrate.

PDEs are classified into 11 families according to their specificity towards cAMP or cGMP, their regulatory features, and sequence homology. For obvious reasons, the PDEs with the broadest expression and highest activity were identified first (and obtained the lowest numbers). PDE5, the photoreceptor PDE6 and PDE9 are cGMP specific; the Ca^{2+}/calmodulin (CaM)-stimulated PDE1, PDEs 2, 10, and 11 and the "cGMP-inhibited" PDE3 degrade both cyclic nucleotides; and the three remaining PDEs (PDE4, 7, 8) are cAMP specific. All PDEs share a common domain structure with regulatory domains in the N-terminal part and the catalytic domain in the C-terminal region. The regulatory domain can confer diverse regulatory features to the enzymes, e.g., activation by Ca^{2+}/CaM, by phosphorylation, or by cyclic nucleotides (see below), and can target the enzyme to specific subcellular compartments. For example, PDE2A1 is cytosolic, PDE2A2 is located in mitochondria, and PDE2A3 is located at synaptic membranes [21, 22]. Five PDE families contain a tandem of regulatory GAF domains (c*G*MP-specific and -stimulated phosphodiesterases, Anabaena *A*denylate cyclases, and *Escherichia coli F*hlA) in their N-terminal halves that can bind cGMP (PDE 2, 5, 6, 11) or cAMP (PDE10) and stimulate the enzymes, although in case of the PDE11 GAF domains the physiological ligand remains to be identified [23].

2.2 Shaping of Spike-Like cGMP Signals by PDE5

As an example, the interplay of NO-GC and PDE5 in the generation of cGMP signals in platelets and smooth muscle will be discussed in detail. Platelets and smooth muscle cells display the most pronounced NO-induced cGMP increases (150-fold in aortic tissue and 30-fold in platelets). Full time courses of these signals can be

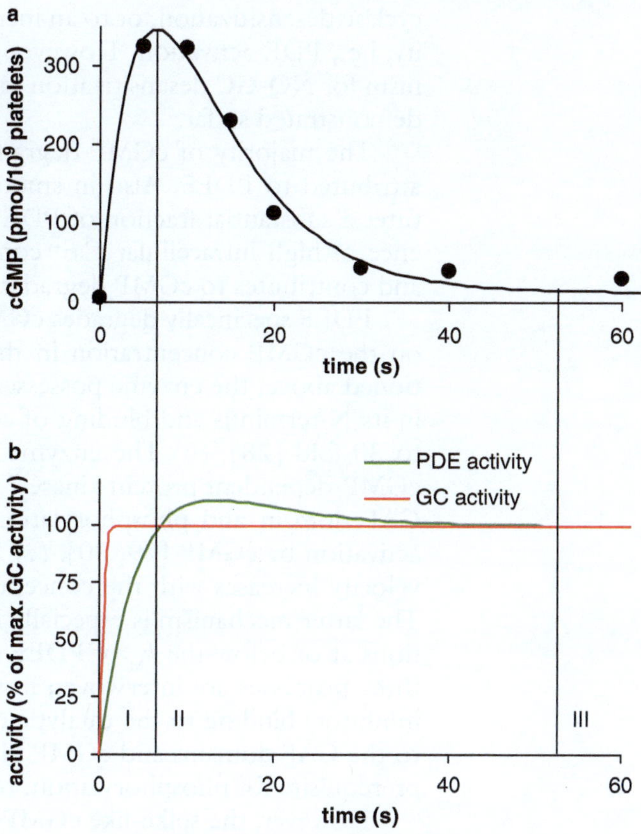

Fig. 1 NO-induced cGMP signals in platelets resulting from joint action of NO-sensitive guanylyl cyclase and phosphodiesterase 5. (**a**) Biphasic, spike-like cGMP signal in human platelets after stimulation with NO. Closed circles are measured values (from 25, the line represents calculated cGMP levels obtained from the time courses of GC and PDE activity depicted in (**b**)). (**b**) Almost immediate activation of NO-GC by NO and delayed activation of PDE5 activity were simulated. Kinetic parameters of this simulation were fitted to obtain the cGMP time course depicted in (**a**). During phase I, GC activity is higher than PDE activity, leading to cGMP accumulation, whereas during phase II, PDE activity outperforms GC activity leading to cGMP degradation. During phase III, PDE activity equals GC activity causing a plateau twofold above the level before stimulation. For further discussion see text

measured by radioimmunoassay as described in Chapter 4 [24], whereas intracellular cGMP indicators reach saturation because of their relatively narrow measuring range. In platelets and smooth muscle cells, a tremendous cGMP increase is followed by a sharp decline in cGMP within seconds to minutes (Fig. 1a), resulting in biphasic, spike-like cGMP signals [25, 26]. Theoretically, the decline of cGMP in the ongoing presence of NO can be attributed either to a reduction in cGMP-forming activity, i.e., guanylyl cyclase desensitization, or to an increase in cGMP-degrading activity, i.e., PDE activation. However a conclusive molecular mechanism for NO-GC desensitization in a physiological setting was not demonstrated so far.

The majority of cGMP-degrading activity in platelets can be attributed to PDE5. Also in smooth muscle cells, PDE5 constitutes a substantial fraction of PDE activity. However, in the presence of high intracellular Ca^{2+} concentrations, PDE1 is activated and contributes to cGMP degradation in smooth muscle [27].

PDE5 specifically degrades cGMP and its activity is dependent on the cGMP concentration in three ways: (a) As already mentioned above, the enzyme possesses cGMP-binding GAF domains in its N-terminus and binding of cGMP activates the enzyme 20- to 30-fold [28]. (b) The enzyme can be phosphorylated by the cGMP-dependent protein kinase I in the vicinity of its N-terminal GAF domain and phosphorylation stabilizes the GAF-mediated activation by cGMP [29, 30]. (c) As with all enzymes, enzymatic velocity increases with the concentration of the substrate, cGMP. The latter mechanism is especially important at cGMP concentrations at or below the k_M of PDE5 of approx. 1 μM cGMP. These three processes are interwoven in several ways, e.g., substrate (or inhibitor) binding to the catalytic center facilitates cGMP binding to the GAF domains and cGMP binding to the GAF domains is a prerequisite for phosphorylation, respectively [31, 32].

However, the spike-like cGMP responses in intact cells cannot be explained by a simple cGMP-dependent activation of PDE5, because instantaneous increases in activity upon cGMP elevation would simply limit the cGMP elevation, resulting in the formation of a plateau. Rather, the activity increase has to be delayed, such that cGMP degradation remains lower than cGMP formation despite the increase in cGMP (Fig. 1b, phase I). Subsequently, PDE activity has to outcompete NO-GC activity leading to the decline in cGMP (*see* Fig. 1b, phase II). Recently, a study on PDE5 activation using reasonably specific GAF domain ligands unravelled that both the GAF domain-dependent activation and the activity increase caused by elevation of the substrate concentration are unusually slow processes, requiring tens of seconds to minutes to complete [30]. Thus, the slow activation kinetics of PDE5 explain the formation of cGMP spikes. The deactivation of PDE5 after reduction of the cGMP concentration is a slow process too [28–30]. Even after cGMP declined, PDE5 remains in the active state with cGMP bound to the GAF domains, a conformation in part stabilized by phosphorylation of the enzyme. Thereby, the enzyme can efficiently dampen subsequent signals providing an explanation for long-term desensitization of cGMP signalling which partly underlies the phenomenon of NO resistance in vascular smooth muscle. Vice versa, low NO generation in the vascular bed sensitizes the NO/cGMP signalling pathway towards NO [33]. In sum, the degree of PDE5 activation adjusts sensitivity of the NO/cGMP cascade to the NO available.

2.3 Biological Responses Mediated by NO/cGMP

NO/cGMP has long been known to be involved in smooth muscle relaxation, platelet aggregation, and synaptic transmission. Now, *knockout mice deficient for NO-GCs* offer the possibility to learn about the protein's function in its natural environment. Mice lines deficient in either NO-GC subunit (α_1, α_2, β_1) have been generated. The knockout of the β_1 subunit results in a complete loss of both NO-GC receptors [34] whereas knockout of an α subunit causes deletion of the respective NO-GC1 and NO-GC2 isoform, respectively [35, 36]. Mice with a complete lack of NO-GC exhibit a drastically reduced life expectancy. As the NO-GC1 and NO-GC2 KO lines show normal life spans, the amounts of cGMP formed by either of the GC isoforms are apparently sufficient to prevent the high mortality found for the complete NO-GC KO (see above) which is also seen in the KO mice of the cGKI [37].

Surprisingly, in the complete NO-GC KO (deletion of the β_1 subunit) no α subunit was detectable on the protein level, and in the NO-GC1 or NO-GC2 KO (deletion of the α_1 or α_2 subunit), expression of the β_1 subunit was reduced. The results are compatible with the earlier observation that one of the GC subunits is not stable without its dimerizing partner [38]; therefore, speculations of altered expression of single NO-GC subunits do not appear to be reasonable.

Analysis of the NO-GC isoform expression identified NO-GC1 as the only NO-GC in platelets and confirmed the dominant expression (>90 %) in lung and aorta whereas in brain, NO-GC1 and NO-GC2 exist in comparable amounts. No compensatory upregulation of the non-deleted subunit was observed in either NO-GC isoform KO.

As the role of NO/cGMP signalling within *blood pressure* was well established [39], the pronounced elevation of systolic blood pressure (26 mmHg) in the mice with a complete NO-GC KO was not completely surprising. Two NO-GC1 KO mice strains exist. Gender-dependent hypertension (147 mmHg vs. 118 mmHg for male and female, respectively) reported in one NO-GC1 KO strain depended on the genetic background and was shown to be caused by a genetic modification of the renin-angiotensin-aldosterone system [36, 40]. The other NO-GC1 strain showed an only mildly elevated systolic blood pressure that vanished after backcrossing ([35] and unpublished). Because more than 90 % of the aortic NO-GC activity was shown to be deleted, only a fraction of the normal NO-GC content (provided by NO-GC2) in the vessel wall is sufficient to maintain normal blood pressure under physiological conditions. Accordingly, a marked blood pressure increase was observed in the NO-GC1 KO mice upon treatment with the NO synthase inhibitor L-NAME. The results emphasize the important role of NO in blood pressure regulation and are in good accordance with results found in the eNOS and cGKI KO mice [37, 41]. Vice versa, application of NO donors decreased blood pressure NO-GC1 KO mice [36, 42] whereas the response to an NO donor was totally abolished in the complete NO-GC KO mice [34] underlining that the blood pressure effects of NO are entirely mediated by the NO receptor GCs.

The importance of NO for *smooth muscle relaxation* has been established for a long time and numerous targets of NO other than NO-GC have been reported. However, in the complete NO-GC KO neither endogenously nor pharmacologically applied NO caused any relaxation. Thus, in the aorta, the relaxing effect of NO is solely conveyed by NO-GC. The results are in good agreement with the abrogated endothelium-induced relaxation in the cGKI [37] and eNOS KO mice [43, 44].

In the NO-GC1 KO, the cGMP-forming capacity is greatly reduced (94 %) but the vessels are still completely relaxed by NO albeit requiring higher concentrations and show a reduction of endothelium-dependent relaxation (50 % of WT). The results demonstrate that cGMP generated by NO-GC2 is able to substitute for the cGMP produced by NO-GC1 and that small increases in cGMP exert a profound effect on vascular tone suggesting that most of NO-GCs act as spare receptors to increase sensitivity towards NO [35]. Despite the finding that cGMP formed by one NO-GC isoform relaxes smooth muscle as does the cGMP generated by the other isoform, the NO-GC isoforms may generate different cGMP pools which may serve different targets. Therefore, methodological approaches allowing the detection of local cGMP pools are of high scientific interest and four chapters (Chapters 5–8) in this book deal with methods to measure cGMP in living cells [45–48].

For the NO-GC1 KO mice invasive hemodynamic catheter measurements yielded *increased cardiac contractility* as well as impaired ventricular relaxation [36].

Penile erection greatly depends on smooth muscle relaxation as blood filling of the corpus cavernosum is the key event which then restricts venous outflow [49]. The important role of NO/cGMP in penile erection is underlined by the efficiency of PDE5 inhibitors such as sildenafil as today's most successful therapy for the treatment of erectile dysfunction. Both eNOS-produced NO from the endothelium and nNOS-produced NO from non-adrenergic and non-cholinergic (NANC) nerves are considered to be of importance in this physiological situation [50]. In accordance with the assumption that NO-GC1 and NO-GC2 mediate NO effects in penile erection, smooth muscle relaxation of corpus cavernosum was greatly reduced in the NO-GC1 KO, while pharmacologically applied NO still caused a substantial relaxation which nevertheless was ODQ sensitive [51].

As has been observed in the cGKI KO before, the NO/cGMP cascade is required for the *motility of the gastrointestinal tract*. The complete NO-GC KOs exhibit an enlarged cecum and an extremely increased whole-gut transient time resulting in fatal gastrointestinal obstruction [34]. Under normal diet, these mice die shortly after weaning which can be almost completely prevented by a fiber-free diet. Again, the NO-GC1-deficient KO mice despite greatly reduced cGMP did neither show functional gastrointestinal impairment nor visible abnormalities indicating that the relatively small amounts of cGMP formed by the NO-GC2 isoform are still sufficient [52].

KOs lacking other members of the NO/cGMP signalling cascade like cGKI or NOSs display a gastrointestinal phenotype similar to that of the complete NO-GC KO. Mice lacking cGKI also show a reduction of life span which is less severe (50 % dead before 6 weeks of age) and show very similar disturbances in the gastrointestinal tract [37, 53]. The comparatively mild gastrointestinal phenotype of the nNOS-KO mice with enlargement of the stomach, hypertrophy of the pyloric sphincter, and delayed gastric emptying may be due to an additional catalytically active splice variant [54]. Even the triple-KO strain in which apparently all three NOS isoforms have been deleted [55] shows a less severe phenotype than the NO-GC KO, but again the additional catalytically active splice variant may still exist in this KO.

Functionally, NO is considered as transmitter of non-adrenergic non-cholinergic (NANC) neurons in the gastrointestinal tract. NO synthesized by nNOS is released from the NANC nerve terminals and is thought to diffuse to gastrointestinal smooth muscle cells where it causes relaxation via stimulation of NO-GC. Nitrergic relaxation was greatly reduced in the NO-GC1 KO mice; relaxation induced by exogenous NO was decreased but still substantial with the remaining relaxation being ODQ sensitive. The results show a role of NO-GC1 in gastric nitrergic relaxation in vitro, but obviously in vivo the cGMP formed by the NO-GC2 isoform is still sufficient to avoid fatal consequences in NO-GC1-deficient KO mice.

In the complete NO-GC KO, the response towards endogenously released or pharmacologically applied NO was completely abrogated again identifying NO-GC as the sole NO target [56]. However, NO-induced relaxation was only slightly reduced in a smooth muscle-specific KO of NO-GC and the mice did show a WT-like whole-gut transition time indicating NO-GC other than in smooth muscle cells is responsible for the NO effects.

The NO/cGMP signalling cascade has been implicated to play a role in *pulmonary vascular tone and remodelling* as the NO-GC content is comparatively high and inhalative NO selectively dilates pulmonary resistance vessels [57]. In NO-GC1-deficient mice,

NO-stimulated cGMP-forming activity was greatly reduced but not completely abrogated again suggesting NO-GC1 as the major isoform and the additional occurrence of NO-GC2 [42]. Pulmonary arterial pressure was not altered in these mice nor did the response to acute hypoxia differ from the one in WT. Under prolonged hypoxia (3 weeks), the increase in right ventricular pressure, right ventricular hypertrophy, and remodelling of pulmonary vessels was significantly greater suggesting that NO-GC1 serves to limit hypoxia-induced pulmonary vascular remodelling. Pulmonary hypertension was also observed in the eNOS KO mice [58], the lack of this alteration in the NO-GC1 KO can most likely be attributed to the remaining NO-GC2 in the NO-GC1 KO.

Similar to smooth muscle relaxation, NO has been postulated to mediate *inhibition of platelet aggregation* via cGMP, although cGMP-independent mechanisms have been proposed as well [59–61]. The inhibitory effect of NO on platelet aggregation induced by various agonists was completely lost in NO-GC KO platelets [34]. Analysis of the NO-GC receptor isoforms revealed NO-GC2 to be completely absent and, thus, NO-GC1 to be the only isoform in platelets. In accordance, platelets from NO-GC1-deficient mice featured the same characteristics as those from the complete NO-GC KO [35]. These data suggest that in platelets NO signals solely via NO-GC1. The results are in good accordance with those from cGKI KO [62] and IRAG KO mice (IP3-receptor-associated cGMP-kinase substrate; [63]).

NO/cGMP has been proposed to contribute to *inflammatory and neuropathic pain*. Indeed the complete NO-GC KO exhibited a considerably reduced nociceptive behavior in models of inflammatory and neuropathic pain but their responses to acute pain were not impaired [64]. Interestingly, immunohistochemical analysis revealed that NO-GC does not colocalize with cGKI indicating the involvement in other pain signalling pathways. A method describing the intrathecal application of drugs has been described in Chapter 14 of this book [65].

In the *central nervous system*, the NO/cGMP cascade has been postulated to modulate *synaptic transmission* and to participate in certain types of *synaptic plasticity* such as long-term potentiation, the use-dependent increase of transmission efficacy at synapses implicated in learning and memory [66–68]. Also in brain, NO is generated by eNOS and nNOS as signalling molecule under physiological conditions. Functionally various effects of NO on neuronal excitability and/or synaptic transmission have been described, which can be generally classified as excitatory or inhibitory [68]. NO formed postsynaptically in response to NMDA receptor activation stands for the archetype of neurotransmitter-induced NO generation and is explained by the physical association of nNOS and NMDA (N-methyl-D-aspartate) receptor subunit NR2B via PSD-95 (postsynaptic density protein 95 kDa; [69]). As a retrograde

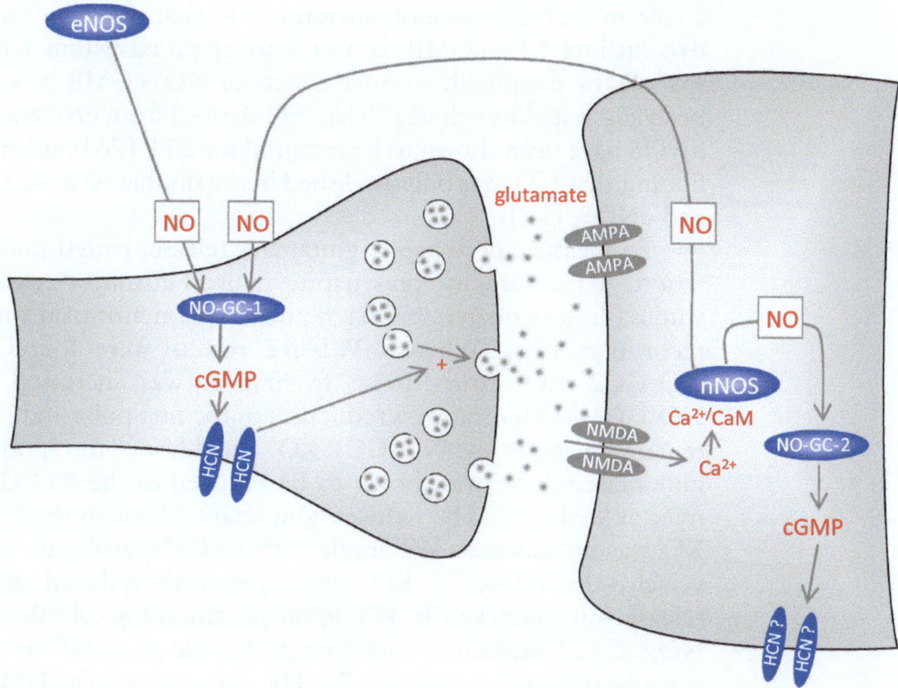

Fig. 2 Simplified schematic diagram of a glutamatergic synapse with the proposed localization of the members of the NO/cGMP signalling cascade. Endothelial NO synthase (eNOS)-derived NO increases the glutamate release via NO-GC1, cGMP, and presynaptic HCN channels. Postsynaptically, NO is generated by the neuronal NO synthase (nNOS) in response to NMDA receptor-mediated calcium. As a retrograde messenger, this NO can activate NO-GC1 presynaptically or it can cause activation of NO-GC2 predicted to occur in the postsynaptic neuron. For further explanation see text. *eNOS* endothelial NO synthase, *nNOS* neuronal NO synthase, *HCN* hyperpolarization-activated nucleotide-gated channel, *GC* guanylyl cyclase, *AMPA* and *NMDA* glutamate receptors

messenger the postsynaptically formed NO has been proposed to cause long-lasting increases in the transmitter release in the presynaptic terminals (Fig. 2; [70–72]). A major impact of NO on long-term potentiation (LTP) has been demonstrated in the hippocampal CA1 region. The finding that only the double NOS KO deficient in eNOS and nNOS showed a complete loss of LTP underlines that not only nNOS is responsible for NO formation within LTP but that also eNOS-derived NO is required.

NO-GC1 and NO-GC2 were found to be expressed in brain in similar amounts. Moreover, *LTP* measured by field potential responses revealed a lack of LTP in both the NO-GC1 *and* the NO-GC2 KO in the hippocampal CA1 region and in the visual cortex indicating that either NO-GC isoform is required for LTP [73, 74]. With one exception (hippocampal LTP in the NO-GC1 KO), the function of the isoforms within LTP could be reconstituted with a cGMP analogue and vice versa the inhibition of NO-GC isoforms with ODQ (1H-[1,2,4]oxadiazolo[4,3-a]quinoxalin-1-one) abrogated LTP in WT. The finding that both NO-GC isoforms play a role in LTP and cannot substitute for each other indicates that two distinct NO/cGMP pathways are required within LTP. Two spatially or temporally distinct effects of NO/cGMP have already been suggested by others [75] as NO derived from eNOS and from nNOS have been shown to be required for LTP [76] explaining the finding that LTP was only abolished in the double NOS KO (nNOS and eNOS; [77]).

To measure presynaptic glutamate release, paired-pulse facilitation, a measure for presynaptic neurotransmitter release, was studied in glutamatergic CA1 hippocampal neurons in single-cell recordings [73]. Whereas WT-like results were found in the NO-GC2 KO, paired-pulse facilitation was increased in the NO-GC1 KO indicating a reduction of the first pulse-induced glutamate release in the NO-GC1 KO. In addition, the spontaneous glutamate release was shown to be reduced in the NO-GC1 KO mice as well [78]. The reduced glutamate release in the NO-GC1 KO was increased to WT levels with a cGMP analogue and conversely, the NO-GC1 KO phenotype with reduced glutamate release was mimicked in WT upon pharmacological inhibition of NO/cGMP signalling underlining *the role of cGMP in the neurotransmitter release* (*see* Fig. 2). The occurrence of a cGMP effect already under basal conditions indicates continuous NO and cGMP formation under non-stimulated conditions. Endothelial NOS- and not nNOS-derived NO was shown to be responsible for the glutamate release under these experimental conditions, i.e., in the presence of an NMDA receptor blocker. Finally, hyperpolarization-activated cyclic nucleotide-gated (HCN) channels were considered as cGMP target that facilitates glutamate release and the hypothesis was experimentally addressed by applying HCN channel blockers (ZD7288 and DK-AH269). Indeed, the HCN blockers reduced glutamate release in WT- to KO-like levels and did not affect glutamate release in NO-GC1 KO suggesting that HCN channels act as executors of cGMP effects.

In sum, increasing evidence suggests that presynaptic cGMP formed by NO-GC1 increases the neurotransmitter release probability by gating HCN channels and that NO-GC2 is responsible for postsynaptic cGMP generation. Thus, two NO/cGMP pathways exist that enhance the strength of synaptic transmission on either side of the synaptic cleft.

3 The Future

Future experiments have to clarify the role of postsynaptically generated cGMP and have to show whether the pre- and postsynaptic localizations of NO-GC1 and NO-GC2 apply to other brain regions as well. In addition, a possible GABA-ergic localization of the NO-GC isoforms has to be addressed and the functional impact of NO/cGMP signalling for the strength of synaptic transmission and the development of synaptic plasticity have to be elucidated.

Outside the central nervous system, local cGMP generation possibly pointing to different cGMP effector molecules should be addressed and the involvement of NO/cGMP in the regulation of additional physiological responses and the modification of pathophysiological processes should be investigated.

References

1. Schultz G, Böhme E, Munske K (1969) Guanyl cyclase. Determination of enzyme activity. Life Sci 8:1323–1332
2. Chrisman TD, Garbers DL, Parks MA, Hardman JG (1975) Characterization of particulate and soluble guanylate cyclases from rat lung. J Biol Chem 250:374–381
3. Palmer RMJ, Ferrige AG, Moncada S (1987) Nitric oxide release accounts for the biological activity of endothelium-derived relaxing factor. Nature 327:524–526
4. Kuno T, Andresen JW, Kamisaki Y, Waldman SA, Chang LY, Saheki S, Leitman DC, Nakane M, Murad F (1986) Co-purification of an atrial natriuretic factor receptor and particulate guanylate cyclase from rat lung. J Biol Chem 261: 5817–5823
5. Kuhn M (2009) Function and dysfunction of mammalian membrane guanylyl cyclase receptors: lessons from genetic mouse models and implications for human diseases. Handb Exp Pharmacol 191:47–69
6. Arnold WP, Mittal CK, Katsuki S, Murad F (1977) Nitric oxide activates guanylate cyclase and increases guanosine 3':5'-cyclic monophosphate levels in various tissue preparations. Proc Natl Acad Sci USA 74:3203–3207
7. Hofmann F, Wegener JW (2013) cGMP dependent protein kinases (cGK). Methods Mol Biol
8. Gerzer R, Hofmann F, Schultz G (1981) Purification of a soluble, sodium-nitroprusside-stimulated guanylate cyclase from bovine lung. Eur J Biochem 116:479–486
9. Kamisaki Y, Saheki S, Nakane M, Palmieri JA, Kuno T, Chang BY, Waldman SA, Murad F (1986) Soluble guanylate cyclase from rat lung exists as a heterodimer. J Biol Chem 261: 7236–7241
10. Gerzer R, Böhme E, Hofmann F, Schultz G (1981) Soluble guanylate cyclase purified from bovine lung contains heme and copper. FEBS Lett 132:71–74
11. Russwurm M, Behrends S, Harteneck C, Koesling D (1998) Functional properties of a naturally occurring isoform of soluble guanylyl cyclase. Biochem J 335:125–130
12. Russwurm M, Wittau N, Koesling D (2001) Guanylyl cyclase/PSD-95 interaction: targeting of the nitric oxide-sensitive alpha2beta1 guanylyl cyclase to synaptic membranes. J Biol Chem 276:44647–44652
13. Mergia E, Russwurm M, Zoidl G, Koesling D (2003) Major occurrence of the new alpha2beta1 isoform of NO-sensitive guanylyl cyclase in brain. Cell Signal 15:189–195
14. Stone JR, Marletta MA (1994) Soluble guanylate cyclase from bovine lung: activation with nitric oxide and carbon monoxide and spectral characterization of the ferrous and ferric states. Biochemistry 33:36–5640
15. Russwurm M, Koesling D (2004) NO activation of guanylyl cyclase. EMBO J 23: 4443–4450
16. Fernhoff NB, Derbyshire ER, Marletta MA (2009) A nitric oxide/cysteine interaction mediates the activation of soluble guanylate cyclase. Proc Natl Acad Sci USA 106: 21602–21607
17. Schmidt HHHW, Schmidt PM, Stasch J-P (2009) NO- and haem-independent soluble guanylate cyclase activators. Handb Exp Pharmacol 191:309–339
18. Stasch J-P, Hobbs AJ (2009) NO-independent, haem-dependent soluble guanylate cyclase stimulators. Handb Exp Pharmacol 191:277–308
19. Russwurm M, Mergia E, Mullershausen F, Koesling D (2002) Inhibition of deactivation of NO-sensitive guanylyl cyclase accounts for the sensitizing effect of YC-1. J Biol Chem 277:24883–24888
20. Schmidt PM, Stasch JP (2013) Receptor binding assay for NO-independent activators of soluble guanylate cyclase. Methods Mol Biol
21. Acin-Perez R, Russwurm M, Günnewig K, Gertz M, Zoidl G, Ramos L, Buck J, Levin LR, Rassow J, Manfredi G, Steegborn C (2011) A phosphodiesterase 2A isoform localized to mitochondria regulates respiration. J Biol Chem 286:30423–30432
22. Russwurm C, Zoidl G, Koesling D, Russwurm M (2009) Dual acylation of PDE2A splice variant 3: targeting to synaptic membranes. J Biol Chem 284:25782–25790

23. Jäger R, Russwurm C, Schwede F, Genieser HG, Koesling D, Russwurm M (2012) Activation of PDE10 and PDE11 phosphodiesterases. J Biol Chem 287:1210–1219
24. Jäger R, Groneberg D, Lies B, Bettaga N, Kümmel M, Friebe A (2013) Radioimmunoassay for the quantification of cGMP levels in cells and tissues. Methods Mol Biol
25. Mullershausen F, Russwurm M, Thompson WJ, Liu L, Koesling D, Friebe A (2001) Rapid nitric oxide-induced desensitization of the cGMP response is caused by increased activity of phosphodiesterase type 5 paralleled by phosphorylation of the enzyme. J Cell Biol 155:271–278
26. Mullershausen F, Lange A, Mergia E, Friebe A, Koesling D (2006) Desensitization of NO/cGMP signaling in smooth muscle: blood vessels versus airways. Mol Pharmacol 69:1969–1974
27. Hagiwara M, Endo T, Hidaka H (1984) Effects of vinpocetine on cyclic nucleotide metabolism in vascular smooth muscle. Biochem Pharmacol 33:453–457
28. Rybalkin SD, Rybalkina IG, Shimizu-Albergine M, Tang XB, Beavo JA (2003) PDE5 is converted to an activated state upon cGMP binding to the GAF A domain. EMBO J 22:469–478
29. Mullershausen F, Friebe A, Feil R, Thompson WJ, Hofmann F, Koesling D (2003) Direct activation of PDE5 by cGMP: long-term effects within NO/cGMP signaling. J Cell Biol 160:719–727
30. Jäger R, Schwede F, Genieser HG, Koesling D, Russwurm M (2010) Activation of PDE2 and PDE5 by specific GAF ligands: delayed activation of PDE5. Br J Pharmacol 161:1645–1660
31. Turko IV, Francis SH, Corbin JD (1998) Binding of cGMP to both allosteric sites of cGMP-binding cGMP-specific phosphodiesterase (PDE5) is required for its phosphorylation. Biochem J 329:505–510
32. Turko IV, Ballard SA, Francis SH, Corbin JD (1999) Inhibition of cyclic GMP-binding cyclic GMP-specific phosphodiesterase (Type 5) by sildenafil and related compounds. Mol Pharmacol 56:124–130
33. Mullershausen F, Russwurm M, Koesling D, Friebe A (2003) The enhanced NO-induced cGMP response induced by long-term L-NAME treatment is not due to enhanced expression of NO-sensitive guanylyl cyclase. Vascul Pharmacol 40:161–165
34. Friebe A, Mergia E, Dangel O, Lange A, Koesling D (2007) Fatal gastrointestinal obstruction and hypertension in mice lacking nitric oxide-sensitive guanylyl cyclase. Proc Natl Acad Sci USA 104:7699–7704
35. Mergia E, Friebe A, Dangel O, Russwurm M, Koesling D (2006) Spare guanylyl cyclase NO receptors ensure high NO sensitivity in the vascular system. J Clin Invest 116:1731–1737
36. Buys ES, Sips P, Vermeersch P, Raher MJ, Rogge E, Ichinose F, Dewerchin M, Bloch KD, Janssens S, Brouckaert P (2008) Gender-specific hypertension and responsiveness to nitric oxide in sGC alpha1 knockout mice. Cardiovasc Res 79:179–186
37. Pfeifer A, Klatt P, Massberg S, Ny L, Sausbier M, Hirneiss C, Wang GX, Korth M, Aszódi A, Andersson KE, Krombach F, Mayerhofer A, Ruth P, Fässler R, Hofmann F (1998) Defective smooth muscle regulation in cGMP kinase I-deficient mice. EMBO J 17:3045–3051
38. Wagner C, Russwurm M, Jäger R, Friebe A, Koesling D (2005) Dimerization of nitric oxide-sensitive guanylyl cyclase requires the alpha 1 N terminus. J Biol Chem 280: 17687–17693
39. Rees DD, Palmer RM, Moncada S (1989) Role of endothelium-derived nitric oxide in the regulation of blood pressure. Proc Natl Acad Sci USA 86:3375–3378
40. Buys ES, Raher MJ, Kirby A, Mohd S, Baron DM, Hayton SR, Tainsh LT, Sips PY, Rauwerdink KM, Yan Q, Tainsh RE, Shakartzi HR, Stevens C, Decaluwé K, Rodrigues-Machado Mda G, Malhotra R, Van de Voorde J, Wang T, Brouckaert P, Daly MJ, Bloch KD (2012) Genetic modifiers of hypertension in soluble guanylate cyclase α1-deficient mice. J Clin Invest 122:2316–2325
41. van Vliet BN, Chafe LL, Montani JP (2003) Characteristics of 24 h telemetered blood pressure in eNOS-knockout and C57Bl/6 J control mice. J Physiol 549:313–325
42. Vermeersch P, Buys E, Pokreisz P, Marsboom G, Ichinose F, Sips P, Pellens M, Gillijns H, Swinnen M, Graveline A, Collen D, Dewerchin M, Brouckaert P, Bloch KD, Janssens S (2007) Soluble guanylate cyclase-alpha1 deficiency selectively inhibits the pulmonary vasodilator response to nitric oxide and increases the pulmonary vascular remodeling response to chronic hypoxia. Circulation 116:936–943
43. Huang PL, Huang Z, Mashimo H, Bloch KD, Moskowitz MA, Bevan JA, Fishman MC (1995) Hypertension in mice lacking the gene for endothelial nitric oxide synthase. Nature 377:239–242
44. Chataigneau T, Félétou M, Huang PL, Fishman MC, Duhault J, Vanhoutte PM (1999) Acetylcholine-induced relaxation in blood vessels from endothelial nitric oxide synthase knockout mice. Br J Pharmacol 126:219–226
45. Rich TC, Britain AL, Stedman T, Leavesley SJ (2013) Hyperspectral imaging of FRET-based cGMP probes. Methods Mol Biol

46. Thunemann M, Fomin N, Krawutschke C, Russwurm M, Feil R (2013) Visualization of cGMP with cGi biosensors. Methods Mol Biol
47. Götz KR, Nikolaev VO (2013) Advances and techniques to measure cGMP in intact cardiomyocytes. Methods Mol Biol
48. Held KF, Dostmann WR (2013) Real-time monitoring the spatio-temporal dynamics of intracellular cGMP in vascular smooth muscle cells. Methods Mol Biol
49. Andersson KE, Wagner G (1995) Physiology of penile erection. Physiol Rev 75:191–236
50. Musicki B, Burnett AL (2006) eNOS function and dysfunction in the penis. Exp Biol Med (Maywood) 231:154–165
51. Nimmegeers S, Sips P, Buys E, Decaluwé K, Brouckaert P, Van de Voorde J (2007) Role of the soluble guanylyl cyclase alpha(1)-subunit in mice corpus cavernosum smooth muscle relaxation. Int J Impot Res 20:278–284
52. Vanneste G, Dhaese I, Sips P, Buys E, Brouckaert P, Lefebvre RA (2007) Gastric motility in soluble guanylate cyclase alpha 1 knock-out mice. J Physiol 584:907–920
53. Ny L, Pfeifer A, Aszòdi A, Ahmad M, Alm P, Hedlund P, Fässler R, Andersson KE (2000) Impaired relaxation of stomach smooth muscle in mice lacking cyclic GMP-dependent protein kinase I. Br J Pharmacol 129:395–401
54. Huang PL, Dawson TM, Bredt DS, Snyder SH, Fishman MC (1993) Targeted disruption of the neuronal nitric oxide synthase gene. Cell 75:1273–1286
55. Morishita T, Tsutsui M, Shimokawa H, Sabanai K, Tasaki H, Suda O, Nakata S, Tanimoto A, Wang KY, Ueta Y, Sasaguri Y, Nakashima Y, Yanagihara N (2005) Nephrogenic diabetes insipidus in mice lacking all nitric oxide synthase isoforms. Proc Natl Acad Sci USA 102:10616–10621
56. Groneberg D, König P, Koesling D, Friebe A (2011) Nitric oxide-sensitive guanylyl cyclase is dispensable for nitrergic signaling and gut motility in mouse intestinal smooth muscle. Gastroenterology 140:1608–1617
57. Ichinose F, Roberts JD Jr, Zapol WM (2004) Inhaled nitric oxide: a selective pulmonary vasodilator: current uses and therapeutic potential. Circulation 109:3106–3111
58. Steudel W, Scherrer-Crosbie M, Bloch KD, Weimann J, Huang PL, Jones RC, Picard MH, Zapol WM (1998) Sustained pulmonary hypertension and right ventricular hypertrophy after chronic hypoxia in mice with congenital deficiency of nitric oxide synthase 3. J Clin Invest 101:2468–2477
59. Moncada S, Radomski MW, Palmer RM (1998) Endothelium-derived relaxing factor. Identification as nitric oxide and role in the control of vascular tone and platelet function. Biochem Pharmacol 37:2495–2501
60. Wanstall JC, Homer KL, Doggrell SA (2005) Evidence for, and importance of, cGMP-independent mechanisms with NO and NO donors on blood vessels and platelets. Curr Vasc Pharmacol 3:41–53
61. Crane MS, Rossi AG, Megson IL (2005) A potential role for extracellular nitric oxide generation in cGMP-independent inhibition of human platelet aggregation: biochemical and pharmacological considerations. Br J Pharmacol 144:849–859
62. Massberg S, Sausbier M, Klatt P, Bauer M, Pfeifer A, Siess W, Fässler R, Ruth P, Krombach F, Hofmann F (1999) Increased adhesion and aggregation of platelets lacking cyclic guanosine 3′,5′-monophosphate kinase I. J Exp Med 189:1255–1264
63. Antl M, von Brühl ML, Eiglsperger C, Werner M, Konrad I, Kocher T, Wilm M, Hofmann F, Massberg S, Schlossmann J (2007) IRAG mediates NO/cGMP-dependent inhibition of platelet aggregation and thrombus formation. Blood 109:552–559
64. Schmidtko A, Gao W, König P, Heine S, Motterlini R, Ruth P, Schlossmann J, Koesling D, Niederberger E, Tegeder I, Friebe A, Geisslinger G (2008) cGMP produced by NO-sensitive guanylyl cyclase essentially contributes to inflammatory and neuropathic pain by using targets different from cGMP-dependent protein kinase I. J Neurosci 28:8568–8576
65. Lu R, Schmidtko A (2013) Direct Intrathecal drug delivery in mice for detecting in vivo effects of cGMP on pain processing. Methods Mol Biol
66. Snyder SH, Bredt DS (1991) Nitric oxide as a neuronal messenger. Trends Pharmacol Sci 12:125–128
67. Boehning D, Snyder SH (2003) Novel neural modulators. Annu Rev Neurosci 26:105–131
68. Garthwaite J (2008) Concepts of neural nitric oxide-mediated transmission. Eur J Neurosci 27:2783–2802
69. Christopherson KS, Hillier BJ, Lim WA, Bredt DS (1999) PSD-95 assembles a ternary complex with the N-methyl-D-aspartic acid receptor and a bivalent neuronal NO synthase PDZ domain. J Biol Chem 274:27467–27473
70. Garthwaite J, Charles SL, Chess-Williams R (1988) Endothelium-derived relaxing factor release on activation of NMDA receptors suggests role as intercellular messenger in the brain. Nature 336:385–388
71. O'Dell TJ, Hawkins RD, Kandel ER, Arancio O (1991) Tests of the roles of two diffusible

substances in long-term potentiation: evidence for nitric oxide as a possible early retrograde messenger. Proc Natl Acad Sci USA 88:11285–11289

72. Schuman EM, Madison DV (1991) A requirement for the intercellular messenger nitric oxide in long-term potentiation. Science 254:1503–1506
73. Taqatqeh F, Mergia E, Neitz A, Eysel UT, Koesling D, Mittmann T (2009) More than a retrograde messenger: nitric oxide needs two cGMP pathways to induce hippocampal long-term potentiation. J Neurosci 29:9344–9350
74. Haghikia A, Mergia E, Friebe A, Eysel UT, Koesling D, Mittmann T (2007) Long-term potentiation in the visual cortex requires both nitric oxide receptor guanylyl cyclases. J Neurosci 27:818–823
75. Bon CL, Garthwaite J (2003) On the role of nitric oxide in hippocampal long-term potentiation. J Neurosci 23:1941–1948
76. Hopper RA, Garthwaite J (2006) Tonic and phasic nitric oxide signals in signals in hippocampal long-term potentiation. J Neurosci 26:11513–11521
77. Son H, Hawkins RD, Martin K, Kiebler M, Huang PL, Fishman MC, Kandel ER (1996) Long-term potentiation is reduced in mice that are doubly mutant in endothelial and neuronal nitric oxide synthase. Cell 13: 1015–1023
78. Neitz A, Mergia E, Eysel UT, Koesling D, Mittmann T (2011) NO/cGMP facilitates glutamate release via HCN channels. Eur J Neurosci 33:1611–1621

Chapter 2

cGMP-Dependent Protein Kinases (cGK)

Franz Hofmann and Jörg W. Wegener

Abstract

cGMP-dependent protein kinases (cGK) are serine/threonine kinases that are widely distributed in eukaryotes. Two genes—*prkg1* and *prkg2*—code for cGKs, namely, cGKI and cGKII. In mammals, two isozymes, cGKIα and cGKIβ, are generated from the *prkg1* gene. The cGKI isozymes are prominent in all types of smooth muscle, platelets, and specific neuronal areas such as cerebellar Purkinje cells, hippocampal neurons, and the lateral amygdala. The cGKII prevails in the secretory epithelium of the small intestine, the juxtaglomerular cells, the adrenal cortex, the chondrocytes, and in the nucleus suprachiasmaticus. Both cGKs are major downstream effectors of many, but not all, signalling events of the NO/cGMP and the ANP/cGMP pathways. cGKI relaxes smooth muscle tone and prevents platelet aggregation, whereas cGKII inhibits renin secretion, chloride/water secretion in the small intestine, the resetting of the clock during early night, and endochondral bone growth. This chapter focuses on the involvement of cGKs in cardiovascular and non-cardiovascular processes including cell growth and metabolism.

Key words cGKs, Cardiovascular, Non-cardiovascular, Cell growth, Metabolism

1 Introduction

NO is generated by three different isozymes (NO synthases (NOS) 1–3). In many cells, NO increases the concentration of cyclic guanosine monophosphate (cGMP) by activation of the soluble guanylyl cyclase (sGC) [1, 2]. cGMP is also generated by membrane-bound particulate guanylyl cyclases (pGCs, e.g., GC-A, GC-B, and GC-C). GC-A and GC-B are major receptors for a family of natriuretic peptides released from the heart and vascular endothelium, like atrial natriuretic peptide (ANP), brain natriuretic peptide (BNP), and C-type natriuretic peptide (CNP), whereas GC-C is the receptor for guanylin, an intestinal peptide involved in intestinal fluid regulation [3]. Further analysis of the cGMP system identified three major intracellular targets for cGMP. For example, cGMP binds to cyclic adenosine monophosphate (cAMP)-specific phosphodiesterases (PDEs) and thereby modulates the concentration of cAMP enabling a cross talk between both cyclic nucleotides pathways [4, 5]. cGMP and cAMP activate cyclic nucleotide-gated

Thomas Krieg and Robert Lukowski (eds.), *Guanylate Cyclase and Cyclic GMP: Methods and Protocols*, Methods in Molecular Biology, vol. 1020, DOI 10.1007/978-1-62703-459-3_2,

(CNG) cation channels that are an important part of the signal transduction pathway in the visual and olfactory system [6, 7]. In addition, most cells contain at least one of three cGMP-dependent protein kinases (cGKs)—cGKIα, cGKIβ, or cGKII [8–10]—that are targeted by their distinct amino termini to different substrates that are involved in the regulation of different cellular functions.

NO not only signals through the cGMP pathway but has several effects that are independent of the cGMP/cGK signalling pathway [11]. Interpretation of cGMP effects should include the same precautions, because (a) cGMP has several effectors that may be used simultaneously in various tissues, (b) cGMP might activate directly or indirectly cAMP-dependent protein kinases (cAK), and (c) some of the effects of "cGK-specific" activators and inhibitors are not mediated by cGKs [12–16].

This chapter will concentrate on recent results obtained by total or tissue-specific deletion or modulation of the cGK genes and of some of their substrates. We will discuss mainly the contribution of the NO/cGMP/cGK signalling system in the cardiovascular organs and on some results for the peripheral tissues. Details on the characteristics of cGKs tissue distribution and substrates will not be discussed in this chapter. The interested reader is referred to the following articles [8, 9, 17, 18].

2 cGK Signalling in the Cardiovascular System

Signalling via NO and natriuretic peptides (NP) is involved in cardiovascular physiology and pathophysiology [7, 19–27]. NO derived from the endothelium relaxes blood vessels and reduces blood pressure [28]. In addition, NO modulates cardiac and vascular remodelling processes that are associated with heart failure and atherosclerosis [24]. Atrial natriuretic peptide (ANP) also relaxes blood vessels [29, 30] and thus controls blood pressure in addition to regulating homeostasis of sodium [31]. These physiological and pathophysiological effects are supposed to be mediated by cGMP acting as intracellular second messenger, since both NO and ANP increase cGMP levels in the heart [32–34]. The prominent mediator of cGMP effects in the heart is cGKI [4, 35–37], being the main isoform of cGK in this tissue [38, 39]. Drugs interfering with cGMP degradation, like the PDE5 inhibitor sildenafil, increase cGKI activity in the heart [40]. However, the molecular mechanisms of cardiovascular NO/NP signalling are not well understood, since the heart consists of various cell types, i.e., cardiomyocytes, smooth muscle cells, fibroblasts, macrophages, and probably blood cells like thrombocytes, which all may differentially contribute to the cardiovascular effects of NO and NP.

2.1 cGKI and Cardiac Contractility

Contraction of the heart is determined by the intrinsic contractile property of the cardiomyocytes. Cardiomyocytes express cGKI, most probably cGKIα, which has been shown by Western blot analysis [41, 42] and immunocytochemistry [43], albeit at about tenfold lower levels as compared to vascular smooth muscle [38, 43, 44]. Several signalling pathways that increase cGMP levels in cardiomyocytes have been proposed to modulate cardiac contractility via cGKI, e.g., the ACh, ANP, and NO signalling pathway.

The most prominent pathway is the negative inotropic effect of ACh on cardiac contractility after β-adrenergic stimulation [45]. Muscarinic stimulation increases cGMP levels in cardiac tissue [46] and cardiomyocytes [47]. However, analysis of conventional and cardiomyocyte-specific cGKI knockout mice revealed that cGKI is not involved in this pathway [39].

Another pathway is the production of NPs (ANP, BNP, and CNP) by the heart which is commonly used in the diagnosis of heart failure [48]. NPs act via activation of a transmembrane guanylyl cyclase-coupled receptor which increases cGMP levels in heart and cardiomyocytes [34, 42, 49, 50]. However, several studies report no effects of ANP on contractility of isolated heart muscle, even at concentrations that were effective for vasorelaxation [42, 51–53]. Only in isolated cardiomyocytes, ANP decreased [54], whereas CNP increased contractility [55], the latter being more pronounced in cardiomyocytes overexpressing cGKI [56]. The dual effects of NPs on cardiac contractility might be due to differences in the amount and duration of cGMP synthesized and subcellular localization of the ANP/cGMP/cGKI versus CNP/cGMP/cGKI signalling pathway [57, 58].

Similar to the vascular system, signalling for NO via cGMP/cGKI has been proposed to regulate cardiac contractility but this signalling pathway has been discussed controversially [22, 59]. The main reasons for this debate are the findings that NO can act via cGMP-dependent and independent mechanisms [60] and that cardiac myoglobin represents a scavenger for NO [32, 61, 62]. In addition, although NO increases cGMP levels in the heart [63], NO-mediated increases in cGMP levels in cardiomyocytes were reported inconsistently [49, 64]. This difference seems to be due to the concentrations used since NO and NO donors show positive and negative inotropic effects which are dependent on concentration [65]. The combined analysis of conventional and cardiomyocyte-specific cGKI knockout mice demonstrated that cGMP/cGKI contributes to the negative inotropic effect of high concentrations of NO in the juvenile as well as in the adult murine heart [39]. However, although eNOS has been suggested to couple muscarinic receptor activation to cGMP-dependent control of $Ca_V1.2$ currents in cardiac myocytes [66, 67], the NO/cGMP/cGKI pathway does not appear to be involved in the negative inotropic action of acetylcholine [39, 68, 69].

The mainly proposed mechanism of cGMP/cGKI signalling in cardiomyocytes is the modulation of cardiac Ca^{2+} channel activity. This view comes from the finding that the cardiac $Ca_V1.2$ channel is phosphorylated by cGKI in vitro [70]. However, a protein identified in vitro as a cGKI substrate may also be a cAK substrate in vivo [71]. Electrophysiological recordings of cardiac $Ca_V1.2$ channel currents revealed inhibitory and stimulatory effects of cGMP/cGKI [44, 72, 73]. Cardiomyocyte-directed overexpression of cGKIα augmented cGMP inhibition but not muscarinic inhibition of L-type Ca^{2+} channel activity [74]. Thus, the dual effects of cGMP/cGKI on $Ca_V1.2$ might represent differences in species and tissue [75] and subcellular localization of cGKI and associated phosphodiesterases [76, 77]. In addition, cGMP in cardiomyocytes may readily reach levels that would activate cAMP-signalling pathways since cAK can also be activated by cGMP under physiological conditions [78]. Further, it cannot be excluded that other or yet unidentified alternative signals mediate in concert with cGMP/cGKI the cardiac effects by either NO or CNP. For example, the cGMP-/cGKI-stimulated phosphorylation of phospholamban seems to be involved in the effects of CNP on cardiac contractility [56].

Recently, evidence has been presented that cGMP/cGKI signalling is involved in regulating passive myocardial contractility by acting directly on the contractile machinery located in the cardiac sarcomere. Especially, titin, the third myofilament of the cardiac sarcomere, has been found to be a substrate for cGMP-/cGKI-mediated phosphorylation [79]. Titin exhibits a springlike function and is the principal determinant of the passive length–tension curve, i.e., stiffness, of the cardiomyocytes [80, 81] as well as of the length-dependent activation of cardiac muscle underlying the Frank–Starling mechanism, which describes an increase in cardiac work output with increased diastolic filling [82, 83]. Skinned cardiomyocytes of patients suffering from heart failure showed an elevated resting tension (stiffness) that can be reduced by cAK and cGKI application indicating that diastolic stiffness of failing myocardium is associated with hypophosphorylation of titin [84]. In line with this view, patients suffering from heart failure with preserved ejection fraction displayed lower myocardial cGMP levels and cGKI activity, concomitant with cardiomyocyte stiffness that was corrected by in vitro cGKI administration [85]. Furthermore, in aged dogs, treatment with sildenafil and/or BNP increased plasma cGMP levels, accompanied by a reduction in cardiomyocyte stiffness and increased titin phosphorylation [86]. However, since both cAK and cGKI are able to phosphorylate titin in the same region (N2B) [81, 87, 88] and, in the latter study, sildenafil and BNP increased also plasma cAMP levels, it is not clear which of the kinases is physiologically dominant in these conditions. Nevertheless, these finding suggest that therapies activating myocardial cGMP/cGKI signalling may provide benefit in the treatment of a reduced diastolic compliance.

2.2 cGKI and Cardiac Remodelling

Heart failure is associated with a complex mixture of structural, functional, and biologic remodelling processes which account for its progressive nature [89]. Several models have been combined to describe heart failure as a cascade of independent and cooperating mechanisms [90]. Such mechanisms include those that regulate cardiomyocyte hypertrophy, calcium homoeostasis, energetics, and cell survival, and processes that take place outside the cardiomyocytes, e.g., in the myocardial vasculature and extracellular matrix [91].

Evidence exists that cGMP/cGKI signalling attenuates these mechanisms suggesting that cGMP/cGKI signalling plays a "brake"-like role, especially in the cardiac hypertrophy remodelling process. This view comes from the findings that signalling pathways that increase cardiac cGMP levels, i.e., the ANP and NO signalling pathways or the reduction of cGMP breakdown by phosphodiesterase inhibition, modulate cardiac hypertrophy. For example, mice lacking ANP or the ANP receptor, GC-A, develop pressure-independent cardiac hypertrophy [92–96]. Mice lacking eNOS and/or nNOS also develop cardiac hypertrophy and dysfunction [97–99] whereas mice with cardiomyocyte-specific overexpression of eNOS showed an attenuated hypertrophy, at least after infarction [100]. In addition, the hypertrophic response to $G\alpha q$ agonists, including angiotensin, endothelin, and the α_1-adrenergic agonist phenylephrine, is suppressed by ANP, NO, or cGMP in cultured cardiomyocytes [50, 101–103]. Further, inhibition of cGMP breakdown by administration of a PDE inhibitor, sildenafil, suppresses the development of cardiac hypertrophy in response to pressure overload [40]. However, in the latter study, although the antihypertrophic effect of sildenafil was linked to an increase in cardiac cGKI activity, myocardial cGMP levels were decreased. Thus, it is unclear whether or not the antihypertrophic effect of cGMP is mediated by cGKI in cardiomyocytes since neither global nor cardiomyocyte-specific ablation of cGKI affected the development of cardiac hypertrophy under basal conditions or in response to pressure overload [43]. Therefore, a causal relationship between cGMP, cGKI, and the antihypertrophic action of cGMP-increasing signalling pathways is not absolutely clear.

It has been suggested that cGKI modulates cardiac remodelling through the control of pathological Ca^{2+} signalling pathways, namely, by attenuation of Ca^{2+} entry mechanisms and downstream signalling cascades that have been shown to be involved in cardiac remodelling. The Ca^{2+} entry mechanisms include the activity of TRPC channels and NHE [104, 105]. Downstream cascades include the activation of the calcineurin/NFAT, the CaMKII, and the AKT/GSK3β pathway [106–108]. TRPC6 channel activation and associated calcineurin/NFAT activation signalling was inhibited by cGKI-dependent channel phosphorylation [109]. cGKI-dependent inhibition of NHE activity by elevations in $[cGMP]_i$ has been suggested to occur indirectly via a cGKI-dependent activation of protein phosphatase 1, which then dephosphorylates critical residues in the C-terminal tail of NHE1 [110, 111]. Adenoviral overexpression of cGKI and elevations in $[cGMP]_i$ inhibit the calcineurin–NFAT pathway and attenuate cardiomyocyte hypertrophy in vitro [112, 113]. However, after knockdown of functional calcineurin, potent cardiac antihypertrophic and functional effects of cGMP/cGKI activation are still observed which are supposed to be related to CaMKII signalling [114]. Treatment of hearts and cardiomyocytes by sildenafil increases cGKI activity which enhances phosphorylation of ERK1/2, and glycogen synthase kinase 3 leading to cardioprotection [115]. Recent studies suggest that RGS 2 mediates cardiac remodelling in response to pressure overload and antihypertrophic effects of PDE5 inhibition and cardiac ANP in mice [116, 117].

The cGMP/cGKI pathway has also been implicated in cardiac protection against apoptosis and cell death-inducing stress, such as ischemic injury [118, 119]. NO, BNP, and 8-Br-cGMP all ameliorate cell death in cultured cardiomyocytes exposed to ischemia-reoxygenation [120]. Mice lacking cGKI develop a larger infarct than controls after ischemia-reperfusion [121]. The underlying mechanisms are complex involving mitochondrial K_{ATP} channel regulation by cGKI [122] and stress responsive signalling cascades. For example, stimulation with NO induced opening of mitochondrial K_{ATP} channels by the cGMP/cGKI pathway [123].

In summary, it is widely accepted that cGMP/cGKI signalling plays a "brake"-like role in the cardiac remodelling processes. However, it is unclear whether this role is solely related to cGMP/cGKI signalling in cardiomyocytes or whether other cardiac cell types are also involved in these processes. For example, activation of cGMP/cGKI signalling by ANP disrupts downstream events mediated by TGF-β in cardiac fibroblasts which play a major role in cardiac remodelling under stress conditions [124].

2.3 cGKI, Smooth Muscle Relaxation, and Blood Pressure

Smooth muscle tone is controlled by a balance between the cellular signalling pathways that mediate vasoconstriction and vasodilation [125]. Vasoconstriction is associated with increases in intracellular Ca^{2+}, activation of myosin light-chain kinase (MLCK), and increases in the phosphorylation of the regulatory myosin light chains leading to enhanced actin-myosin cross-bridge cycling. Vasodilatation is related to decrease in intracellular $[Ca^{2+}]$, activation of myosin light-chain phosphatase (MLCP), and decreases in the phosphorylation of the regulatory myosin light chains leading to less actin-myosin cross-bridge cycling. Several signalling pathways exist that modulate Ca^{2+} mobilization and Ca^{2+} sensitivity of the contractile machinery that secondarily regulates the contractile status of vascular smooth muscle cells (VSMC). Among these pathways, the cGMP/cGKI system is considered as the main cascade that mediates vasodilation under physiological conditions. Evidence for this view comes from mice being deficient for cGKI (cGKI$^{-/-}$ mice). cGMP or other cGMP-elevating agents like NO did not relax several smooth muscles from these mice, e.g., aorta, cremaster arterioles, arteria tibialis, fundus muscle, and intestinal smooth muscle, that were contracted either by hormone stimulation or by depolarization [126–131].

Smooth muscle cells express both cGKIα and cGKIβ [130, 132–135] albeit at different concentrations. For example, the dominant form of cGKI in lung is cGKIα whereas that in aorta is cGKIβ [134]. Both isoforms are supposed to mediate different intracellular mechanisms of relaxation (see below), although reconstitution experiments with cGKI-deficient mice indicated that each isoform can functionally substitute each other [130].

At least five distinct mechanisms are currently thought to be involved in the vasodilator effect of cGMP/cGKI: (1) the modulation of receptor signalling through modulating the activity of G proteins and their associated partners; (2) the decrease in cytosolic [Ca^{2+}] that can be achieved through inhibition of Ca^{2+} release from the SR, decreased extracellular Ca^{2+} influx, activation of Ca^{2+} uptake by the SR, and/or increased intracellular Ca^{2+} efflux; (3) the hyperpolarization of the smooth muscle cell membrane potential through activation of outward potassium channels leading to inactivation of voltage-dependent Ca^{2+} channels; (4) the reduction in the sensitivity of the contractile machinery by decreasing the Ca^{2+} sensitivity of myosin light-chain phosphorylation due to a decrease of the MLCK activity and/or to an increase of the MLCP activity; and (5) the reduction in the sensitivity of the contractile machinery by uncoupling contraction from myosin light-chain phosphorylation via a thin-filament regulatory process [136]. All these mechanisms seem to be expressed in a given type of smooth muscle even though the contribution of each of these mechanisms depends on the smooth muscle type, species, and contractile stimulus and is determined by the relative expression and/or co-localization of the different proteins [137]. For example, cGMP/cGKI signalling differentially relaxed colon and jejunum smooth muscle, i.e., with and without changing intracellular [Ca^{2+}], respectively [126]. Further, ACh-induced relaxation was impaired in aorta but not in cremaster arterioles from cGKI knockout mice [127, 128]. Another example is that cGMP/cGKI signalling via the regulation of BK_{Ca} channels can largely differ in large conduit arteries (aorta) compared to small arteries (mesenteric artery) [138].

2.3.1 Receptor Signalling

A number of contractile agonists, e.g., adrenergic hormones and prostaglandins, activate Gαq-dependent phospholipase C (PLC) through binding to their respective receptors leading to the generation of the second messengers IP_3 and DAG. The agonist-induced activation of phospholipase C and the consequent IP_3 generation in smooth muscle were shown to be inhibited by cGMP and cAMP, presumably through activation of cGKI and cAK, respectively [139]. It was reported that cGKI can directly phosphorylate smooth muscle PLC-β2 and PLC-β3 in vitro and in vivo and thus inhibits their activity [140]. However, these results have not been repeated so far. In addition, the interaction between phospholipase C and guanine nucleotide regulatory proteins is attenuated by cGMP/cGKI [141]. These proteins, known as regulators of G-protein signalling (RGS), regulate the strength and duration of Gαq signalling [142]. Two members of the RGS family, RGS2 and RGS4, are phosphorylated by cGKI [143–145]. In mice with a deleted RGS2 protein, cGMP analogues fail to inhibit hormone-triggered Ca^{2+} transients in smooth muscle cells even though cGKIα and cGKIβ are expressed and activated normally [146]. Interestingly, the cGKIα isoform selectively binds, phosphorylates, and activates RGS2 which terminates signalling by Gq-coupled receptors for contractile agonists [143]. This finding may explain why cGMP selectively reduced noradrenaline-triggered Ca^{2+} signals in cultured smooth muscle cells transfected with cGKIα [133]. A direct in vivo phosphorylation of thromboxane receptor by cyclic GMP-dependent protein kinase has been shown for thrombocytes [147]. In smooth muscle, the thromboxane receptor mainly signals through $G\alpha_{12/13}$ and induces Ca^{2+}-independent relaxation [148, 149].

2.3.2 Decrease in Cytosolic [Ca^{2+}]

A number of studies have established that cGMP/cGKI signalling decreases agonist- and depolarization-induced Ca^{2+} signals in smooth muscle cells [126, 128, 131, 133]. The Ca^{2+} transients are originated by Ca^{2+} release and Ca^{2+} influx. Inhibition of agonist-mediated Ca^{2+} release is thought to be mediated either by direct phosphorylation of the IP_3R in response to cGMP-generating vasodilators or cell-penetrating cGMP analogues [150, 151] or by phosphorylation of the IP_3R associated substrate for cGKI (IRAG) which has been identified in a complex with the smooth muscle IP_3 receptor type 1 and cGKIβ [152]. Phosphorylation of IRAG by cGKIβ inhibits IP_3-induced Ca^{2+} release from intracellular stores in transfected COS cells and in smooth muscle cells [131, 152, 153]. In mice expressing a mutated IRAG protein which does not interact with the IP_3 receptor [131] or mice being deficient in IRAG [154], cGMP was unable to fully suppress hormone-induced increases in $[Ca^{2+}]_i$ and contractility, indicating that the cGKIβ/IRAG/IP_3 receptor pathway inhibits hormone receptor-triggered intracellular Ca^{2+} release and contraction in vivo. Thus, the cGKI isoforms act via different signalling pathways, e.g., cGKIα inhibits preferentially RGS2, terminates second messenger generation, and activates Ca^{2+}-independent MLC phosphatase, whereas cGKIβ inhibits preferentially IP_3R-mediated Ca^{2+} release.

Smooth muscle Ca^{2+} influx mechanisms include the activation of L-type Ca^{2+} channels, probably of the $Ca_V1.2b$ type [155], and store- or receptor-activated channels, which are part of the TRP channel family. Inhibition of Ca_V channels by cGMP/cGKI lowers calcium influx and cell excitability and therefore causes vasodilation [156, 157]. Indeed, cGKI has been shown to phosphorylate cardiac $Ca_V1.2$ and the corresponding β subunit in vitro [70]. Inhibition of transient receptor potential canonical channels (TRPC), namely, TRPC1 and TRPC3, by cGKI contributes also to NO-mediated vasorelaxation [158]. Further, cGKI phosphorylates the TRPC6 channel at Thr-70 and Ser-322 which inactivates the associated inward calcium current [109] that is thought to be an essential component of noradrenaline-mediated Ca^{2+} influx in smooth muscle [159].

Besides controlling Ca^{2+} release and Ca^{2+} influx, cGKI has been also connected to Ca^{2+} efflux mechanisms mediated by sarcoplasmic/endoplasmic reticulum Ca^{2+} ATPase (SERCA), plasmalemmal Ca^{2+} ATPase (PMCA), and NCX. The cGKI may activate SERCA by phosphorylation of the SERCA regulator phospholamban (PLB) [160–162] which promotes Ca^{2+} reuptake into the sarcoplasmic/endoplasmic reticulum. Phosphorylation of PLB by cGKI at the Ser-16 residue relieves its inhibitory action on SERCA, thus increasing ATPase activity and the rate of Ca^{2+} uptake by the SR [163]. Further, results obtained in cultured rat aortic smooth muscle cells suggest a possible mechanism of action for cGMP in mediating decreases in cytosolic $[Ca^{2+}]$ through activation of a PMCA and the subsequent removal of Ca^{2+} from the cell [164, 165]. cGKI, but not cAK, stimulates the PMCA of pig aorta smooth muscle indirectly by increasing phosphatidylinositol phosphate, probably via the phosphorylation of a phosphatidylinositol kinase that co-purified with the Ca^{2+} ATPase [166]. Furthermore, a cGMP-dependent phosphorylation enhancement of the affinity for Ca^{2+} and the activity of the PMCA pump has been observed in the solubilized plasmalemmal Ca^{2+} pump purified by calmodulin affinity chromatography [167]. ANP or 8-bromo-cGMP enhanced Na^+/Ca^{2+} exchange in primary cultures of VSM cells prepared from rat aorta [168].

2.3.3 Hyperpolarization

Membrane depolarization and elevation of $[Ca^{2+}]_i$ increase the activity of BK_{Ca} channels. These channels are involved in the relaxant response induced by many endogenous vasodilators (e.g., adenosine, prostacyclin, NO) and play an important role as a negative feedback mechanism to limit membrane depolarization and, hence, vasoconstriction. BK_{Ca} activation by cGMP/PKG was reported in several blood vessels [129, 169, 170], and functional studies confirm that such activation can contribute to vasorelaxation. The phosphorylation of the alpha subunit of BK_{Ca} channels by cGKI and its activity modulation was shown in *Xenopus* oocytes co-expressing both proteins [171]. Controversial evidence is given by a report that NO activates BK_{Ca} channels directly [172]. BK_{Ca} channels are activated by ANP either through direct phosphorylation by cGKI [171, 173] or through phosphorylation of protein phosphatase 2A [174, 175].

2.3.4 Ca^{2+} Sensitization

The most recent work suggests that cGKI activates MLCP, thereby inhibiting MLC20 phosphorylation and contraction [149, 176–178]. The mechanism of cGMP-dependent activation of MLCP may involve direct phosphorylation of the regulatory subunit (MYPT1) of MLCP at Ser-695 [179]. Several studies demonstrated that phosphorylation of Ser-695 by cGKI prevents phosphorylation of MYPT1 at Thr-696 by several kinases and therefore inhibition of MLCP [180]. Thus, phosphorylation of MYPT1 by cGKI increases MLCP activity resulting in the decrease in MLC20 phosphorylation and smooth muscle relaxation. It is worth mentioning that Thr-696 phosphorylation decreased by only 50 % with 8-Br-cGMP stimulation, whereas the tension of phenylephrine-pre-contracted arteries and regulatory light-chain phosphorylation fell to baseline, presumably indicating that Ser-695 phosphorylation is not the sole explanation of cGKI-induced Ca^{2+} desensitization. In fact, activation of cGKI can mediate other processes resulting in changes in contractility, such as cGKI phosphorylation of telokin [181] or other similar mediator proteins. A further candidate is the calponin homology-associated smooth muscle protein (CHASM), which activates MLCP activity, independently of changes in MYPT1 phosphorylation, through an unknown mechanism [182]. On the other hand, using yeast two-hybrid system screening, it was also shown that cGKI is targeted to the smooth muscle cell contractile apparatus by a leucine zipper interaction with the myosin-binding subunit (MBS) of MLCP. MBS is the regulatory subunit of MLCP that confers the specificity for MLC and is the site of regulation by Rho-kinase. Uncoupling of the cGKI–MBS interaction prevents cGMP-dependent dephosphorylation of MLC, demonstrating that this interaction is essential to the regulation of vascular smooth muscle cell tone [176].

cGMP-induced inhibition of Ca^{2+} sensitization can be directly mediated by RhoA/Rho-kinase signalling [183]. RhoA is a member of the Ras family of small GTP-binding proteins. RhoA is translocated to the plasma membrane and activated by $G\alpha_{12/13}$. Rho-kinase (ROCK) can be partially translocated from the cytosol to the membrane and activated by active RhoA. ROCK is believed to be the most important modulator of Ca^{2+} sensitivity in smooth muscle [184]. Myosin light-chain phosphatase activity is controlled by ROCK. Phosphorylation of the myosin-binding subunit by

ROCKs leads to the inhibition of MLCP activity, which results in a decreased MLC dephosphorylation and hence to an increased Ca^{2+} sensitivity and smooth muscle contractility [185]. In addition, ROCK targets other substrates that are important for smooth muscle contraction such as CPI-17, calponin, or MLC [184]. It was shown that cGKI-mediated phosphorylation of RhoA at Ser-188 prevents its binding to ROCK, and this phosphorylation inhibits RhoA-induced Ca^{2+} sensitization and actin cytoskeleton organization [183, 186].

2.3.5 Uncoupling

There are data indicating that increases in intracellular cGMP concentrations can lead to relaxation of tonic smooth muscle by mechanisms independent of changes in intracellular [Ca^{2+}] or of the state of MLC20 phosphorylation. Inhibition of contraction was associated with increases in the small 20-kDa heat shock protein (HSP20) phosphorylation [187]. In fact, there are several studies indicating that relaxation induced by cGMP is associated with increases in the phosphorylation of the small heat shock-related protein HSP20 at Ser-16 by cGKI [188, 189].

The presented results show that cGKI inhibits receptor-induced vascular smooth muscle contraction by multiple mechanisms including the cGKIβ/IRAG, the cGKIα/RGS2, and the cGKIα/MLCP signalling pathway. The mechanism by which cGKI interferes with depolarization-induced contraction remains controversial. cGKI reduced cardiac Ca^{2+} influx [44], most likely by phosphorylation of the calcium channel β_{2a} subunit [70], and may contribute thereby to vascular relaxation, because Ca^{2+} influx via the L-type $Ca_V1.2$ calcium channel is essential for sustained contraction of various smooth muscles [190–192]. It is appreciated that the individual contribution of each individual pathway to cGKI-mediated relaxation presumably varies with the type, the physiological function, and the pathological state of the smooth muscle.

2.3.6 Blood Pressure

The effect of cGMP/cGKI signalling on blood pressure is quite complex. Global cGKI deficiency results in a slight hypertension in young animals, whereas in adult mice, the basal blood pressure does not differ between wild-type and cGKI knockout animals [128]. Interestingly, global deletion of eNOS in mice leads to hypertension [193], whereas the smooth muscle-specific deletion of the ANP receptor GC-A suggests that the vascular GC-A is dispensable in the chronic but crucial in the acute regulation of blood pressure by ANP [194]. Deficiency in single elements of the smooth muscle contractile machinery, like inactivation of the BK_{Ca} gene [195] or the phospholamban gene [162], affected the blood pressure only marginally. One reason for this outcome may be that compensatory mechanisms substitute the respective deficient pathways. Results have been presented that sometimes mice carrying a disrupted gene appear phenotypically normal, suggesting the presence of several redundant pathways [196]. Evidence for such "backup" systems comes, for example, from mice that lack the IP receptor for prostacyclin. These mice are normotensive, but more susceptible to vasoconstrictor stimuli than their wild-type litter mates [197]. In eNOS-deficient mice, the neuronal isoform of NO synthase may substitute for eNOS in mediating responses of cerebral arterioles to acetylcholine [198]. Individual inactivation of the soluble guanylyl cyclase (sGC) has moderate to minimal effects on blood pressure and intestinal activity, presumably since the other sGC compensates the deletion of one sGC enzyme [199]. However, deletion of the common β subunit of sGC that results in inactivation of both sGC isozymes results in intestinal dysmotility, a pronounced increase in blood pressure, and a defect in NO-dependent modulation of aortic smooth muscle contraction and platelet aggregation [199]. These studies suggest that ANP- and NO-dependent activation of the cGMP/cGKI system mediates vasorelaxation and is essential to the regulation of basal, chronic blood pressure.

Alternative models of blood pressure regulation by the redox state of the VSMC have been proposed. The two cGKI isozymes share eight cysteines outside of the amino terminus. The cGKIα contains additional cysteines at aa 43 and aa 118. Oxidation of the bovine cGKIα enzyme by Cu^{2+} activates the enzyme through formation of two disulfide bonds between Cys-118 and Cys-196 and Cys-313 and Cys-519 [200, 201]. The biological significance of this activation is not obvious, because oxidation of the enzyme by transition metals is unlikely to occur in vivo. The cGKIα forms a disulfide bridge at Cys-43 linking its two subunits in cells exposed to exogenous hydrogen peroxide [202]. Cys-43 is only present in the Iα isoform. Oxidation of Cys-43 activates the kinase independent of cGMP and has been suggested as an alternative activation mechanism of cGKI in vascular smooth muscle under oxidative stress [203] leading to relaxation.

2.3.7 cGKI and Vascular Remodelling

In addition to smooth muscle relaxation, NO/cGMP signalling has been reported to be involved in the development of vasculoproliferative disorders, such as restenosis and atherosclerosis. The analysis of transgenic mice showed that NO can both promote [204–209] and inhibit [210–215] pathological vascular remodelling. These findings might explain why NO-generating drugs have not been reported to limit the progression of atherosclerosis or restenosis in humans [216]. The opposing effects of NO on vascular remodelling might depend on the spatiotemporal profile of its production (cellular source, time, and quantity) and are probably mediated by different cellular and molecular mechanisms [217].

A key process in vascular remodelling is the phenotypic modulation of VSMCs from contractile to synthetic phenotype [218–221]. High concentrations of NO inhibit VSMC growth in vitro, even in the absence of cGKI [222], whereas the "growth-promoting" effect of cGMP was absent in primary VSMCs isolated from cGKI-deficient mice [223]. Recent evidence shows that the "growth-promoting" effect of cGKI in primary smooth muscle cells is caused by an increased adhesion of the wild-type VSMCs to the substrate during the establishment of the cell culture [223]. These investigations demonstrated also that cGMP slightly inhibited the growth of cGKI-positive but not cGKI-deficient VSMCs, if the cells were passaged several times.

Most likely, the cGKI-dependent regulation of "cell growth" is mediated by affecting gene expression [224–226]. VSMCs can reversibly change their phenotype from a differentiated, "contractile" phenotype with high levels of smooth muscle (SM)-specific gene expression to a dedifferentiated, "synthetic" phenotype with reduced levels of SM-specific gene expression [220, 227]. This phenotypic switching plays an important role in the development of vascular diseases: in acutely injured blood vessels, e.g., after balloon angioplasty, VSMCs proliferate and migrate from the medial layer of the vessel wall to a "neointimal" layer. The majority of SM-like cells found in atherosclerotic plaques appear to represent dedifferentiated VSMCs originating from the medial layer [220, 227]. The regulation of VSMC phenotypic switching is complex and mediated by multiple factors, but it is clear that dedifferentiated VSMCs are a major cell type responsible for the generation of vascular lesions [228]. Primary VSMCs cultured in vitro undergo changes similar to those observed in neointimal smooth muscle-like cells, including phenotypic dedifferentiation, decreased expression of SM-specific genes, and loss of cGKI [227, 229, 230]. When these dedifferentiated, cGKI-deficient VSMCs are transfected with expression vectors encoding cGKI to restore physiologic levels of cGKI activity, the cells develop a more contractile phenotype; increase expression of SM-specific genes such as SM-myosin heavy chain (SM-MHC), SM-α-actin (SMA), and calponin; and reduce production of extracellular matrix proteins and growth-related genes. In line with this view, cGKI maintains the differentiated contractile phenotype in pulmonary VSMC in normoxia conditions, whereas acute hypoxia decreases cGKI activity leading to proliferation [231]. These results suggest that cGKI may contribute to the switch from synthetic/proliferative to contractile VSMCs.

Another process involved in vascular remodelling is the expression of matrix-degrading metalloproteinases (MMPs) by VSMC [232]. Two major forms of the MMP family, MMP-2 and MMP-9, are found in atherosclerotic lesions, suggesting a role for these two proteases in vascular wall remodelling [233]. It has been suggested that cGKI inhibits SMC proliferation and migration by suppressing MMP-2 expression and increasing the expression of an MMP inhibitor, namely, TIMP-2 [234].

Most SM-specific promoters, including the SM-MHC, SMA, and calponin promoter, contain multiple CArG (CC(AT)6GG) elements recognized by the ubiquitously expressed serum response factor (SRF) [235]. Expression of these genes depends on the interaction of SRF with multiple cofactors including the cysteine-rich LIM-only proteins CRP1 and CRP2/smLIM (smooth muscle LIM protein) [235]. Recently, a new member of the CRP family was identified through a yeast two-hybrid screen that used cGKIβ as bait [236]. This protein was independently cloned from a rat brain and human intestinal cDNA library [237, 238] and was named "CRP2" and recently been renamed to "CRP4" [239]. CRP4 is phosphorylated by cGKI in vitro and in vivo [239, 240]. cGKI was required for maintaining SM-specific gene expression in several differentiated smooth muscle cell lines, and enhanced SRF- and GATA6-induced differentiation of pluripotent embryonal cells into smooth muscle cells [239]. CRP4 was associated with SM-specific promoters and mediated positive transcriptional effects of cGKI on SM-specific gene expression. These results establish for the first time a plausible link for the effect of cGKI on gene expression.

The above-discussed results led to the suggestion that cGMP/cGKI signalling might also contribute to the phenotypic modulation and growth of VSMCs during pathological vascular remodelling in vivo [239, 241]. To test this hypothesis, the consequences of postnatal smooth muscle-specific cGKI deletion were studied in wild-type and apoE-deficient mice. Smooth muscle-specific deletion of cGKI did not affect restenosis in response to carotid ligation in normo-lipidemic wild-type mice nor in apoE-deficient mice [242]. Continuous elevation of cGMP by sildenafil had no effect on neointima proliferation after carotid ligation. Furthermore, VSMC-specific deletion of cGKI had no effect on vascular remodelling after wire-induced removal of the endothelial cell layer [242]. These studies are consistent with the conclusion that the role of the smooth muscle cGMP/cGKI pathway is negligible in vascular remodelling in vivo under normo-lipidemic condition. An identical conclusion has been published by Sinnaeve and coauthors [241]. Gene transfer of the wild-type cGKIβ isoform had no effect on restenosis in vivo. These results contrast those reported above and suggest that cGMP/cGKI signalling is not critically involved in vascular restenosis in vivo. Further experiments with different cGKI-deficient mouse models and cultured VSMCs are required to decipher the exact effect of cGMP/cGKI on vascular smooth muscle differentiation/dedifferentiation.

3 The Non-cardiovascular Effects of cGKs

3.1 Neuronal Effects of cGKI and cGKII

cGKI and cGKII, signal transducer of the NO/cGMP cascade, are widely distributed throughout the mammalian brain including regions thought to be related to emotional behavior, e.g., cortex, hippocampus, the amygdala, midbrain raphe nuclei, the nucleus tractus solitarius, and cerebellum [243–246]. Both kinase are apparently involved in the control of anxiety-like behavior and fear memory [243, 247]. Wild-type and cGKII-deficient mice showed marked differences in their ethanol consumption in a free-choice test without general differences in place preference [243]. When given free access to two bottles filled with water and ethanol, respectively, $cGKII^{-/-}$ mice consumed significantly more ethanol over the course of 6 days, and this effect, again, was observed for two different genetic backgrounds [243].

Many behavioral responses of animals (e.g., feeding, drinking, and locomotor activity) and the underlying neurohumoral activities are organized in circadian rhythms. Both cGKI and cGKII are expressed in subpopulations of SCN neurons [244, 248, 249]. Mice with an inactivated cGKII gene displayed a virtually normal spontaneous circadian rhythm and unaltered expression patterns of the clock genes mPer1 and mPer2, supporting the view that the circadian clock is still intact in the absence of cGKII [248]. Disruption of the cGKII gene, however, resulted in an impaired light-induced entrainment of the circadian clock within a definite time window. Compared to wild-type mice, the phase delay of the clock induced by a light pulse applied at circadian time (CT) corresponding to early night, CT14, was reduced by nearly 50 %. The phase advance of the clock induced by a light pulse applied at CT22, corresponding to late night, was not affected in cGKII knockout mice [248]. At the molecular level, cGKII knockout mice displayed marked differences to wild-type mice in light induction of two clock genes during the early period of the night: induction of mPer1 was enhanced and induction of mPer2 was strongly reduced. The absence of cGKII did not affect light induction of these genes during late night. In contrast to the results obtained with cGKII-deficient mice, pharmacological studies have placed the cGKs into the signalling pathway for phase advances [250, 251], and a recent in vitro study reported an essential role of cGKII in the progression of the circadian cycle [252]. The reason for these discrepancies are not clear, but may be explained in part by effects of inhibitors not related to cGKII inhibition as well as by differences in the experimental systems used, e.g., slice cultures versus whole animals. Furthermore, a function of cGKI, which is also expressed in the SCN [249], cannot be excluded. Conditional knockout mice lacking cGKI in the SCN did not reveal a significant effect on circadian rhythmicity but had an altered wake/sleep rhythm [253].

cGKI is highly expressed in the cerebellar Purkinje nerve cells [254–256] and hippocampus [245, 257, 258]. Inactivation of the cGKI gene in Purkinje cells reduced LTP and did not affect general behavior and motor performance. The mutants exhibited an impaired adaptation of the vestibulo-ocular reflex (VOR) [256]. These results indicate that cGKI in Purkinje cells is dispensable for general motor coordination, but that it is required for cerebellar LTD and specific forms of motor learning, namely, the adaptation of the VOR.

Studies with cultured hippocampal neurons suggested that cGKI is involved in LTP mediating the retrograde effects of NO on the release of glutamate [259]. Studies using global cGKI$^{-/-}$ mice were unable to confirm these results. Even conventional double knockout mice (cGKI$^{-/-}$ x cGKII$^{-/-}$) showed normal LTP [257]. A potential caveat to these results was that due to the severe phenotype, only very young mice (age 2–4 weeks) were used. This problem was circumvented by the use of mice with a hippocampus-specific deletion of cGKI [258]. Unlike conventional cGKI knockout mice, hippocampus-specific cGKI$^{-/-}$ mice lack the gastrointestinal and cardiovascular phenotype and have a normal life expectancy. Hippocampal LTP after repetitive episodes of theta burst stimulation was impaired in adult (12–14 weeks of age) but not in juvenile (3–4 weeks of age) hippocampus-specific cGKI$^{-/-}$ mice. The LTP difference between adult control and hippocampus-specific cGKI$^{-/-}$ mice was abolished by the protein synthesis inhibitor anisomycin, suggestive of a defect in late-phase LTP [258]. Despite their deficit in LTP, adult hippocampus-specific cGKI$^{-/-}$ mutants showed normal performance in a discriminatory water maze and had intact contextual fear conditioning, suggesting that hippocampal cGKI supports an age- and protein synthesis-dependent form of hippocampal LTP.

Previous in vitro studies suggested that cyclic nucleotides modulate growth cone direction [260]. cGKIα is predominantly expressed in sensory axons during developmental stages. Analysis of the trajectories of axons within the spinal cord showed a longitudinal guidance defect of sensory axons within the developing dorsal root entry zone in the absence of cGKI [261]. Further analysis demonstrated that cGMP is generated by the particulate guanylyl cyclase B, NPR2 [262]. These axon guidance defects in cGKI-deficient mice lead to a substantial impairment in nociceptive flexion reflexes. Further investigations supported the notion that dorsal root ganglion cGKI is involved in pain sensation [263]. Deletion of cGKI reduced inflammatory hyperalgesia. Further in-depth studies with neuron-specific deletion of cGKIα supported these findings and showed that cGKIα is a key component of spinal synaptic potentiation and pain hypersensitivity [264]. These effects are most likely mediated by phosphorylation of the neuronal IP_3 receptor I [264].

3.2 Nonneuronal Effects of cGKI and cGKII

3.2.1 Intestine

cGKI is an important regulator of intestinal motility [128], whereas cGKII modulates gastrointestinal secretion of chloride and water [265]. cGKII is located in the secretory epithelium of the small intestine and stimulates chloride and water secretion possibly through phosphorylation of CFTR [266]. cGKII increased Na^+ absorption in the small intestine by inhibition of the Na^+/H^+ exchanger 3 (NHE3) through interaction with the G kinase-anchoring protein NHERF2 [267, 268]. Stimulation of the cGMP signalling cascade by toxins causes diarrhea as the *Escherichia coli* heat-stable toxin (STa) and guanylin activate the GC-C and thereby increase water secretion in the small intestine. Correspondingly, STa did not induce diarrhea in cGKII knockout mice [265]. The motility defect induced by inactivation of cGKI is severe [128], but is corrected by the selective expression of cGKIα or cGKIβ in all smooth muscles [130]. The conventional cGKI$^{-/-}$ mice develop a bleeding duodenal ulcer [246]. Ulcer development is also observed in mice that expressed selectively cGKIα or cGKIβ in all smooth muscle cells, indicating that development of a bleeding ulcer is not connected to the motility imbalance in cGKI$^{-/-}$ mice. Bleeding leads to anemia [246] that may be further aggravated by a defect in erythrocyte stability [269]. The duodenal ulcer is caused by the inability of the duodenal secretory epithelium to produce bicarbonate in response to an acid challenge [246]. Surprisingly, acid-induced bicarbonate secretion depends on an intact vagal reflex that requires the presence of cGKI in the nucleus tractus solitarius as shown by the specific inactivation of the neuronal cGKI gene [246].

3.2.2 Kidney and Adrenal Gland

cGMP affects blood pressure not only by directly regulating the vascular smooth muscle tone but also by regulation of the renin and aldosterone secretion [270–272]. Renin secretion is enhanced by NO through cGMP-dependent inhibition of cAMP hydrolysis mediated by PDE 3 [272]. In contrast, ACTH-dependent aldosterone secretion is inhibited by ANP through cGMP-dependent stimulation of cAMP hydrolysis mediated by PDE2 [270, 271]. In addition, it was reported that cGMP analogues reduced renin secretion from isolated kidney or juxtaglomerular cells [273, 274]. Kidney expresses both isozymes of cGK. cGKII is localized together with storage granules in juxtaglomerular cells [275]. Renin secretion from juxtaglomerular cells was increased in cGKII-deficient mice [276], suggesting that cGMP has a dual role in the regulation of renin secretion. cGKII has been detected in rat and murine zona glomerulosa cells of the adrenal gland [277] and might regulate aldosterone secretion. The overexpression of cGKII in rat zona glomerulosa cells enhanced the production of aldosterone [277]. Likewise, a low-salt diet activating the aldosterone system enhanced expression of cGKII [277]. In contrast, cGKII deletion did not alter basal or low-salt-stimulated aldosterone secretion in mice and had only a mild effect on ANP-regulated aldosterone secretion [278]. So far, cGKII deletion has not affected resting blood pressure. Therefore, it is unlikely that the above-reported findings are relevant for the overall blood pressure regulation in resting mice and other animals.

Both isozymes of cGKI are present in arterioles, mesangium, and within the cortical interstitium of the kidney. In contrast to cGKIα, the β isoform was not detected in the juxtaglomerular apparatus and medullary fibroblasts. cGKIα has an antifibrotic effect on renal fibrosis that was induced by unilateral ureter obstruction (UUO) [279]. It is unclear at present if this antifibrotic effect is specific for renal fibroblasts or can be extended to other tissues.

3.2.3 Bone Growth

The endochondral ossification of bones is stimulated by the C-type natriuretic peptide (CNP) [280]. Overexpression of CNP rescued achondroplasia which was induced by a defect in fibroblast growth factor receptor 3 signalling [281]. cGKII knockout mice are dwarfs which develop short bones caused by a defect in the endochondral ossification at the endochondral plate [265, 282]. This defect was not rescued by CNP overexpression, indicating that cGKII is essential for endochondral bone development [283]. Interestingly, the regulation of body size was also impaired in the cGMP kinase-deficient nematode *egl4* [284]. Furthermore, cGKII-deficient rat exhibited an expanded growth plate and impaired bone healing. The observed accumulation of postmitotic but nonhypertrophic cells in these rats might be caused by induction of the transcription factor Sox9 and decreased phosphorylation of GSK3β leading to hypertrophic differentiation of chondrocytes [285–287].

NO stimulates osteoclast motility that is mediated by cGKI and phosphorylation of VASP [288]. Interestingly, attachment of osteoclasts was regulated through cGKIβ-dependent phosphorylation of IRAG [289], a regulator of IP_3 receptor-mediated calcium release [152]. In contrast to osteoclasts, CNP, an activator of particulate guanylyl cyclase, enhances the proliferation of osteoblasts [290]. Indirect evidence supports the notion that these effects are partially mediated by cGKII [291]. Furthermore, cGKI and cGKII appear to be necessary to mediate the antiapoptotic effect of 17β-estradiol in osteocyctes [292]. In general, these results are in line with the hypothesis that the NO–cGMP–cGK pathway mediates anti-osteoporosis effects [293].

3.2.4 Metabolism and Fat Cells

Controversial data have been reported for the physiological significance of the nitric oxide (NO) pathways in pancreatic islets. Both types of constitutive NOS (eNOS, nNOS) isozymes have been identified in islets [294–298]. The role of cGKI for the regulation of glucose homeostasis has been studied using conventional $cGKI^{-/-}$ mice [128] and smooth muscle-specific cGKI rescue mice [130]. The lack of cGKI affected severely the secretion of glucagon and increased the fasting glucose level [299], whereas insulin secretion was unaffected. These results were supplemented by Lutz et al. [300]. These researchers reported that genetic deletion of cGKI in nonneuronal cells resulted in a complex metabolic phenotype including liver inflammation and fasting hyperglycemia. Loss of cGKI in hepatic stellate cells may have affected liver metabolism via a paracrine mechanism that involves enhanced macrophage infiltration and IL-6 signalling.

Brown fat cells contain cGKI [301], although it was reported that they lack this enzyme [300], the finding of which may contribute to the complex metabolic phenotype. Haas and colleagues showed that cGKI is essential for brown fat cell (BAT) differentiation [301]. Interestingly, cGKI interfered with insulin signalling in BAT by interfering with Rho/Rho-kinase activity. These findings point to a possible way by which the number of BAT cells and the uncoupling protein I can be increased in adult mice and man. However, no in vivo data are available for this notion.

Acknowledgements

Research in the author's laboratories was supported by German Research Foundation (DFG) and Fond der Chemischen Industrie.

References

1. Friebe A, Koesling D (2003) Regulation of nitric oxide-sensitive guanylyl cyclase. Circ Res 93(2):96–105
2. Katsuki S, Arnold W, Mittal C, Murad F (1977) Stimulation of guanylate cyclase by sodium nitroprusside, nitroglycerin and nitric oxide in various tissue preparations and comparison to the effects of sodium azide and hydroxylamine. J Cyclic Nucleotide Res 3(1):23–35
3. Garbers DL (1992) Guanylyl cyclase receptors and their endocrine, paracrine, and autocrine ligands. Cell 71(1):1–4
4. Rybalkin SD, Yan C, Bornfeldt KE, Beavo JA (2003) Cyclic GMP phosphodiesterases and regulation of smooth muscle function. Circ Res 93(4):280–291
5. Bender AT, Beavo JA (2006) Cyclic nucleotide phosphodiesterases: molecular regulation to clinical use. Pharmacol Rev 58(3):488–520
6. Biel M, Zong X, Ludwig A, Sautter A, Hofmann F (1999) Structure and function of cyclic nucleotide-gated channels. Rev Physiol Biochem Pharmacol 135:151–171
7. Hofmann F, Biel M, Feil R, Kleppisch T (2004) Mouse models of NO/natriuretic peptide/cGMP kinase signaling. In: Offermanns S (ed) Handbook of experimental pharmacology, vol 159, Transgenic models in pharmacology. Elsevier, Amsterdam, pp 95–130
8. Hofmann F (2005) The biology of cyclic GMP-dependent protein kinases. J Biol Chem 280(1):1–4
9. Pfeifer A, Ruth P, Dostmann W, Sausbier M, Klatt P, Hofmann F (1999) Structure and function of cGMP-dependent protein kinases. Rev Physiol Biochem Pharmacol 135:105–149
10. Feil R, Lohmann SM, de Jonge H, Walter U, Hofmann F (2003) Cyclic GMP-dependent protein kinases and the cardiovascular system: insights from genetically modified mice. Circ Res 93(10):907–916
11. Hess DT, Matsumoto A, Kim SO, Marshall HE, Stamler JS (2005) Protein S-nitrosylation: purview and parameters. Nat Rev Mol Cell Biol 6(2):150–166
12. Burkhardt M, Glazova M, Gambaryan S, Vollkommer T, Butt E, Bader B, Heermeier K, Lincoln TM, Walter U, Palmetshofer A (2000) KT5823 inhibits cGMP-dependent protein kinase activity in vitro but not in intact human platelets and rat mesangial cells. J Biol Chem 275(43):33536–33541
13. Marshall SJ, Senis YA, Auger JM, Feil R, Hofmann F, Salmon G, Peterson JT, Burslem F, Watson SP (2004) GPIb-dependent platelet activation is dependent on Src kinases but not MAP kinase or cGMP-dependent kinase. Blood 103(7):2601–2609
14. Wyatt TA, Pryzwansky KB, Lincoln TM (1991) KT5823 activates human neutrophils and fails

to inhibit cGMP-dependent protein kinase phosphorylation of vimentin. Res Commun Chem Pathol Pharmacol 74(1):3–14

15. Daugirdas JT, Zhou HL, Tamulaitis VV, Nutting CW, Fiscus RR (1991) Effect of H-8, an isoquinolinesulfonamide inhibitor of cyclic nucleotide-dependent protein kinase, on cAMP- and cGMP-mediated vasorelaxation. Blood Vessels 28(5):366–371
16. Valtcheva N, Nestorov P, Beck A, Russwurm M, Hillenbrand M, Weinmeister P, Feil R (2009) The commonly used cGMP-dependent protein kinase type I (cGKI) inhibitor Rp-8-Br-PET-cGMPS can activate cGKI in vitro and in intact cells. J Biol Chem 284(1):556–562. doi:10.1074/jbc.M806161200
17. Hofmann F, Bernhard D, Lukowski R, Weinmeister P (2009) cGMP regulated protein kinases (cGK). Handb Exp Pharmacol 191:137–162
18. Francis SH, Busch JL, Corbin JD, Sibley D (2010) cGMP-dependent protein kinases and cGMP phosphodiesterases in nitric oxide and cGMP action. Pharmacol Rev 62(3):525–563
19. D'Souza SP, Davis M, Baxter GF (2004) Autocrine and paracrine actions of natriuretic peptides in the heart. Pharmacol Ther 101(2):113–129
20. Garbers DL, Dubois SK (1999) The molecular basis of hypertension. Annu Rev Biochem 68:127–155
21. Lloyd-Jones DM, Bloch KD (1996) The vascular biology of nitric oxide and its role in atherogenesis. Annu Rev Med 47:365–375
22. Massion PB, Feron O, Dessy C, Balligand JL (2003) Nitric oxide and cardiac function: ten years after, and continuing. Circ Res 93(5):388–398
23. Ignarro LJ (2002) Nitric oxide as a unique signaling molecule in the vascular system: a historical overview. J Physiol Pharmacol 53(4 Pt 1):503–514
24. Shimokawa H, Tsutsui M (2010) Nitric oxide synthases in the pathogenesis of cardiovascular disease: lessons from genetically modified mice. Pflugers Arch 459(6):959–967. doi:10.1007/s00424-010-0796-2
25. Ziolo MT, Kohr MJ, Wang H (2008) Nitric oxide signaling and the regulation of myocardial function. J Mol Cell Cardiol 45(5):625–632. doi:10.1016/j.yjmcc.2008.07.015
26. Abassi Z, Karram T, Ellaham S, Winaver J, Hoffman A (2004) Implications of the natriuretic peptide system in the pathogenesis of heart failure: diagnostic and therapeutic importance. Pharmacol Ther 102(3):223–241. doi:10.1016/j.pharmthera.2004.04.004
27. Roy LF, Ogilvie RI, Larochelle P, Hamet P, Leenen FH (1989) Cardiac and vascular effects of atrial natriuretic factor and sodium nitroprusside in healthy men. Circulation 79(2):383–392
28. Moncada S, Higgs A (1993) The l-arginine-nitric oxide pathway. N Engl J Med 329(27):2002–2012. doi:10.1056/NEJM199312303292706
29. Matsumoto T, Wada A, Tsutamoto T, Omura T, Yokohama H, Ohnishi M, Nakae I, Takahashi M, Kinoshita M (1999) Vasorelaxing effects of atrial and brain natriuretic peptides on coronary circulation in heart failure. Am J Physiol 276(6 Pt 2):H1935–H1942
30. Winquist RJ, Faison EP, Waldman SA, Schwartz K, Murad F, Rapoport RM (1984) Atrial natriuretic factor elicits an endothelium-independent relaxation and activates particulate guanylate cyclase in vascular smooth muscle. Proc Natl Acad Sci USA 81(23):7661–7664
31. Inagami T (1994) Atrial natriuretic factor as a volume regulator. J Clin Pharmacol 34(5):424–426
32. Ishibashi T, Hamaguchi M, Kato K, Kawada T, Ohta H, Sasage H, Imai S (1993) Relationship between myoglobin contents and increases in cyclic GMP produced by glyceryl trinitrate and nitric oxide in rabbit aorta, right atrium and papillary muscle. Naunyn Schmiedebergs Arch Pharmacol 347(5):553–561
33. Yanagisawa T, Hashimoto H, Taira N (1988) The negative inotropic effect of nicorandil is independent of cyclic GMP changes: a comparison with pinacidil and cromakalim in canine atrial muscle. Br J Pharmacol 95(2):393–398
34. Cramb G, Banks R, Rugg EL, Aiton JF (1987) Actions of atrial natriuretic peptide (ANP) on cyclic nucleotide concentrations and phosphatidylinositol turnover in ventricular myocytes. Biochem Biophys Res Commun 148(3):962–970
35. Mullershausen F, Friebe A, Feil R, Thompson WJ, Hofmann F, Koesling D (2003) Direct activation of PDE5 by cGMP: long-term effects within NO/cGMP signaling. J Cell Biol 160(5):719–727
36. Kuhn M (2003) Structure, regulation, and function of mammalian membrane guanylyl cyclase receptors, with a focus on guanylyl cyclase-A. Circ Res 93(8):700–709
37. Francis SH, Busch JL, Corbin JD, Sibley D (2010) cGMP-dependent protein kinases and cGMP phosphodiesterases in nitric oxide and cGMP action. Pharmacol Rev 62(3):525–563. doi:10.1124/pr.110.002907

38. Ecker T, Gobel C, Hullin R, Rettig R, Seitz G, Hofmann F (1989) Decreased cardiac concentration of cGMP kinase in hypertensive animals. An index for cardiac vascularization? Circ Res 65(5):1361–1369
39. Wegener JW, Nawrath H, Wolfsgruber W, Kuhbandner S, Werner C, Hofmann F, Feil R (2002) cGMP-dependent protein kinase I mediates the negative inotropic effect of cGMP in the murine myocardium. Circ Res 90(1):18–20
40. Takimoto E, Champion HC, Li M, Belardi D, Ren S, Rodriguez ER, Bedja D, Gabrielson KL, Wang Y, Kass DA (2005) Chronic inhibition of cyclic GMP phosphodiesterase 5A prevents and reverses cardiac hypertrophy. Nat Med 11(2):214–222
41. MacDonell KL, Diamond J (1997) Cyclic GMP-dependent protein kinase activation in the absence of negative inotropic effects in the rat ventricle. Br J Pharmacol 122(7):1425–1435. doi:10.1038/sj.bjp.0701492
42. Pierkes M, Gambaryan S, Boknik P, Lohmann SM, Schmitz W, Potthast R, Holtwick R, Kuhn M (2002) Increased effects of C-type natriuretic peptide on cardiac ventricular contractility and relaxation in guanylyl cyclase A-deficient mice. Cardiovasc Res 53(4):852–861
43. Lukowski R, Rybalkin SD, Loga F, Leiss V, Beavo JA, Hofmann F (2010) Cardiac hypertrophy is not amplified by deletion of cGMP-dependent protein kinase I in cardiomyocytes. Proc Natl Acad Sci USA 107(12):5646–5651. doi:10.1073/pnas.1001360107
44. Mery PF, Lohmann SM, Walter U, Fischmeister R (1991) Ca^{2+} current is regulated by cyclic GMP-dependent protein kinase in mammalian cardiac myocytes. Proc Natl Acad Sci USA 88(4):1197–1201
45. George WJ, Polson JB, O'Toole AG, Goldberg ND (1970) Elevation of guanosine 3′,5′-cyclic phosphate in rat heart after perfusion with acetylcholine. Proc Natl Acad Sci USA 66(2):398–403
46. Watanabe AM, Besch HR Jr (1975) Interaction between cyclic adenosine monophosphate and cyclic gunaosine monophosphate in guinea pig ventricular myocardium. Circ Res 37(3):309–317
47. MacDonell KL, Tibbits GF, Diamond J (1995) cGMP elevation does not mediate muscarinic agonist-induced negative inotropy in rat ventricular cardiomyocytes. Am J Physiol 269(6 Pt 2):H1905–H1912
48. Hayek S, Nemer M (2011) Cardiac natriuretic peptides: from basic discovery to clinical practice. Cardiovasc Ther 29(6):362–376. doi:10.1111/j.1755-5922.2010.00152.x
49. Wegener JW, Gath I, Forstermann U, Nawrath H (1997) Activation of soluble guanylyl cyclase by YC-1 in aortic smooth muscle but not in ventricular myocardium from rat. Br J Pharmacol 122(7):1523–1529. doi:10.1038/sj.bjp.0701542
50. Rosenkranz AC, Woods RL, Dusting GJ, Ritchie RH (2003) Antihypertrophic actions of the natriuretic peptides in adult rat cardiomyocytes: importance of cyclic GMP. Cardiovasc Res 57(2):515–522
51. Hiwatari M, Satoh K, Angus JA, Johnston CI (1986) No effect of atrial natriuretic factor on cardiac rate, force and transmitter release. Clin Exp Pharmacol Physiol 13(2):163–168
52. Bergey JL, Kotler D (1985) Effects of atriopeptins I, II and III on atrial contractility, sinus nodal rate (guinea pig) and agonist-induced tension in rabbit aortic strips. Eur J Pharmacol 110(2):277–281
53. Bohm M, Diet F, Pieske B, Erdmann E (1988) h-ANF does not play a role in the regulation of myocardial force of contraction. Life Sci 43(16):1261–1267
54. Neyses L, Vetter H (1989) Action of atrial natriuretic peptide and angiotensin II on the myocardium: studies in isolated rat ventricular cardiomyocytes. Biochem Biophys Res Commun 163(3):1435–1443
55. Brusq JM, Mayoux E, Guigui L, Kirilovsky J (1999) Effects of C-type natriuretic peptide on rat cardiac contractility. Br J Pharmacol 128(1):206–212. doi:10.1038/sj.bjp.0702766
56. Wollert KC, Yurukova S, Kilic A, Begrow F, Fiedler B, Gambaryan S, Walter U, Lohmann SM, Kuhn M (2003) Increased effects of C-type natriuretic peptide on contractility and calcium regulation in murine hearts overexpressing cyclic GMP-dependent protein kinase I. Br J Pharmacol 140(7):1227–1236
57. Castro LR, Schittl J, Fischmeister R (2010) Feedback control through cGMP-dependent protein kinase contributes to differential regulation and compartmentation of cGMP in rat cardiac myocytes. Circ Res 107(10):1232–1240. doi:10.1161/CIRCRESAHA.110.226712
58. Piggott LA, Hassell KA, Berkova Z, Morris AP, Silberbach M, Rich TC (2006) Natriuretic peptides and nitric oxide stimulate cGMP synthesis in different cellular compartments. J Gen Physiol 128(1):3–14. doi:10.1085/jgp.200509403
59. Hare JM, Stamler JS (2005) NO/redox disequilibrium in the failing heart and cardiovascular system. J Clin Invest 115(3):509–517
60. Ziolo MT (2008) The fork in the nitric oxide road: cyclic GMP or nitrosylation? Nitric Oxide 18(3):153–156. doi:10.1016/j.niox.2008.01.008

61. Hendgen-Cotta UB, Flogel U, Kelm M, Rassaf T (2010) Unmasking the Janus face of myoglobin in health and disease. J Exp Biol 213(Pt 16):2734–2740. doi:10.1242/jeb.041178
62. Wegener JW, Godecke A, Schrader J, Nawrath H (2002) Effects of nitric oxide donors on cardiac contractility in wild-type and myoglobin-deficient mice. Br J Pharmacol 136(3):415–420. doi:10.1038/sj.bjp.0704740
63. Maulik N, Engelman DT, Watanabe M, Engelman RM, Maulik G, Cordis GA, Das DK (1995) Nitric oxide signaling in ischemic heart. Cardiovasc Res 30(4):593–601
64. Kojda G, Kottenberg K, Nix P, Schluter KD, Piper HM, Noack E (1996) Low increase in cGMP induced by organic nitrates and nitrovasodilators improves contractile response of rat ventricular myocytes. Circ Res 78(1):91–101
65. Kohr MJ, Wang H, Wheeler DG, Velayutham M, Zweier JL, Ziolo MT (2008) Biphasic effect of SIN-1 is reliant upon cardiomyocyte contractile state. Free Radic Biol Med 45(1):73–80. doi:10.1016/j.freeradbiomed.2008.03.019
66. Han X, Kubota I, Feron O, Opel DJ, Arstall MA, Zhao YY, Huang P, Fishman MC, Michel T, Kelly RA (1998) Muscarinic cholinergic regulation of cardiac myocyte ICa-L is absent in mice with targeted disruption of endothelial nitric oxide synthase. Proc Natl Acad Sci USA 95(11):6510–6515
67. Hare JM, Colucci WS (1995) Role of nitric oxide in the regulation of myocardial function. Prog Cardiovasc Dis 38(2):155–166
68. Vandecasteele G, Eschenhagen T, Scholz H, Stein B, Verde I, Fischmeister R (1999) Muscarinic and beta-adrenergic regulation of heart rate, force of contraction and calcium current is preserved in mice lacking endothelial nitric oxide synthase. Nat Med 5(3):331–334
69. Godecke A, Heinicke T, Kamkin A, Kiseleva I, Strasser RH, Decking UK, Stumpe T, Isenberg G, Schrader J (2001) Inotropic response to beta-adrenergic receptor stimulation and anti- adrenergic effect of ACh in endothelial NO synthase-deficient mouse hearts. J Physiol 532(Pt 1):195–204
70. Yang L, Liu G, Zakharov SI, Bellinger AM, Mongillo M, Marx SO (2007) Protein kinase G phosphorylates Cav1.2 alpha1c and beta2 subunits. Circ Res 101(5):465–474
71. Rodriguez-Pascual F, Ferrero R, Miras-Portugal MT, Torres M (1999) Phosphorylation of tyrosine hydroxylase by cGMP-dependent protein kinase in intact bovine chromaffin cells. Arch Biochem Biophys 366(2):207–214. doi:10.1006/abbi.1999.1199
72. Wang Y, Wagner MB, Joyner RW, Kumar R (2000) cGMP-dependent protein kinase mediates stimulation of l-type calcium current by cGMP in rabbit atrial cells. Cardiovasc Res 48(2):310–322
73. Wahler GM, Dollinger SJ (1995) Nitric oxide donor SIN-1 inhibits mammalian cardiac calcium current through cGMP-dependent protein kinase. Am J Physiol 268(1 Pt 1):C45–C54
74. Schroder F, Klein G, Fiedler B, Bastein M, Schnasse N, Hillmer A, Ames S, Gambaryan S, Drexler H, Walter U, Lohmann SM, Wollert KC (2003) Single l-type $Ca^{(2+)}$ channel regulation by cGMP-dependent protein kinase type I in adult cardiomyocytes from PKG I transgenic mice. Cardiovasc Res 60(2):268–277
75. Fischmeister R, Castro L, Abi-Gerges A, Rochais F, Vandecasteele G (2005) Species- and tissue-dependent effects of NO and cyclic GMP on cardiac ion channels. Comp Biochem Physiol A Mol Integr Physiol 142(2):136–143. doi:10.1016/j.cbpb.2005.04.012
76. Fischmeister R, Castro LR, Abi-Gerges A, Rochais F, Jurevicius J, Leroy J, Vandecasteele G (2006) Compartmentation of cyclic nucleotide signaling in the heart: the role of cyclic nucleotide phosphodiesterases. Circ Res 99(8):816–828. doi:10.1161/01.RES.0000246118.98832.04
77. Zhang M, Takimoto E, Lee DI, Santos CX, Nakamura T, Hsu S, Jiang A, Nagayama T, Bedja D, Yuan Y, Eaton P, Shah AM, Kass DA (2012) Pathological cardiac hypertrophy alters intracellular targeting of phosphodiesterase type 5 from nitric oxide synthase-3 to natriuretic Peptide signaling. Circulation 126(8):942–951
78. Jiang H, Colbran JL, Francis SH, Corbin JD (1992) Direct evidence for cross-activation of cGMP-dependent protein kinase by cAMP in pig coronary arteries. J Biol Chem 267(2):1015–1019
79. Kruger M, Kotter S, Grutzner A, Lang P, Andresen C, Redfield MM, Butt E, dos Remedios CG, Linke WA (2009) Protein kinase G modulates human myocardial passive stiffness by phosphorylation of the titin springs. Circ Res 104(1):87–94. doi:10.1161/CIRCRESAHA.108.184408
80. LeWinter MM, Granzier H (2010) Cardiac titin: a multifunctional giant. Circulation 121(19):2137–2145. doi:10.1161/CIRCULATIONAHA.109.860171
81. Voelkel T, Linke WA (2011) Conformation-regulated mechanosensory control via titin

domains in cardiac muscle. Pflugers Arch 462(1):143–154. doi:10.1007/s00424-011-0938-1

82. Lee EJ, Peng J, Radke M, Gotthardt M, Granzier HL (2010) Calcium sensitivity and the Frank-Starling mechanism of the heart are increased in titin N2B region-deficient mice. J Mol Cell Cardiol 49(3):449–458. doi:10.1016/j.yjmcc.2010.05.006
83. Terui T, Sodnomtseren M, Matsuba D, Udaka J, Ishiwata S, Ohtsuki I, Kurihara S, Fukuda N (2008) Troponin and titin coordinately regulate length-dependent activation in skinned porcine ventricular muscle. J Gen Physiol 131(3):275–283. doi:10.1085/jgp.200709895
84. Borbely A, Falcao-Pires I, van Heerebeek L, Hamdani N, Edes I, Gavina C, Leite-Moreira AF, Bronzwaer JG, Papp Z, van der Velden J, Stienen GJ, Paulus WJ (2009) Hypophosphorylation of the Stiff N2B titin isoform raises cardiomyocyte resting tension in failing human myocardium. Circ Res 104(6):780–786. doi:10.1161/CIRCRESAHA.108.193326
85. van Heerebeek L, Hamdani N, Falcao-Pires I, Leite-Moreira AF, Begieneman MP, Bronzwaer JG, van der Velden J, Stienen GJ, Laarman GJ, Somsen A, Verheugt FW, Niessen HW, Paulus WJ (2012) Low myocardial protein kinase g activity in heart failure with preserved ejection fraction. Circulation 126(7):830–839. doi:10.1161/CIRCULATIONAHA.111.076075
86. Bishu K, Hamdani N, Mohammed SF, Kruger M, Ohtani T, Ogut O, Brozovich FV, Burnett JC Jr, Linke WA, Redfield MM (2011) Sildenafil and B-type natriuretic peptide acutely phosphorylate titin and improve diastolic distensibility in vivo. Circulation 124(25):2882–2891. doi:10.1161/CIRCULATIONAHA.111.048520
87. Kruger M, Linke WA (2006) Protein kinase-A phosphorylates titin in human heart muscle and reduces myofibrillar passive tension. J Muscle Res Cell Motil 27(5–7):435–444. doi:10.1007/s10974-006-9090-5
88. Yamasaki R, Wu Y, McNabb M, Greaser M, Labeit S, Granzier H (2002) Protein kinase A phosphorylates titin's cardiac-specific N2B domain and reduces passive tension in rat cardiac myocytes. Circ Res 90(11):1181–1188
89. Jessup M, Brozena S (2003) Heart failure. N Engl J Med 348(20):2007–2018. doi:10.1056/NEJMra021498
90. Zamanillo D, Sprengel R, Hvalby O, Jensen V, Burnashev N, Rozov A, Kaiser KM, Koster HJ, Borchardt T, Worley P, Lubke J, Frotscher M, Kelly PH, Sommer B, Andersen P, Seeburg PH, Sakmann B (1999) Importance of AMPA receptors for hippocampal synaptic plasticity but not for spatial learning. Science 284(5421):1805–1811
91. Shah AM, Mann DL (2011) In search of new therapeutic targets and strategies for heart failure: recent advances in basic science. Lancet 378(9792):704–712. doi:10.1016/S0140-6736(11)60894-5
92. John SW, Krege JH, Oliver PM, Hagaman JR, Hodgin JB, Pang SC, Flynn TG, Smithies O (1995) Genetic decreases in atrial natriuretic peptide and salt-sensitive hypertension. Science 267(5198):679–681
93. Oliver PM, Fox JE, Kim R, Rockman HA, Kim HS, Reddick RL, Pandey KN, Milgram SL, Smithies O, Maeda N (1997) Hypertension, cardiac hypertrophy, and sudden death in mice lacking natriuretic peptide receptor A. Proc Natl Acad Sci USA 94(26):14730–14735
94. Kishimoto I, Rossi K, Garbers DL (2001) A genetic model provides evidence that the receptor for atrial natriuretic peptide (guanylyl cyclase-A) inhibits cardiac ventricular myocyte hypertrophy. Proc Natl Acad Sci USA 98(5):2703–2706
95. Knowles JW, Esposito G, Mao L, Hagaman JR, Fox JE, Smithies O, Rockman HA, Maeda N (2001) Pressure-independent enhancement of cardiac hypertrophy in natriuretic peptide receptor A-deficient mice. J Clin Invest 107(8):975–984
96. Holtwick R, Van Eickels M, Skryabin BV, Baba HA, Bubikat A, Begrow F, Schneider MD, Garbers DL, Kuhn M (2003) Pressure-independent cardiac hypertrophy in mice with cardiomyocyte-restricted inactivation of the atrial natriuretic peptide receptor guanylyl cyclase-A. J Clin Invest 111(9):1399–1407
97. Ruetten H, Dimmeler S, Gehring D, Ihling C, Zeiher AM (2005) Concentric left ventricular remodeling in endothelial nitric oxide synthase knockout mice by chronic pressure overload. Cardiovasc Res 66(3):444–453. doi:10.1016/j.cardiores.2005.01.021
98. Wenzel S, Rohde C, Wingerning S, Roth J, Kojda G, Schluter KD (2007) Lack of endothelial nitric oxide synthase-derived nitric oxide formation favors hypertrophy in adult ventricular cardiomyocytes. Hypertension 49(1):193–200. doi:10.1161/01.HYP.0000250468.02084.ce
99. Flaherty MP, Brown M, Grupp IL, Schultz JE, Murphree SS, Jones WK (2007) eNOS deficient mice develop progressive cardiac hypertrophy with altered cytokine and calcium handling protein expression. Cardiovasc Toxicol 7(3):165–177. doi:10.1007/s12012-007-0028-y

100. Janssens S, Pokreisz P, Schoonjans L, Pellens M, Vermeersch P, Tjwa M, Jans P, Scherrer-Crosbie M, Picard MH, Szelid Z, Gillijns H, Van de Werf F, Collen D, Bloch KD (2004) Cardiomyocyte-specific overexpression of nitric oxide synthase 3 improves left ventricular performance and reduces compensatory hypertrophy after myocardial infarction. Circ Res 94(9):1256–1262. doi:10.1161/01.RES.0000126497.38281.23

101. Calderone A, Thaik CM, Takahashi N, Chang DL, Colucci WS (1998) Nitric oxide, atrial natriuretic peptide, and cyclic GMP inhibit the growth-promoting effects of norepinephrine in cardiac myocytes and fibroblasts. J Clin Invest 101(4):812–818

102. Horio T, Nishikimi T, Yoshihara F, Matsuo H, Takishita S, Kangawa K (2000) Inhibitory regulation of hypertrophy by endogenous atrial natriuretic peptide in cultured cardiac myocytes. Hypertension 35(1 Pt 1):19–24

103. Laskowski A, Woodman OL, Cao AH, Drummond GR, Marshall T, Kaye DM, Ritchie RH (2006) Antioxidant actions contribute to the antihypertrophic effects of atrial natriuretic peptide in neonatal rat cardiomyocytes. Cardiovasc Res 72(1):112–123. doi:10.1016/j.cardiores.2006.07.006

104. Nakamura TY, Iwata Y, Arai Y, Komamura K, Wakabayashi S (2008) Activation of Na^+/H^+ exchanger 1 is sufficient to generate Ca^{2+} signals that induce cardiac hypertrophy and heart failure. Circ Res 103(8):891–899. doi:10.1161/CIRCRESAHA.108.175141

105. Nakayama H, Wilkin BJ, Bodi I, Molkentin JD (2006) Calcineurin-dependent cardiomyopathy is activated by TRPC in the adult mouse heart. FASEB J 20(10):1660–1670. doi:10.1096/fj.05-5560com

106. De Windt LJ, Lim HW, Bueno OF, Liang Q, Delling U, Braz JC, Glascock BJ, Kimball TF, del Monte F, Hajjar RJ, Molkentin JD (2001) Targeted inhibition of calcineurin attenuates cardiac hypertrophy in vivo. Proc Natl Acad Sci USA 98(6):3322–3327. doi:10.1073/pnas.031371998

107. Backs J, Backs T, Neef S, Kreusser MM, Lehmann LH, Patrick DM, Grueter CE, Qi X, Richardson JA, Hill JA, Katus HA, Bassel-Duby R, Maier LS, Olson EN (2009) The delta isoform of CaM kinase II is required for pathological cardiac hypertrophy and remodeling after pressure overload. Proc Natl Acad Sci USA 106(7):2342–2347. doi:10.1073/pnas.0813013106

108. Condorelli G, Drusco A, Stassi G, Bellacosa A, Roncarati R, Iaccarino G, Russo MA, Gu Y, Dalton N, Chung C, Latronico MV, Napoli C, Sadoshima J, Croce CM, Ross J Jr (2002) Akt induces enhanced myocardial contractility and cell size in vivo in transgenic mice. Proc Natl Acad Sci USA 99(19):12333–12338. doi:10.1073/pnas.172376399

109. Koitabashi N, Aiba T, Hesketh GG, Rowell J, Zhang M, Takimoto E, Tomaselli GF, Kass DA (2010) Cyclic GMP/PKG-dependent inhibition of TRPC6 channel activity and expression negatively regulates cardiomyocyte NFAT activation Novel mechanism of cardiac stress modulation by PDE5 inhibition. J Mol Cell Cardiol 48(4):713–724. doi:10.1016/j.yjmcc.2009.11.015

110. Perez NG, Piaggio MR, Ennis IL, Garciarena CD, Morales C, Escudero EM, Cingolani OH, Chiappe de Cingolani G, Yang XP, Cingolani HE (2007) Phosphodiesterase 5A inhibition induces Na^+/H^+ exchanger blockade and protection against myocardial infarction. Hypertension49(5):1095–1103.doi:10.1161/HYPERTENSIONAHA.107.087759

111. Yeves AM, Garciarena CD, Nolly MB, Chiappe de Cingolani GE, Cingolani HE, Ennis IL (2010) Decreased activity of the Na^+/H^+ exchanger by phosphodiesterase 5A inhibition is attributed to an increase in protein phosphatase activity. Hypertension 56(4):690–695. doi:10.1161/HYPERTENSIONAHA.110.151324

112. Wollert KC, Fiedler B, Gambaryan S, Smolenski A, Heineke J, Butt E, Trautwein C, Lohmann SM, Drexler H (2002) Gene transfer of cGMP-dependent protein kinase I enhances the antihypertrophic effects of nitric oxide in cardiomyocytes. Hypertension 39(1):87–92

113. Fiedler B, Lohmann SM, Smolenski A, Linnemuller S, Pieske B, Schroder F, Molkentin JD, Drexler H, Wollert KC (2002) Inhibition of calcineurin-NFAT hypertrophy signaling by cGMP-dependent protein kinase type I in cardiac myocytes. Proc Natl Acad Sci USA 99(17):11363–11368

114. Hsu S, Nagayama T, Koitabashi N, Zhang M, Zhou L, Bedja D, Gabrielson KL, Molkentin JD, Kass DA, Takimoto E (2009) Phosphodiesterase 5 inhibition blocks pressure overload-induced cardiac hypertrophy independent of the calcineurin pathway. Cardiovasc Res 81(2):301–309. doi:10.1093/cvr/cvn324

115. Das A, Xi L, Kukreja RC (2008) Protein kinase G-dependent cardioprotective mechanism of phosphodiesterase-5 inhibition involves phosphorylation of ERK and GSK3beta. J Biol Chem 283(43):29572–29585. doi:10.1074/jbc.M801547200

116. Takimoto E, Koitabashi N, Hsu S, Ketner EA, Zhang M, Nagayama T, Bedja D, Gabrielson KL, Blanton R, Siderovski DP,

Mendelsohn ME, Kass DA (2009) Regulator of G protein signaling 2 mediates cardiac compensation to pressure overload and antihypertrophic effects of PDE5 inhibition in mice. J Clin Invest 119(2):408–420. doi:10.1172/JCI35620

117. Klaiber M, Kruse M, Volker K, Schroter J, Feil R, Freichel M, Gerling A, Feil S, Dietrich A, Londono JE, Baba HA, Abramowitz J, Birnbaumer L, Penninger JM, Pongs O, Kuhn M (2010) Novel insights into the mechanisms mediating the local antihypertrophic effects of cardiac atrial natriuretic peptide: role of cGMP-dependent protein kinase and RGS2. Basic Res Cardiol 105(5):583–595. doi:10.1007/s00395-010-0098-z

118. Garcia-Dorado D, Agullo L, Sartorio CL, Ruiz-Meana M (2009) Myocardial protection against reperfusion injury: the cGMP pathway. Thromb Haemost 101(4):635–642

119. Costa AD, Pierre SV, Cohen MV, Downey JM, Garlid KD (2008) cGMP signalling in pre- and post-conditioning: the role of mitochondria. Cardiovasc Res 77(2):344–352. doi:10.1093/cvr/cvm050

120. Gorbe A, Giricz Z, Szunyog A, Csont T, Burley DS, Baxter GF, Ferdinandy P (2010) Role of cGMP-PKG signaling in the protection of neonatal rat cardiac myocytes subjected to simulated ischemia/reoxygenation. Basic Res Cardiol 105(5):643–650. doi:10.1007/s00395-010-0097-0

121. Fiedler B, Feil R, Hofmann F, Willenbockel C, Drexler H, Smolenski A, Lohmann SM, Wollert KC (2006) cGMP-dependent protein kinase type I inhibits TAB1-p38 mitogen-activated protein kinase apoptosis signaling in cardiacmyocytes. J Biol Chem 281(43):32831–32840. doi:10.1074/jbc.M603416200

122. Costa AD, Garlid KD, West IC, Lincoln TM, Downey JM, Cohen MV, Critz SD (2005) Protein kinase G transmits the cardioprotective signal from cytosol to mitochondria. Circ Res 97(4):329–336. doi:10.1161/01.RES.0000178451.08719.5b

123. Jang Y, Wang H, Xi J, Mueller RA, Norfleet EA, Xu Z (2007) NO mobilizes intracellular Zn^{2+} via cGMP/PKG signaling pathway and prevents mitochondrial oxidant damage in cardiomyocytes. Cardiovasc Res 75(2):426–433. doi:10.1016/j.cardiores.2007.05.015

124. Li P, Wang D, Lucas J, Oparil S, Xing D, Cao X, Novak L, Renfrow MB, Chen YF (2008) Atrial natriuretic peptide inhibits transforming growth factor beta-induced Smad signaling and myofibroblast transformation in mouse cardiac fibroblasts. Circ Res 102(2):185–192. doi:10.1161/CIRCRESAHA.107.157677

125. Murthy KS (2006) Signaling for contraction and relaxation in smooth muscle of the gut. Annu Rev Physiol 68:345–374. doi:10.1146/annurev.physiol.68.040504.094707

126. Frei E, Huster M, Smital P, Schlossmann J, Hofmann F, Wegener JW (2009) Calcium-dependent and calcium-independent inhibition of contraction by cGMP/cGKI in intestinal smooth muscle. Am J Physiol Gastrointest Liver Physiol 297(4):G834–G839. doi:10.1152/ajpgi.00095.2009

127. Koeppen M, Feil R, Siegl D, Feil S, Hofmann F, Pohl U, de Wit C (2004) cGMP-dependent protein kinase mediates NO- but not acetylcholine-induced dilations in resistance vessels in vivo. Hypertension 44(6):952–955

128. Pfeifer A, Klatt P, Massberg S, Ny L, Sausbier M, Hirneiss C, Wang GX, Korth M, Aszodi A, Andersson KE, Krombach F, Mayerhofer A, Ruth P, Fassler R, Hofmann F (1998) Defective smooth muscle regulation in cGMP kinase I-deficient mice. EMBO J 17(11):3045–3051

129. Sausbier M, Schubert R, Voigt V, Hirneiss C, Pfeifer A, Korth M, Kleppisch T, Ruth P, Hofmann F (2000) Mechanisms of NO/cGMP-dependent vasorelaxation. Circ Res 87(9):825–830

130. Weber S, Bernhard D, Lukowski R, Weinmeister P, Worner R, Wegener JW, Valtcheva N, Feil S, Schlossmann J, Hofmann F, Feil R (2007) Rescue of cGMP kinase I knockout mice by smooth muscle specific expression of either isozyme. Circ Res 101(11):1096–1103

131. Geiselhöringer A, Werner M, Sigl K, Smital P, Worner R, Acheo L, Stieber J, Weinmeister P, Feil R, Feil S, Wegener J, Hofmann F, Schlossmann J (2004) IRAG is essential for relaxation of receptor-triggered smooth muscle contraction by cGMP kinase. EMBO J 23(21):4222–4231

132. Hofmann F, Ammendola A, Schlossmann J (2000) Rising behind NO: cGMP-dependent protein kinases. J Cell Sci 113(Pt 10):1671–1676

133. Feil R, Gappa N, Rutz M, Schlossmann J, Rose CR, Konnerth A, Brummer S, Kuhbandner S, Hofmann F (2002) Functional reconstitution of vascular smooth muscle cells with cGMP-dependent protein kinase I isoforms. Circ Res 90(10):1080–1086

134. Geiselhöringer A, Gaisa M, Hofmann F, Schlossmann J (2004) Distribution of IRAG and cGKI-isoforms in murine tissues. FEBS Lett 575(1–3):19–22

135. Wolfe L, Corbin JD, Francis SH (1989) Characterization of a novel isozyme of

cGMP-dependent protein kinase from bovine aorta. J Biol Chem 264(13):7734–7741

136. Morgado M, Cairrao E, Santos-Silva AJ, Verde I (2012) Cyclic nucleotide-dependent relaxation pathways in vascular smooth muscle. Cell Mol Life Sci 69(2):247–266. doi:10.1007/s00018-011-0815-2

137. Vaandrager AB, de Jonge HR (1996) Signalling by cGMP-dependent protein kinases. Mol Cell Biochem 157(1–2):23–30

138. Tanaka Y, Tang G, Takizawa K, Otsuka K, Eghbali M, Song M, Nishimaru K, Shigenobu K, Koike K, Stefani E, Toro L (2006) Kv channels contribute to nitric oxide- and atrial natriuretic peptide-induced relaxation of a rat conduit artery. J Pharmacol Exp Ther 317(1):341–354. doi:10.1124/jpet.105.096115

139. Rapoport RM (1986) Cyclic guanosine monophosphate inhibition of contraction may be mediated through inhibition of phosphatidylinositol hydrolysis in rat aorta. Circ Res 58(3):407–410

140. Xia C, Bao Z, Yue C, Sanborn BM, Liu M (2001) Phosphorylation and regulation of G-protein-activated phospholipase C-beta 3 by cGMP-dependent protein kinases. J Biol Chem 276(23):19770–19777

141. Hirata M, Kohse KP, Chang CH, Ikebe T, Murad F (1990) Mechanism of cyclic GMP inhibition of inositol phosphate formation in rat aorta segments and cultured bovine aortic smooth muscle cells. J Biol Chem 265(3):1268–1273

142. Hepler JR (1999) Emerging roles for RGS proteins in cell signalling. Trends Pharmacol Sci 20(9):376–382

143. Tang KM, Wang GR, Lu P, Karas RH, Aronovitz M, Heximer SP, Kaltenbronn KM, Blumer KJ, Siderovski DP, Zhu Y, Mendelsohn ME (2003) Regulator of G-protein signaling-2 mediates vascular smooth muscle relaxation and blood pressure. Nat Med 9(12):1506–1512

144. Huang J, Zhou H, Mahavadi S, Sriwai W, Murthy KS (2007) Inhibition of Galphaq-dependent PLC-beta1 activity by PKG and PKA is mediated by phosphorylation of RGS4 and GRK2. Am J Physiol Cell Physiol 292(1):C200–C208. doi:10.1152/ajpcell.00103.2006

145. Pedram A, Razandi M, Kehrl J, Levin ER (2000) Natriuretic peptides inhibit G protein activation. Mediation through cross-talk between cyclic GMP-dependent protein kinase and regulators of G protein-signaling proteins. J Biol Chem 275(10):7365–7372

146. Sun X, Kaltenbronn KM, Steinberg TH, Blumer KJ (2005) RGS2 is a mediator of nitric oxide action on blood pressure and vasoconstrictor signaling. Mol Pharmacol 67(3):631–639

147. Wang GR, Zhu Y, Halushka PV, Lincoln TM, Mendelsohn ME (1998) Mechanism of platelet inhibition by nitric oxide: in vivo phosphorylation of thromboxane receptor by cyclic GMP-dependent protein kinase. Proc Natl Acad Sci USA 95(9):4888–4893

148. Klages B, Brandt U, Simon MI, Schultz G, Offermanns S (1999) Activation of G12/G13 results in shape change and Rho/Rho-kinase-mediated myosin light chain phosphorylation in mouse platelets. J Cell Biol 144(4):745–754

149. Somlyo AP, Somlyo AV (2003) Ca^{2+} sensitivity of smooth muscle and nonmuscle myosin II: modulated by G proteins, kinases, and myosin phosphatase. Physiol Rev 83(4):1325–1358

150. Komalavilas P, Lincoln TM (1994) Phosphorylation of the inositol 1,4,5-trisphosphate receptor by cyclic GMP-dependent protein kinase. J Biol Chem 269(12):8701–8707

151. Komalavilas P, Lincoln TM (1996) Phosphorylation of the inositol 1,4,5-trisphosphate receptor. Cyclic GMP-dependent protein kinase mediates cAMP and cGMP dependent phosphorylation in the intact rat aorta. J Biol Chem 271(36):21933–21938

152. Schlossmann J, Ammendola A, Ashman K, Zong X, Huber A, Neubauer G, Wang GX, Allescher HD, Korth M, Wilm M, Hofmann F, Ruth P (2000) Regulation of intracellular calcium by a signalling complex of IRAG, IP3 receptor and cGMP kinase Ibeta. Nature 404(6774):197–201

153. Ammendola A, Geiselhoringer A, Hofmann F, Schlossmann J (2001) Molecular determinants of the interaction between the inositol 1,4,5-trisphosphate receptor-associated cGMP kinase substrate (IRAG) and cGMP kinase Ibeta. J Biol Chem 276(26):24153–24159

154. Desch M, Sigl K, Hieke B, Salb K, Kees F, Bernhard D, Jochim A, Spiessberger B, Hocherl K, Feil R, Feil S, Lukowski R, Wegener JW, Hofmann F, Schlossmann J (2010) IRAG determines nitric oxide- and atrial natriuretic peptide-mediated smooth muscle relaxation. Cardiovasc Res 86(3):496–505. doi:10.1093/cvr/cvq008

155. Welling A, Ludwig A, Zimmer S, Klugbauer N, Flockerzi V, Hofmann F (1997) Alternatively spliced IS6 segments of the alpha 1C gene determine the tissue-specific

dihydropyridine sensitivity of cardiac and vascular smooth muscle l-type Ca^{2+} channels. Circ Res 81(4):526–532

156. Lorenz JN, Bielefeld DR, Sperelakis N (1994) Regulation of calcium channel current in A7r5 vascular smooth muscle cells by cyclic nucleotides. Am J Physiol 266(6 Pt 1): C1656–C1663

157. Ruiz-Velasco V, Zhong J, Hume JR, Keef KD (1998) Modulation of Ca^{2+} channels by cyclic nucleotide cross activation of opposing protein kinases in rabbit portal vein. Circ Res 82(5):557–565

158. Chen J, Crossland RF, Noorani MM, Marrelli SP (2009) Inhibition of TRPC1/TRPC3 by PKG contributes to NO-mediated vasorelaxation. Am J Physiol Heart Circ Physiol 297(1):H417–H424. doi:10.1152/ajpheart.01130.2008

159. Inoue R, Okada T, Onoue H, Hara Y, Shimizu S, Naitoh S, Ito Y, Mori Y (2001) The transient receptor potential protein homologue TRP6 is the essential component of vascular alpha(1)-adrenoceptor-activated Ca($^{2+}$)-permeable cation channel. Circ Res 88(3):325–332

160. Koller A, Schlossmann J, Ashman K, Uttenweiler-Joseph S, Ruth P, Hofmann F (2003) Association of phospholamban with a cGMP kinase signaling complex. Biochem Biophys Res Commun 300(1):155–160

161. Raeymaekers L, Hofmann F, Casteels R (1988) Cyclic GMP-dependent protein kinase phosphorylates phospholamban in isolated sarcoplasmic reticulum from cardiac and smooth muscle. Biochem J 252(1):269–273

162. Lalli MJ, Shimizu S, Sutliff RL, Kranias EG, Paul RJ (1999) [Ca2+]i homeostasis and cyclic nucleotide relaxation in aorta of phospholamban-deficient mice. Am J Physiol 277(3 Pt 2):H963–H970

163. Cornwell TL, Pryzwansky KB, Wyatt TA, Lincoln TM (1991) Regulation of sarcoplasmic reticulum protein phosphorylation by localized cyclic GMP-dependent protein kinase in vascular smooth muscle cells. Mol Pharmacol 40(6):923–931

164. Kobayashi S, Kanaide H, Nakamura M (1985) Cytosolic-free calcium transients in cultured vascular smooth muscle cells: microfluorometric measurements. Science 229(4713):553–556

165. Rashatwar SS, Cornwell TL, Lincoln TM (1987) Effects of 8-bromo-cGMP on Ca^{2+} levels in vascular smooth muscle cells: possible regulation of Ca^{2+}-ATPase by cGMP-dependent protein kinase. Proc Natl Acad Sci USA 84(16):5685–5689

166. Vrolix M, Raeymaekers L, Wuytack F, Hofmann F, Casteels R (1988) Cyclic GMP-dependent protein kinase stimulates the plasmalemmal Ca^{2+} pump of smooth muscle via phosphorylation of phosphatidylinositol. Biochem J 255(3):855–863

167. Furukawa K, Nakamura H (1987) Cyclic GMP regulation of the plasma membrane (Ca^{2+}–Mg^{2+})ATPase in vascular smooth muscle. J Biochem 101(1):287–290

168. Furukawa K, Ohshima N, Tawada-Iwata Y, Shigekawa M (1991) Cyclic GMP stimulates Na^{+}/Ca^{2+} exchange in vascular smooth muscle cells in primary culture. J Biol Chem 266(19):12337–12341

169. Taniguchi J, Furukawa KI, Shigekawa M (1993) Maxi K^{+} channels are stimulated by cyclic guanosine monophosphate-dependent protein kinase in canine coronary artery smooth muscle cells. Pflugers Arch 423(3–4):167–172

170. Robertson BE, Schubert R, Hescheler J, Nelson MT (1993) cGMP-dependent protein kinase activates Ca-activated K channels in cerebral artery smooth muscle cells. Am J Physiol 265(1 Pt 1):C299–C303

171. Alioua A, Tanaka Y, Wallner M, Hofmann F, Ruth P, Meera P, Toro L (1998) The large conductance, voltage-dependent, and calcium-sensitive K^{+} channel, Hslo, is a target of cGMP-dependent protein kinase phosphorylation in vivo. J Biol Chem 273(49): 32950–32956

172. Bolotina VM, Najibi S, Palacino JJ, Pagano PJ, Cohen RA (1994) Nitric oxide directly activates calcium-dependent potassium channels in vascular smooth muscle. Nature 368(6474):850–853. doi:10.1038/368850a0

173. Fukao M, Mason HS, Britton FC, Kenyon JL, Horowitz B, Keef KD (1999) Cyclic GMP-dependent protein kinase activates cloned BKCa channels expressed in mammalian cells by direct phosphorylation at serine 1072. J Biol Chem 274(16):10927–10935

174. Zhou XB, Ruth P, Schlossmann J, Hofmann F, Korth M (1996) Protein phosphatase 2A is essential for the activation of Ca^{2+}-activated K^{+} currents by cGMP-dependent protein kinase in tracheal smooth muscle and Chinese hamster ovary cells. J Biol Chem 271(33):19760–19767

175. White RE, Lee AB, Shcherbatko AD, Lincoln TM, Schonbrunn A, Armstrong DL (1993) Potassium channel stimulation by natriuretic peptides through cGMP-dependent dephosphorylation. Nature 361(6409):263–266

176. Surks HK, Mochizuki N, Kasai Y, Georgescu SP, Tang KM, Ito M, Lincoln TM, Mendelsohn ME (1999) Regulation of myosin phosphatase by a specific interaction with cGMP-dependent protein kinase Ialpha. Science 286(5444):1583–1587
177. Wooldridge AA, MacDonald JA, Erdodi F, Ma C, Borman MA, Hartshorne DJ, Haystead TA (2004) Smooth muscle phosphatase is regulated in vivo by exclusion of phosphorylation of threonine 696 of MYPT1 by phosphorylation of Serine 695 in response to cyclic nucleotides. J Biol Chem 279(33):34496–34504
178. Lee MR, Li L, Kitazawa T (1997) Cyclic GMP causes Ca^{2+} desensitization in vascular smooth muscle by activating the myosin light chain phosphatase. J Biol Chem 272(8):5063–5068
179. Nakamura K, Koga Y, Sakai H, Homma K, Ikebe M (2007) cGMP-dependent relaxation of smooth muscle is coupled with the change in the phosphorylation of myosin phosphatase. Circ Res 101(7):712–722. doi:10.1161/CIRCRESAHA.107.153981
180. Somlyo AV (2007) Cyclic GMP regulation of myosin phosphatase: a new piece for the puzzle? Circ Res 101(7):645–647. doi:10.1161/CIRCRESAHA.107.161893
181. Khromov AS, Wang H, Choudhury N, McDuffie M, Herring BP, Nakamoto R, Owens GK, Somlyo AP, Somlyo AV (2006) Smooth muscle of telokin-deficient mice exhibits increased sensitivity to Ca^{2+} and decreased cGMP-induced relaxation. Proc Natl Acad Sci USA 103(7):2440–2445. doi:10.1073/pnas.0508566103
182. Borman MA, MacDonald JA, Haystead TA (2004) Modulation of smooth muscle contractility by CHASM, a novel member of the smoothelin family of proteins. FEBS Lett 573(1–3):207–213. doi:10.1016/j.febslet.2004.08.002
183. Sauzeau V, Le Jeune H, Cario-Toumaniantz C, Smolenski A, Lohmann SM, Bertoglio J, Chardin P, Pacaud P, Loirand G (2000) Cyclic GMP-dependent protein kinase signaling pathway inhibits RhoA-induced Ca^{2+} sensitization of contraction in vascular smooth muscle. J Biol Chem 275(28):21722–21729
184. Loirand G, Guerin P, Pacaud P (2006) Rho kinases in cardiovascular physiology and pathophysiology. Circ Res 98(3):322–334. doi:10.1161/01.RES.0000201960.04223.3c
185. Sakurada S, Takuwa N, Sugimoto N, Wang Y, Seto M, Sasaki Y, Takuwa Y (2003) Ca2+-dependent activation of Rho and Rho kinase in membrane depolarization-induced and receptor stimulation-induced vascular smooth muscle contraction. Circ Res 93(6):548–556. doi:10.1161/01.RES.0000090998.08629.60
186. Bonnevier J, Arner A (2004) Actions downstream of cyclic GMP/protein kinase G can reverse protein kinase C-mediated phosphorylation of CPI-17 and Ca(2+) sensitization in smooth muscle. J Biol Chem 279(28):28998–29003
187. Woodrum DA, Brophy CM, Wingard CJ, Beall A, Rasmussen H (1999) Phosphorylation events associated with cyclic nucleotide-dependent inhibition of smooth muscle contraction. Am J Physiol 277(3 Pt 2): H931–H939
188. Rembold CM, Foster DB, Strauss JD, Wingard CJ, Eyk JE (2000) cGMP-mediated phosphorylation of heat shock protein 20 may cause smooth muscle relaxation without myosin light chain dephosphorylation in swine carotid artery. J Physiol 524(Pt 3): 865–878
189. Woodrum D, Pipkin W, Tessier D, Komalavilas P, Brophy CM (2003) Phosphorylation of the heat shock-related protein, HSP20, mediates cyclic nucleotide-dependent relaxation. J Vasc Surg 37(4):874–881. doi:10.1067/mva.2003.153
190. Essin K, Welling A, Hofmann F, Luft FC, Gollasch M, Moosmang S (2007) Indirect coupling between Cav1.2 channels and ryanodine receptors to generate Ca^{2+} sparks in murine arterial smooth muscle cells. J Physiol 584(Pt 1):205–219
191. Moosmang S, Schulla V, Welling A, Feil R, Feil S, Wegener JW, Hofmann F, Klugbauer N (2003) Dominant role of smooth muscle l-type calcium channel Cav1.2 for blood pressure regulation. EMBO J 22(22):6027–6034
192. Wegener JW, Schulla V, Lee TS, Koller A, Feil S, Feil R, Kleppisch T, Klugbauer N, Moosmang S, Welling A, Hofmann F (2004) An essential role of Cav1.2l-type calcium channel for urinary bladder function. FASEB J 18(10):1159–1161
193. Huang PL, Huang Z, Mashimo H, Bloch KD, Moskowitz MA, Bevan JA, Fishman MC (1995) Hypertension in mice lacking the gene for endothelial nitric oxide synthase. Nature 377(6546):239–242
194. Kuhn M (2005) Cardiac and intestinal natriuretic peptides: insights from genetically modified mice. Peptides 26(6):1078–1085
195. Sausbier M, Arntz C, Bucurenciu I, Zhao H, Zhou XB, Sausbier U, Feil S, Kamm S, Essin K, Sailer CA, Abdullah U, Krippeit-Drews P, Feil R, Hofmann F, Knaus HG, Kenyon C, Shipston MJ, Storm JF, Neuhuber W, Korth M, Schubert

R, Gollasch M, Ruth P (2005) Elevated blood pressure linked to primary hyperaldosteronism and impaired vasodilation in BK channel-deficient mice. Circulation 112(1):60–68. doi:10.1161/01.CIR.0000156448.74296.FE

196. Faraci FM, Sigmund CD (1999) Vascular biology in genetically altered mice: smaller vessels, bigger insight. Circ Res 85(12):1214–1225
197. Murata T, Ushikubi F, Matsuoka T, Hirata M, Yamasaki A, Sugimoto Y, Ichikawa A, Aze Y, Tanaka T, Yoshida N, Ueno A, Oh-ishi S, Narumiya S (1997) Altered pain perception and inflammatory response in mice lacking prostacyclin receptor. Nature 388(6643): 678–682. doi:10.1038/41780
198. Meng W, Ayata C, Waeber C, Huang PL, Moskowitz MA (1998) Neuronal NOS-cGMP-dependent ACh-induced relaxation in pial arterioles of endothelial NOS knockout mice. Am J Physiol 274(2 Pt 2): H411–H415
199. Friebe A, Mergia E, Dangel O, Lange A, Koesling D (2007) Fatal gastrointestinal obstruction and hypertension in mice lacking nitric oxide-sensitive guanylyl cyclase. Proc Natl Acad Sci USA 104(18):7699–7704
200. Landgraf W, Regulla S, Meyer HE, Hofmann F (1991) Oxidation of cysteines activates cGMP-dependent protein kinase. J Biol Chem 266(25):16305–16311
201. Osborne BW, Wu J, McFarland CJ, Nickl CK, Sankaran B, Casteel DE, Woods VL Jr, Kornev AP, Taylor SS, Dostmann WR (2011) Crystal structure of cGMP-dependent protein kinase reveals novel site of interchain communication. Structure 19(9):1317–1327
202. Burgoyne JR, Madhani M, Cuello F, Charles RL, Brennan JP, Schroder E, Browning DD, Eaton P (2007) Cysteine redox sensor in PKGIa enables oxidant-induced activation. Science 317(5843):1393–1397
203. Prysyazhna O, Rudyk O, Eaton P (2012) Single atom substitution in mouse protein kinase G eliminates oxidant sensing to cause hypertension. Nat Med 18(2):286–290
204. Ozaki M, Kawashima S, Yamashita T, Hirase T, Namiki M, Inoue N, Hirata K, Yasui H, Sakurai H, Yoshida Y, Masada M, Yokoyama M (2002) Overexpression of endothelial nitric oxide synthase accelerates atherosclerotic lesion formation in apoE-deficient mice. J Clin Invest 110(3):331–340
205. Kuhlencordt PJ, Chen J, Han F, Astern J, Huang PL (2001) Genetic deficiency of inducible nitric oxide synthase reduces atherosclerosis and lowers plasma lipid peroxides in apolipoprotein E-knockout mice. Circulation 103(25):3099–3104
206. Chyu KY, Dimayuga P, Zhu J, Nilsson J, Kaul S, Shah PK, Cercek B (1999) Decreased neointimal thickening after arterial wall injury in inducible nitric oxide synthase knockout mice. Circ Res 85(12):1192–1198
207. Tolbert T, Thompson JA, Bouchard P, Oparil S (2001) Estrogen-induced vasoprotection is independent of inducible nitric oxide synthase expression: evidence from the mouse carotid artery ligation model. Circulation 104(22):2740–2745
208. Detmers PA, Hernandez M, Mudgett J, Hassing H, Burton C, Mundt S, Chun S, Fletcher D, Card DJ, Lisnock J, Weikel R, Bergstrom JD, Shevell DE, Hermanowski-Vosatka A, Sparrow CP, Chao YS, Rader DJ, Wright SD, Pure E (2000) Deficiency in inducible nitric oxide synthase results in reduced atherosclerosis in apolipoprotein E-deficient mice. J Immunol 165(6):3430–3435
209. Sennlaub F, Courtois Y, Goureau O (2001) Inducible nitric oxide synthase mediates the change from retinal to vitreal neovascularization in ischemic retinopathy. J Clin Invest 107(6):717–725
210. Knowles JW, Reddick RL, Jennette JC, Shesely EG, Smithies O, Maeda N (2000) Enhanced atherosclerosis and kidney dysfunction in eNOS(−/−)Apoe(−/−) mice are ameliorated by enalapril treatment. J Clin Invest 105(4):451–458
211. Kuhlencordt PJ, Gyurko R, Han F, Scherrer-Crosbie M, Aretz TH, Hajjar R, Picard MH, Huang PL (2001) Accelerated atherosclerosis, aortic aneurysm formation, and ischemic heart disease in apolipoprotein E/endothelial nitric oxide synthase double-knockout mice. Circulation 104(4):448–454
212. Moroi M, Zhang L, Yasuda T, Virmani R, Gold HK, Fishman MC, Huang PL (1998) Interaction of genetic deficiency of endothelial nitric oxide, gender, and pregnancy in vascular response to injury in mice. J Clin Invest 101(6):1225–1232
213. Rudic RD, Shesely EG, Maeda N, Smithies O, Segal SS, Sessa WC (1998) Direct evidence for the importance of endothelium-derived nitric oxide in vascular remodeling. J Clin Invest 101(4):731–736
214. Chen J, Kuhlencordt PJ, Astern J, Gyurko R, Huang PL (2001) Hypertension does not account for the accelerated atherosclerosis and development of aneurysms in male apolipoprotein e/endothelial nitric oxide synthase double knockout mice. Circulation 104(20): 2391–2394
215. Koglin J, Glysing-Jensen T, Mudgett JS, Russell ME (1998) Exacerbated transplant arterioscle-

rosis in inducible nitric oxide- deficient mice. Circulation 97(20):2059–2065

216. Lablanche JM, Grollier G, Lusson JR, Bassand JP, Drobinski G, Bertrand B, Battaglia S, Desveaux B, Juilliere Y, Juliard JM, Metzger JP, Coste P, Quiret JC, Dubois-Rande JL, Crochet PD, Letac B, Boschat J, Virot P, Finet G, Le Breton H, Livarek B, Leclercq F, Beard T, Giraud T, Bertrand ME et al (1997) Effect of the direct nitric oxide donors linsidomine and molsidomine on angiographic restenosis after coronary balloon angioplasty. The ACCORD study. angioplastic coronaire corvasal diltiazem. Circulation 95(1):83–89
217. Poon BY, Raharjo E, Patel KD, Tavener S, Kubes P (2003) Complexity of inducible nitric oxide synthase: cellular source determines benefit versus toxicity. Circulation 108(9):1107–1112
218. Berk BC (2001) Vascular smooth muscle growth: autocrine growth mechanisms. Physiol Rev 81(3):999–1030
219. Dzau VJ, Braun-Dullaeus RC, Sedding DG (2002) Vascular proliferation and atherosclerosis: new perspectives and therapeutic strategies. Nat Med 8(11):1249–1256
220. Owens GK, Kumar MS, Wamhoff BR (2004) Molecular regulation of vascular smooth muscle cell differentiation in development and disease. Physiol Rev 84(3):767–801
221. Schwartz SM, deBlois D, O'Brien ER (1995) The intima. Soil for atherosclerosis and restenosis. Circ Res 77(3):445–465
222. Hofmann F, Feil R, Kleppisch T, Schlossmann J (2006) Function of cGMP-dependent protein kinases as revealed by gene deletion. Physiol Rev 86(1):1–23
223. Weinmeister P, Lukowski R, Linder S, Traidl-Hoffmann C, Hengst L, Hofmann F, Feil R (2008) Cyclic guanosine monophosphate-dependent protein kinase I promotes adhesion of primary vascular smooth muscle cells. Mol Biol Cell 19(10):4434–4441. doi:10.1091/mbc.E08-04-0370
224. Eigenthaler M, Lohmann SM, Walter U, Pilz RB (1999) Signal transduction by cGMP-dependent protein kinases and their emerging roles in the regulation of cell adhesion and gene expression. Rev Physiol Biochem Pharmacol 135:173–209
225. Lincoln TM, Dey N, Sellak H (2001) Invited review: cGMP-dependent protein kinase signaling mechanisms in smooth muscle: from the regulation of tone to gene expression. J Appl Physiol 91(3):1421–1430
226. Pilz RB, Casteel DE (2003) Regulation of gene expression by cyclic GMP. Circ Res 93(11):1034–1046
227. Lincoln TM, Wu X, Sellak H, Dey N, Choi CS (2006) Regulation of vascular smooth muscle cell phenotype by cyclic GMP and cyclic GMP-dependent protein kinase. Front Biosci 11:356–367
228. Kawai-Kowase K, Owens GK (2007) Multiple repressor pathways contribute to phenotypic switching of vascular smooth muscle cells. Am J Physiol Cell Physiol 292(1):C59–C69
229. Cornwell TL, Arnold E, Boerth NJ, Lincoln TM (1994) Inhibition of smooth muscle cell growth by nitric oxide and activation of cAMP-dependent protein kinase by cGMP. Am J Physiol 267(5 Pt 1):C1405–C1413
230. Boerth NJ, Dey NB, Cornwell TL, Lincoln TM (1997) Cyclic GMP-dependent protein kinase regulates vascular smooth muscle cell phenotype. J Vasc Res 34(4):245–259
231. Negash S, Narasimhan SR, Zhou W, Liu J, Wei FL, Tian J, Raj JU (2009) Role of cGMP-dependent protein kinase in regulation of pulmonary vascular smooth muscle cell adhesion and migration: effect of hypoxia. Am J Physiol Heart Circ Physiol 297(1):H304–H312. doi:10.1152/ajpheart.00077.2008
232. Pauly RR, Passaniti A, Bilato C, Monticone R, Cheng L, Papadopoulos N, Gluzband YA, Smith L, Weinstein C, Lakatta EG et al (1994) Migration of cultured vascular smooth muscle cells through a basement membrane barrier requires type IV collagenase activity and is inhibited by cellular differentiation. Circ Res 75(1):41–54
233. Bendeck MP, Irvin C, Reidy MA (1996) Inhibition of matrix metalloproteinase activity inhibits smooth muscle cell migration but not neointimal thickening after arterial injury. Circ Res 78(1):38–43
234. Dey NB, Lincoln TM (2012) Possible involvement of cyclic-GMP-dependent protein kinase on matrix metalloproteinase-2 expression in rat aortic smooth muscle cells. Mol Cell Biochem 368(1–2):27–35. doi:10.1007/s11010-012-1339-2
235. Liu N, Olson EN (2006) Coactivator control of cardiovascular growth and remodeling. Curr Opin Cell Biol 18(6):715–722
236. Huber A, Neuhuber WL, Klugbauer N, Ruth P, Allescher HD (2000) Cysteine-rich protein 2, a novel substrate for cGMP kinase I in enteric neurons and intestinal smooth muscle. J Biol Chem 275(8):5504–5511
237. Karim MA, Ohta K, Egashira M, Jinno Y, Niikawa N, Matsuda I, Indo Y (1996) Human ESP1/CRP2, a member of the LIM domain protein family: characterization of the cDNA and assignment of the gene locus to chromosome 14q32.3. Genomics 31(2):167–176

238. Okano I, Yamamoto T, Kaji A, Kimura T, Mizuno K, Nakamura T (1993) Cloning of CRP2, a novel member of the cysteine-rich protein family with two repeats of an unusual LIM/double zinc-finger motif. FEBS Lett 333(1–2):51–55

239. Zhang T, Zhuang S, Casteel DE, Looney DJ, Boss GR, Pilz RB (2007) A cysteine-rich LIM-only protein mediates regulation of smooth muscle-specific gene expression by cGMP-dependent protein kinase. J Biol Chem 282(46):33367–33380

240. Schmidtko A, Gao W, Sausbier M, Rauhmeier I, Sausbier U, Niederberger E, Scholich K, Huber A, Neuhuber W, Allescher HD, Hofmann F, Tegeder I, Ruth P, Geisslinger G (2008) Cysteine-rich protein 2, a novel downstream effector of cGMP/cGMP-dependent protein kinase I-mediated persistent inflammatory pain. J Neurosci 28(6):1320–1330. doi:10.1523/JNEUROSCI.5037-07.2008

241. Sinnaeve P, Chiche JD, Gillijns H, Van Pelt N, Wirthlin D, Van De Werf F, Collen D, Bloch KD, Janssens S (2002) Overexpression of a constitutively active protein kinase G mutant reduces neointima formation and in-stent restenosis. Circulation 105(24):2911–2916

242. Lukowski R, Weinmeister P, Bernhard D, Feil S, Gotthardt M, Herz J, Massberg S, Zernecke A, Weber C, Hofmann F, Feil R (2008) Role of smooth muscle cGMP/cGKI signaling in murine vascular restenosis. Arterioscler Thromb Vasc Biol 28(7):1244–1250. doi:10.1161/ATVBAHA.108.166405

243. Werner C, Raivich G, Cowen M, Strekalova T, Sillaber I, Buters JT, Spanagel R, Hofmann F (2004) Importance of NO/cGMP signalling via cGMP-dependent protein kinase II for controlling emotionality and neurobehavioural effects of alcohol. Eur J Neurosci 20(12):3498–3506

244. el-Husseini AE, Bladen C, Vincent SR (1995) Molecular characterization of a type II cyclic GMP-dependent protein kinase expressed in the rat brain. J Neurochem 64(6):2814–2817

245. Feil S, Zimmermann P, Knorn A, Brummer S, Schlossmann J, Hofmann F, Feil R (2005) Distribution of cGMP-dependent protein kinase type I and its isoforms in the mouse brain and retina. Neuroscience 135(3):863–868

246. Singh AK, Spiessberger B, Zheng W, Xiao F, Lukowski R, Wegener JW, Weinmeister P, Saur D, Klein S, Schemann M, Krueger D, Seidler U, Hofmann F (2012) Neuronal cGMP kinase I is essential for stimulation of duodenal bicarbonate secretion by luminal acid. FASEB J 26(4):1745–1754

247. Paul C, Schoberl F, Weinmeister P, Micale V, Wotjak CT, Hofmann F, Kleppisch T (2008) Signaling through cGMP-dependent protein kinase I in the amygdala is critical for auditory-cued fear memory and long-term potentiation. J Neurosci 28(52):14202–14212

248. Oster H, Werner C, Magnone MC, Mayser H, Feil R, Seeliger MW, Hofmann F, Albrecht U (2003) cGMP-dependent protein kinase II modulates mPer1 and mPer2 gene induction and influences phase shifts of the circadian clock. Curr Biol 13(9):725–733

249. Revermann M, Maronde E, Ruth P, Korf HW (2002) Protein kinase G I immunoreaction is colocalized with arginine-vasopressin immunoreaction in the rat suprachiasmatic nucleus. Neurosci Lett 334(2):119–122

250. Weber ET, Gannon RL, Rea MA (1995) cGMP-dependent protein kinase inhibitor blocks light-induced phase advances of circadian rhythms in vivo. Neurosci Lett 197(3):227–230

251. Mathur A, Golombek DA, Ralph MR (1996) cGMP-dependent protein kinase inhibitors block light-induced phase advances of circadian rhythms in vivo. Am J Physiol 270 (5 Pt 2):R1031–R1036

252. Tischkau SA, Mitchell JW, Pace LA, Barnes JW, Barnes JA, Gillette MU (2004) Protein kinase G type II is required for night-to-day progression of the mammalian circadian clock. Neuron 43(4):539–549

253. Langmesser S, Franken P, Feil S, Emmenegger Y, Albrecht U, Feil R (2009) cGMP-dependent protein kinase type I is implicated in the regulation of the timing and quality of sleep and wakefulness. PLoS One 4(1):e4238

254. Sold G, Hofmann F (1974) Evidence for a guanosine-3′:5′-monophosphate-binding protein from rat cerebellum. Eur J Biochem 44(1):143–149

255. Lohmann SM, Walter U, Miller PE, Greengard P, De Camilli P (1981) Immunohistochemical localization of cyclic GMP-dependent protein kinase in mammalian brain. Proc Natl Acad Sci USA 78(1):653–657

256. Li Z, Xi X, Gu M, Feil R, Ye RD, Eigenthaler M, Hofmann F, Du X (2003) A stimulatory role for cGMP-dependent protein kinase in platelet activation. Cell 112(1):77–86

257. Kleppisch T, Pfeifer A, Klatt P, Ruth P, Montkowski A, Fassler R, Hofmann F (1999) Long-term potentiation in the hippocampal CA1 region of mice lacking cGMP-dependent kinases is normal and susceptible to inhibition of nitric oxide synthase. J Neurosci 19(1):48–55

258. Kleppisch T, Wolfsgruber W, Feil S, Allmann R, Wotjak CT, Goebbels S, Nave KA, Hofmann F, Feil R (2003) Hippocampal cGMP-dependent protein kinase I supports an age- and protein synthesis-dependent component of long-term potentiation but is not essential for spatial reference and contextual memory. J Neurosci 23(14):6005–6012

259. Arancio O, Antonova I, Gambaryan S, Lohmann SM, Wood JS, Lawrence DS, Hawkins RD (2001) Presynaptic role of cGMP-dependent protein kinase during long-lasting potentiation. J Neurosci 21(1): 143–149

260. Song H, Ming G, He Z, Lehmann M, McKerracher L, Tessier-Lavigne M, Poo M (1998) Conversion of neuronal growth cone responses from repulsion to attraction by cyclic nucleotides. Science 281(5382): 1515–1518

261. Schmidt H, Werner M, Heppenstall PA, Henning M, More MI, Kuhbandner S, Lewin GR, Hofmann F, Feil R, Rathjen FG (2002) cGMP-mediated signaling via cGKIalpha is required for the guidance and connectivity of sensory axons. J Cell Biol 159(3):489–498

262. Schmidt H, Stonkute A, Juttner R, Schaffer S, Buttgereit J, Feil R, Hofmann F, Rathjen FG (2007) The receptor guanylyl cyclase Npr2 is essential for sensory axon bifurcation within the spinal cord. J Cell Biol 179(2):331–340

263. Tegeder I, Del Turco D, Schmidtko A, Sausbier M, Feil R, Hofmann F, Deller T, Ruth P, Geisslinger G (2004) Reduced inflammatory hyperalgesia with preservation of acute thermal nociception in mice lacking cGMP-dependent protein kinase I. Proc Natl Acad Sci USA 101(9):3253–3257

264. Luo C, Gangadharan V, Bali KK, Xie RG, Agarwal N, Kurejova M, Tappe-Theodor A, Tegeder I, Feil S, Lewin G, Polgar E, Todd AJ, Schlossmann J, Hofmann F, Liu DL, Hu SJ, Feil R, Kuner T, Kuner R (2012) Presynaptically localized cyclic GMP-dependent protein kinase 1 is a key determinant of spinal synaptic potentiation and pain hypersensitivity. PLoS Biol 10(3):e1001283

265. Pfeifer A, Aszodi A, Seidler U, Ruth P, Hofmann F, Fassler R (1996) Intestinal secretory defects and dwarfism in mice lacking cGMP- dependent protein kinase II. Science 274(5295):2082–2086

266. Vaandrager AB, Smolenski A, Tilly BC, Houtsmuller AB, Ehlert EM, Bot AG, Edixhoven M, Boomaars WE, Lohmann SM, de Jonge HR (1998) Membrane targeting of cGMP-dependent protein kinase is required for cystic fibrosis transmembrane conductance regulator Cl- channel activation. Proc Natl Acad Sci USA 95(4):1466–1471

267. Vaandrager AB, Bot AG, Ruth P, Pfeifer A, Hofmann F, De Jonge HR (2000) Differential role of cyclic GMP-dependent protein kinase II in ion transport in murine small intestine and colon. Gastroenterology 118(1):108–114

268. Cha B, Kim JH, Hut H, Hogema BM, Nadarja J, Zizak M, Cavet M, Lee-Kwon W, Lohmann SM, Smolenski A, Tse CM, Yun C, de Jonge HR, Donowitz M (2005) cGMP inhibition of Na+/H+ antiporter 3 (NHE3) requires PDZ domain adapter NHERF2, a broad specificity protein kinase G-anchoring protein. J Biol Chem 280(17):16642–50

269. Foller M, Feil S, Ghoreschi K, Koka S, Gerling A, Thunemann M, Hofmann F, Schuler B, Vogel J, Pichler B, Kasinathan RS, Nicolay JP, Huber SM, Lang F, Feil R (2008) Anemia and splenomegaly in cGKI-deficient mice. Proc Natl Acad Sci USA 105(18):6771–6776. doi:10.1073/pnas.0708940105

270. Nikolaev VO, Gambaryan S, Engelhardt S, Walter U, Lohse MJ (2005) Real-time monitoring of the PDE2 activity of live cells: hormone-stimulated cAMP hydrolysis is faster than hormone-stimulated cAMP synthesis. J Biol Chem 280(3):1716–1719

271. MacFarland RT, Zelus BD, Beavo JA (1991) High concentrations of a cGMP-stimulated phosphodiesterase mediate ANP-induced decreases in cAMP and steroidogenesis in adrenal glomerulosa cells. J Biol Chem 266(1):136–142

272. Kurtz A, Gotz KH, Hamann M, Wagner C (1998) Stimulation of renin secretion by nitric oxide is mediated by phosphodiesterase 3. Proc Natl Acad Sci USA 95(8):4743–4747

273. Henrich WL, McAllister EA, Smith PB, Campbell WB (1988) Guanosine 3′,5′-cyclic monophosphate as a mediator of inhibition of renin release. Am J Physiol 255(3 Pt 2): F474–F478

274. Schricker K, Kurtz A (1993) Liberators of NO exert a dual effect on renin secretion from isolated mouse renal juxtaglomerular cells. Am J Physiol 265(2 Pt 2):F180–F186

275. Gambaryan S, Hausler C, Markert T, Pohler D, Jarchau T, Walter U, Haase W, Kurtz A, Lohmann SM (1996) Expression of type II cGMP-dependent protein kinase in rat kidney is regulated by dehydration and correlated with renin gene expression. J Clin Invest 98(3):662–670

276. Wagner C, Pfeifer A, Ruth P, Hofmann F, Kurtz A (1998) Role of cGMP-kinase II in

the control of renin secretion and renin expression. J Clin Invest 102(8):1576–1582

277. Gambaryan S, Butt E, Marcus K, Glazova M, Palmetshofer A, Guillon G, Smolenski A (2003) cGMP-dependent protein kinase type II regulates basal level of aldosterone production by zona glomerulosa cells without increasing expression of the steroidogenic acute regulatory protein gene. J Biol Chem 278(32):29640–29648

278. Spiessberger B, Bernhard D, Herrmann S, Feil S, Werner C, Luppa PB, Hofmann F (2009) cGMP-dependent protein kinase II and aldosterone secretion. FEBS J 276(4): 1007–1013

279. Schinner E, Hofmann F, Schlossmann J (2012) Role of cGMP-dependent protein kinase I for kidney fibrosis. N-Schmied Arch Pharmacol 385(Suppl1):S81

280. Chusho H, Tamura N, Ogawa Y, Yasoda A, Suda M, Miyazawa T, Nakamura K, Nakao K, Kurihara T, Komatsu Y, Itoh H, Tanaka K, Saito Y, Katsuki M (2001) Dwarfism and early death in mice lacking C-type natriuretic peptide. Proc Natl Acad Sci USA 98(7):4016–4021

281. Yasoda A, Komatsu Y, Chusho H, Miyazawa T, Ozasa A, Miura M, Kurihara T, Rogi T, Tanaka S, Suda M, Tamura N, Ogawa Y, Nakao K (2004) Overexpression of CNP in chondrocytes rescues achondroplasia through a MAPK-dependent pathway. Nat Med 10(1):80–86

282. Talts JF, Pfeifer A, Hofmann F, Hunziker EB, Zhou XH, Aszodi A, Fassler R (1998) Endochondral ossification is dependent on the mechanical properties of cartilage tissue and on intracellular signals in chondrocytes. Ann N Y Acad Sci 857:74–85

283. Miyazawa T, Ogawa Y, Chusho H, Yasoda A, Tamura N, Komatsu Y, Pfeifer A, Hofmann F, Nakao K (2002) Cyclic GMP-dependent protein kinase II plays a critical role in C-type natriuretic peptide-mediated endochondral ossification. Endocrinology 143(9):3604–3610

284. Nakano Y, Nagamatsu Y, Ohshima Y (2004) cGMP and a germ-line signal control body size in C. elegans through cGMP-dependent protein kinase EGL-4. Genes Cells 9(9):773–779

285. Chikuda H, Kugimiya F, Hoshi K, Ikeda T, Ogasawara T, Shimoaka T, Kawano H, Kamekura S, Tsuchida A, Yokoi N, Nakamura K, Komeda K, Chung UI, Kawaguchi H (2004) Cyclic GMP-dependent protein kinase II is a molecular switch from proliferation to hypertrophic differentiation of chondrocytes. Genes Dev 18(19):2418–2429

286. Zhao X, Zhuang S, Chen Y, Boss GR, Pilz RB (2005) Cyclic GMP-dependent protein kinase regulates CCAAT enhancer-binding protein beta functions through inhibition of glycogen synthase kinase-3. J Biol Chem 280(38): 32683–32692

287. Kawasaki Y, Kugimiya F, Chikuda H, Kamekura S, Ikeda T, Kawamura N, Saito T, Shinoda Y, Higashikawa A, Yano F, Ogasawara T, Ogata N, Hoshi K, Hofmann F, Woodgett JR, Nakamura K, Chung UI, Kawaguchi H (2008) Phosphorylation of GSK-3beta by cGMP-dependent protein kinase II promotes hypertrophic differentiation of murine chondrocytes. J Clin Invest 118(7):2506–2515

288. Yaroslavskiy BB, Zhang Y, Kalla SE, Garcia Palacios V, Sharrow AC, Li Y, Zaidi M, Wu C, Blair HC (2005) NO-dependent osteoclast motility: reliance on cGMP-dependent protein kinase I and VASP. J Cell Sci 118(Pt 23):5479–5487

289. Yaroslavskiy BB, Turkova I, Wang Y, Robinson LJ, Blair HC (2010) Functional osteoclast attachment requires inositol-1,4,5-trisphosphate receptor-associated cGMP-dependent kinase substrate. Lab Invest 90(10): 1533–1542

290. Lenz A, Bennett M, Skelton WP, Vesely DL (2010) Vessel dilator and C-type natriuretic peptide enhance the proliferation of human osteoblasts. Pediatr Res 68(5):405–408

291. Rangaswami H, Schwappacher R, Marathe N, Zhuang S, Casteel DE, Haas B, Chen Y, Pfeifer A, Kato H, Shattil S, Boss GR, Pilz RB (2010) Cyclic GMP and protein kinase G control a Src-containing mechanosome in osteoblasts. Sci Signal 3(153):ra91

292. Marathe N, Rangaswami H, Zhuang S, Boss GR, Pilz RB (2012) Pro-survival effects of 17beta-estradiol on osteocytes are mediated by nitric oxide/cGMP via differential actions of cGMP-dependent protein kinases I and II. J Biol Chem 287(2):978–988

293. Chow JW, Fox SW, Lean JM, Chambers TJ (1998) Role of nitric oxide and prostaglandins in mechanically induced bone formation. J Bone Miner Res 13(6):1039–1044

294. Spinas GA, Laffranchi R, Francoys I, David I, Richter C, Reinecke M (1998) The early phase of glucose-stimulated insulin secretion requires nitric oxide. Diabetologia 41(3): 292–299

295. Panagiotidis G, Alm P, Lundquist I (1992) Inhibition of islet nitric oxide synthase increases arginine-induced insulin release. Eur J Pharmacol 229(2–3):277–278

296. Alm P, Ekstrom P, Henningsson R, Lundquist I (1999) Morphological evidence for the existence of nitric oxide and carbon monoxide pathways in the rat islets of Langerhans:

an immunocytochemical and confocal microscopical study. Diabetologia 42(8):978–986

297. Schmidt HH, Warner TD, Ishii K, Sheng H, Murad F (1992) Insulin secretion from pancreatic B cells caused by L-arginine-derived nitrogen oxides. Science 255(5045): 721–723

298. Salehi A, Carlberg M, Henningson R, Lundquist I (1996) Islet constitutive nitric oxide synthase: biochemical determination and regulatory function. Am J Physiol 270(6 Pt 1):C1634–C1641

299. Leiss V, Friebe A, Welling A, Hofmann F, Lukowski R (2011) Cyclic GMP kinase I modulates glucagon release from pancreatic alpha-cells. Diabetes 60(1):148–156

300. Lutz SZ, Hennige AM, Feil S, Peter A, Gerling A, Machann J, Krober SM, Rath M, Schurmann A, Weigert C, Haring HU, Feil R (2011) Genetic ablation of cGMP-dependent protein kinase type I causes liver inflammation and fasting hyperglycemia. Diabetes 60(5):1566–1576

301. Haas B, Mayer P, Jennissen K, Scholz D, Berriel Diaz M, Bloch W, Herzig S, Fassler R, Pfeifer A (2009) Protein kinase G controls brown fat cell differentiation and mitochondrial biogenesis. Sci Signal 2(99):ra78. doi:10.1126/scisignal.2000511

Chapter 3

Enzyme Assays for cGMP Hydrolyzing Phosphodiesterases

Sergei D. Rybalkin, Thomas R. Hinds, and Joseph A. Beavo

Abstract

Cyclic nucleotides (cAMP and cGMP) as second messengers regulate a wide variety of biological processes such as cellular growth, secretary signaling, and neuroplasticity. These processes can be regulated by increasing the synthesis of cyclic nucleotides (cyclases), by regulation of cAMP and cGMP effector proteins such as cAMP- and cGMP-dependent protein kinases, or by regulation of cyclic nucleotide degradation via cyclic nucleotide phosphodiestases (PDEs). At present PDEs are classified into 11 gene families, each containing several different isoforms and splice variants. All PDEs share considerable homology in their catalytic domains but substantially differ in their N-terminal regions, that contain different types of regulatory. The different PDEs show complex substrate specificity. PDE5, PDE6, and PDE9 are considered to be cGMP specific, while PDE1, PDE2, PDE3, PDE10, and PDE11 can hydrolyze both cGMP and cAMP. PDE4, PDE7, and PDE8 use mainly cAMP as their substrates at physiological substrate levels. Here we describe two methods designed for measuring cGMP (cAMP) hydrolytic activities. The first one is a traditional method using radioactive substrates and the second one is a recently developed nonradioactive method based on Isothermal Titration Calorimetry.

Key words Phosphodiesterases, cGMP, Hydrolysis, Radioactivity, Isothermal titration calorimetry

1 Introduction

Cyclic nucleotide phosphodiesterases provide precise control of intracellular levels of second messengers, cAMP and cGMP by hydrolyzing them to their corresponding inactive 5′-nucleotide [1, 2]. Since 11 PDE families differ in their substrate specificity, regulation, and pharmacological properties [3], a number of different PDE assays have been developed. The original methods for determining PDE activity were fixed-time assays that all used a coupled enzyme assay that ended up measuring the release of inorganic phosphate. Usually in these reactions the coupled step was conversion of the product of the PDE reaction (either 5′AMP or 5′GMP) to the corresponding nucleoside and phosphate by an excess of some form of 5′nucleotidase (most commonly *Crotalus atrox* snake venom). This assay had the advantage of simplicity, but the disadvantage of requiring very sensitive phosphate measurement

Thomas Krieg and Robert Lukowski (eds.), *Guanylate Cyclase and Cyclic GMP: Methods and Protocols*, Methods in Molecular Biology, vol. 1020, DOI 10.1007/978-1-62703-459-3_3, © Springer Science+Business Media, LLC 2013

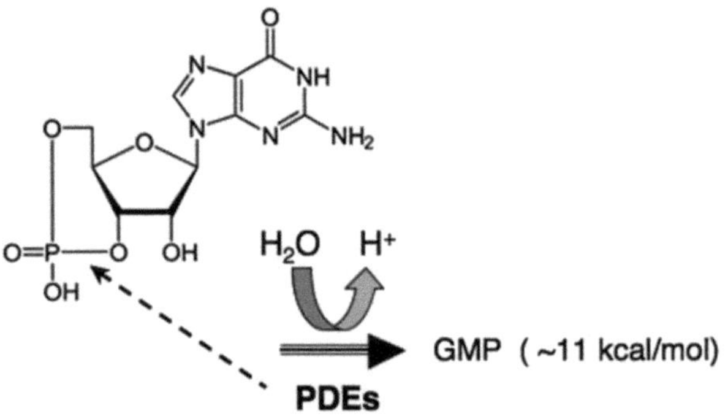

Fig. 1 A schematic illustration of cGMP PDE reaction, showing the 3′ cyclic phosphate bond of cGMP, hydrolyzed by PDEs

procedures to be applicable for most of the high-affinity low-K_m PDEs. However, an adaptation of this assay is still occasionally used in high throughput format for purified preparations of several of the higher K_m/V_{max} PDEs, such as PDE6 [4].

As a result of the sensitivity issues and the requirement for purified preparations of enzyme, the inorganic phosphate release assays were largely surplanted by methods for measuring PDE activities that utilized radioactive labeled cyclic nucleotides as substrates, such as [^{3}H], [^{14}C] or [^{32}P] cAMP/or cGMP, again followed by a second reaction to produce adenosine/guanosine, and their separation from unhydrolyzed substrates by ion-exchange resin [5, 6]. This assay is the only one presently available that can efficiently measure the activity of the very low K_m PDEs (e.g., PDE3, PDE7, PDE8, and PDE9). This basic assay is described below (*see* Subheading 3.1).

Another assay, the pH based PDE assay [7], makes use of the fact that a proton is released during the hydrolysis of cNMP to 5′NMP (Fig. 1). This assay has the advantage that with an appropriately sensitive pH meter, PDE activity can be measured in real time. The pH assay has been mostly applied for measuring activity of the light-activated, cGMP-specific PDE6 since this PDE isozyme has such high catalytic activity. However, in principle, it can be used for any of the higher V_{max} PDEs where substrate concentrations in the mM range can be utilized. This PDE assay has the real advantage of allowing real time measurement of activity and is not limited to selected fixed-time responses. It can also be used with most cNMP analogs if they are a reasonable substrate.

More recently, higher throughput PDE assays such as a luminescence based PDE method [8] and method using fluorescein-labeled cyclic nucleotides for fluorescence polarization [9] have been developed. In general, all of these protocols to measure PDE hydrolytic activities utilize either cAMP or cGMP or their fluorescent analogs as substrates.

Finally, there has been a need to determine the 3′–5′ cyclic nucleotide phosphodiesterase activity of the various PDEs using various commonly used cyclic nucleotide analogs as substrates. Since many of these analogs are extensively used in intact cells as tools to elucidate cyclic nucleotide regulated pathways, researchers have needed to know if they were substrates for, inhibitors of, or activators of the various PDEs present in the cells. As most of these compounds are not available in a radioactive form and many of them are rather poor substrates, this has been a difficult task. Therefore, in the second part of this chapter we also describe the use of a nonradioactive method for direct determination of PDE enzyme activity by use of Isothermal Titration Calorimetry (ITC) (*see* Subheading 3.2).

ITC is a useful method for the characterization of the thermodynamics of binding and kinetic parameters of the reaction. The ITC PDE assay is based on the fact that the free energy of hydrolysis of 3′ the phosphate bonds of cAMP and cGMP are comparable with the hydrolysis of ATP and found to be 14 kcal/mol and 11 kcal/mol, respectively [10]. This method was originally applied for cyclic nucleotide hydrolysis by measuring the kinetic parameters of PDEs and also their hydrolysis of cyclic nucleotide analogs [11].

2 Materials

1. DEAE-Sephadex A-25: make a slurry with the resin and deionized water (10 g in 100 mL of water) and after a couple of hours use approximately 5 mL for filling the columns. Packed volume of column is 0.25 mL.
2. Snake venom: 5′ nucleotidase (from *C. atrox* venom), dilute to stock concentration of 2.5 mg/mL in water, store in the −20 °C.
3. Low-salt buffer: 20 mM Tris–HCl, pH 6.8 (Room Temperature).
4. High-salt buffer: 20 mM Tris–HCl, pH 6.8 + 500 mM NaCl.
5. Columns for PDE assays: 5″ Polypropylene Chromatography column, 120 μm filter (Evegreen Scientific, LA).
6. Reaction buffer A (5×) for PDE assay: 100 mM Tris–HCl, pH 7.5, 4 mM EGTA, 1.0 mg/mL BSA.
7. Reaction buffer B (5×) for PDE assay: 100 mM Tris–HCl, pH 7.5, 75 mM MgAcetate, 100,000 cpm [^{3}H]-cAMP or [^{3}H]-cGMP and unlabeled substrates.
8. Reaction buffer C (5×) to assay calmodulin-dependent PDE (PDE1) in the presence of calmodulin:100 mM Tris–HCl, pH 7.5, 100 mM Imidazole, 15 mM $MgCl_2$, 1.0 mg/mL BSA, 20 μg/mL calmodulin, 0.2 mM $CaCl_2$.

9. Reaction buffer D (5×) to assay calmodulin-dependent PDE (PDE1) in the presence of EGTA: the same as buffer C, except 10 mM EGTA replaces calmodulin.
10. Scintillation fluid (Ultima Gold, PerkinElmer).
11. Liquid scintillation counter (PerkinElmer).
12. The reaction buffer for the calorimetric assays: 40 mM Mops, pH 7.5, 1 mM $MgCl_2$.
13. Purified recombinant PDEs, used for the calorimetric assays, are diluted in 40 mM MOPS pH 7.5 (concentrations: ~0.1–10 nM).
14. Calorimeter: VP-ITC MicroCalorimeter (Microcal Inc., Northampton, MA).

3 Methods

3.1 PDE Assay Using [^{3}H]-cGMP (or cAMP) as Substrates

This Phosphodiesterase Assay is a modification of previously published methods [12, 13]. The assay reaction involves a two steps procedure. In the first step, the phosphodiesterase hydrolyzes the cyclic nucleotide and produces a 5′-monoposphate product. This reaction is terminated by boiling the sample for 1 min and then cooling it. In the second step, *C. atrox* snake venom, which contains a 5′ phosphatase activity but little or no PDE activity, is added to the sample (*see* **Note 1a**). Since both cyclic nucleotide and 5′-derivative contain a negative charge while the nucleoside does not, the radiolabeled nucleoside can be easily separated from each using any of several different ion exchange resins.

1. Pipette 50 μL buffer A into 12×75 mm disposable glass tubes.
2. Add buffer B (*see* **Note 1b**) and inhibitors (if required). Add enough dH_2O to bring reaction volume to 200 μL.
3. Initiate reaction by addition of 50 μL of PDE sample (cell or tussues extract or purified PDE). Total reaction volume should be 250 μL. Vortex and incubate at 30 °C for desired time (usually 5–10 min).
4. To analyze calmoduli-dependent PDE1 activity use buffer C and buffer D to measure basal non-stimulated activity of PDE1.
5. Stop the reaction by placing the tube in a boiling water bath for 1 min. After allowing the tube to cool, add 10 μL of 2.5 mg/mL snake venom. Incubate for 5 min at 30 °C.
6. Dilute the assay with 250 μL of Low-salt buffer and transfer the entire sample to an ion exchange resin column, which has been previously washed with 8 mL High-salt buffer followed

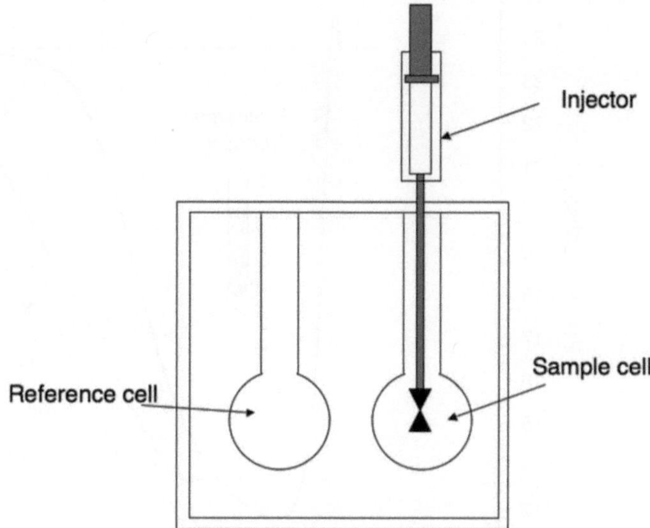

Fig. 2 The VP-ITC unit. A spinning syringe is used for injections and mixing of reaction components. Temperature differences between the reference cells and the sample cells are recorded and used for calculation of PDE activities

by 8 mL Low-salt buffer. The packed volume of the exchange resin is 0.6 mL (*see* **Note 1c**).

7. Elute ^{3}H-nucleoside from resin with 4 × 0.5 mL washes of Low-salt buffer, collecting eluates in a scintillation vial (*see* **Note 1d**).
8. Add 3.0 mL scintillation fluid and mix thoroughly. Count on a liquid scintillation counter after allowing any luminescence to subside. Use an Excel program that can calculate a specific activity (*see* **Note 1e**).
9. Specific Activity is calculated as moles cyclic nucleotides hydrolyzed per min-mL:

   (assay cpm − background cpm){(cGMP, mol)/net cpm per assay}(vol, mL) − 1(time, min) − 1.

3.2 PDE Assays Using Isothermal Titration Calorimetry

The VP-ITC MicroCalorimeter has two cells (1.42 mL each): one a reference that is filled with buffer and the other a measuring cell, containing the reaction mixture with enzyme (Fig. 2). The calorimeter measures the current, applied to each cell to maintain them at the same temperature. As the enzymatic hydrolysis proceeds heat is generated and there is less current applied to the measuring cell; the difference in the amount of current applied to the two cells is measured as differential power (current times heater resistance) (Fig. 3).

The initial downward phase is where the substrate is mixed into the reaction media while the enzyme is also hydrolyzing the substrate (Fig. 4). The next phase is where the uniformly mixed

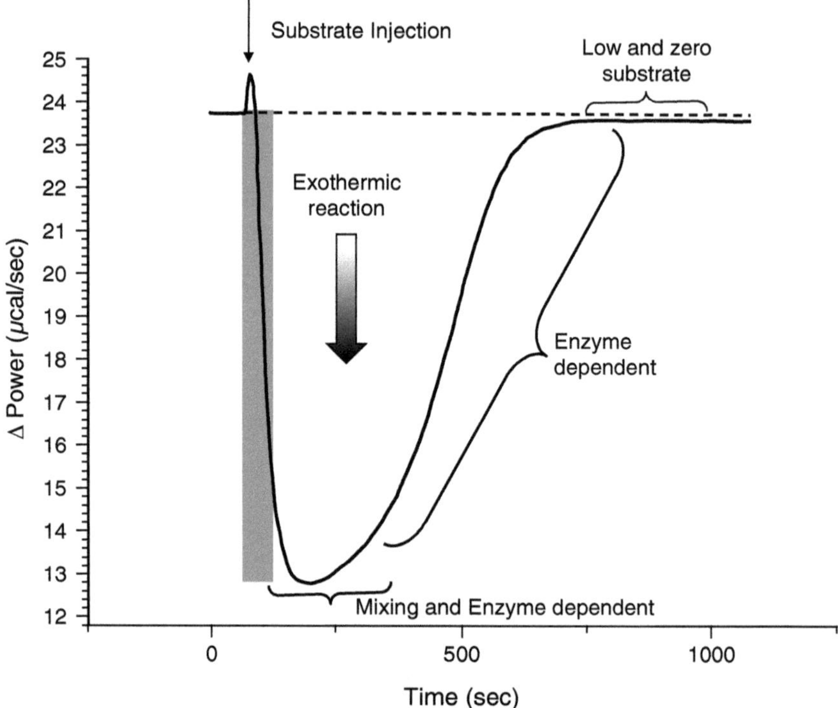

Fig. 3 Typical thermograph of cyclic nucleotide hydrolysis. cGMP (3–20 μL) is injected after 60 s of stable baseline. Reaction cell contains 1.42 mL of PDE2 enzyme in 40 μM Mops, 1.0 mM $MgCl_2$

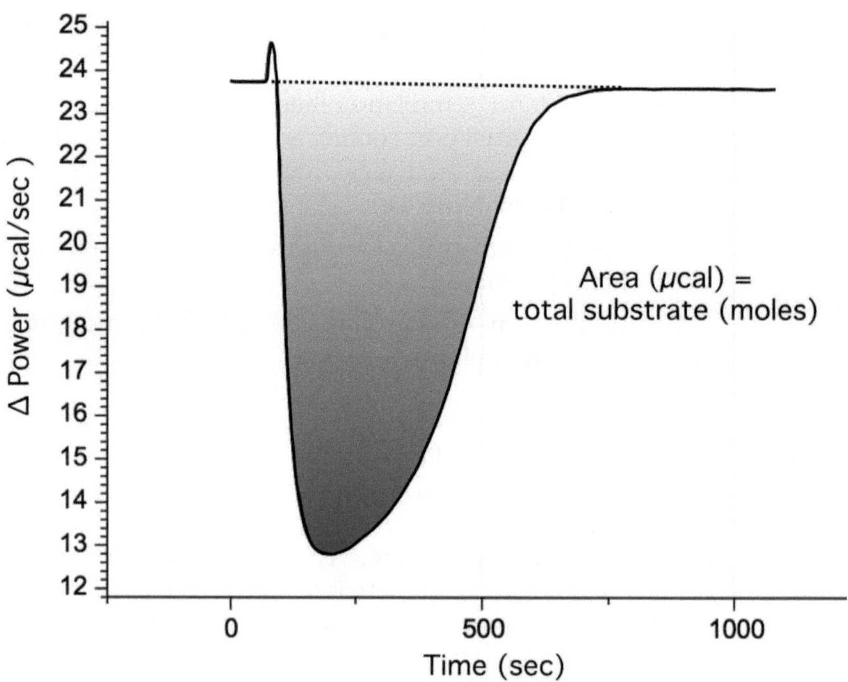

Fig. 4 Determination of the amount of heat generated during the PDE reaction. Total area (minus the injection peak) is equal to the total amount of heat produced by the complete hydrolysis of cGMP. The conversion factor (calories per moles) is calculated from this plot as the molar heat of hydrolysis

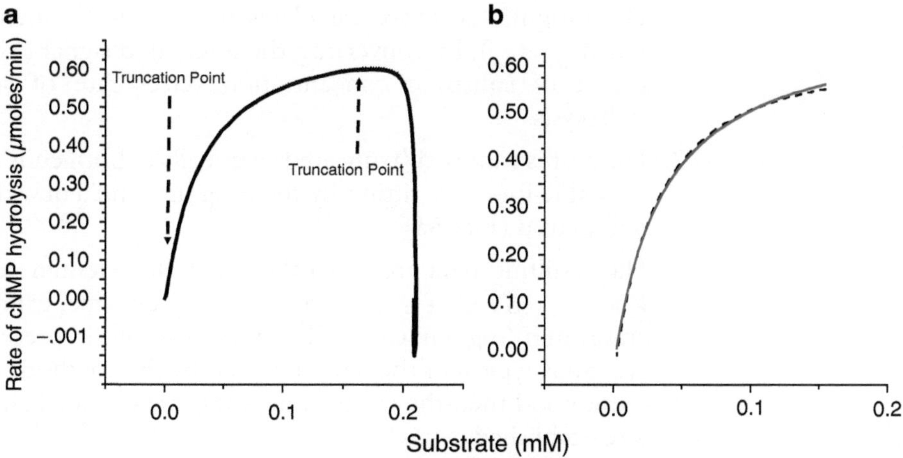

Fig. 5 Calculation of PDE activity data from the ITC data. (**a**) Replotting of data as hydrolysis rate versus substrate by the calorimeter program. Very small and high end numbers are removed due to mixing artifacts. (**b**) Determination of kinetic parameters of PDEs. The truncated data from **a** is nonlinearly fit to the Michaelis–Menten equation: $V = V_{max} \times (S / K_m + S)$, where K_m—the Michaelis–Menten constant, V_{max}—the maximum enzyme activity, S—substrate concentration. The *dashed line* corresponds to data obtained during measurements, the *continuous line* corresponds to fit curve. All reactions were done in triplicate or more

substrate is getting enzymatically depleted while generating heat. The final phase is where the reaction is completed and the signal returns to the baseline. The total area of heat production (μcal) is proportional to the total amount of substrate (μmol). The Origin™ software provided by the calorimeter manufacturer calculates the heat of hydrolysis and converts these values into substrate concentrations and rate of hydrolysis (Fig. 5).

3.2.1 PDE Kinetic Analysis (Using PDE2 as a Sample)

1. The calorimeter requires thermal stability at 30 °C before injection of the substrate.
2. 10 μL of cGMP at a concentration of 30 mM is injected into the reaction cell (1.42 mL) with continuous monitoring of heat production at 2-s intervals (*see* **Note 2a**).
3. The injection syringe, used as a stirring device at 310 rpm, is able to efficiently mix the substrate (cGMP) with PDE samples (recombinant PDE2). The concentration of PDE2 in the 1.42 mL reaction cell is at 2 nM (*see* **Note 2b**).
4. Total reaction times are usually within 15–40 min (*see* **Note 2c**).
5. The tested compounds, such as cyclic nucleotide analogs at concentrations from 2 to 30 mM, are injected into the reaction cells when needed.

3.2.2 Calculation of PDE Kinetic Parameters Obtained by ITC

Todd and Gomez described a method where the kinetic parameters, K_m (the Michaelis–Menten constant), K_{cat} (the catalytic rate constant of an enzyme), and V_{max} (the maximum enzyme activity) can be calculated from the data obtained from the ITC [14] (*see* **Note 2d**).

1. The Origin™ software calculates the kinetic parameters, K_m and K_{cat} (V_{max}), by converting the integrated signal (Fig. 4) to a plot of substrate concentration verses rate of substrate hydrolysis.
2. The plot is trimmed (truncated) manually at both ends to make it useable for curve fitting by the program, and a new data plot is displayed (Fig. 5a).
3. The resulting data are fit to the Michaelis–Menten equation; $V = V_{max} \times (S / K_m + S)$, where $V_{max} = K_{cat}$/enzyme, expressed as μmol/min/mg enzyme. The goodness of fit presented in Fig. 5b is typical of the data obtained by this method. If the fit is not good then the truncation points are changed and the fit is reestablished.
4. Identification of inhibition type and kinetic formula are obtained as described in Enzyme Kinetics Segel [15].
   (a) K_m and V_{max} values of inhibited reactions are compared to uninhibited K_m and V_{max} reactions. A two-tailed T test ($P<0.05$) is used to determine if the average values are different.
   (b) In case of using cyclic nucleotide analogs different types of kinetic reactions could be detected. For example, some analogs show no effect on the activity of the PDEs, while some can be inhibitors or substrates. In total comparing substrates and analogs on the same day and same enzyme preparation, three types of inhibition can be identified:
   - Competitive; if the V_{max} is the same for inhibited and uninhibited reaction, but the K_m for the inhibited reaction is greater than the K_m for uninhibited reaction;
   - Non-competitive; if the inhibited V_{max} is less than the uninhibited V_{max}, but the K_m is the same for both reactions;
   - Linear mixed; if the inhibited V_{max} is less than the uninhibited V_{max}, but the K_m for the Inhibited reaction is greater than the K_m for uninhibited reaction.

4 Notes

1. Notes for PDE assay (*see* Subheading 3.1).
   (a) The PDE assay has the advantage of being sufficiently sensitive to allow measurement of lower K_m/V_{max} PDEs even in crude homogenates without removal of phosphates and organic phosphate containing substances. It still has the disadvantage of being a fixed-time assay and therefore provides only an estimate of basic enzymatic rates, especially at lower substrate levels (i.e., below the K_m). In practice,

however, the assay is reasonably accurate as long as total hydrolysis is kept below about 30 % even for basic kinetic analysis.

(b) Substrate and Conditions for PDE isozymes are always determined by the type of PDE isoforms present in the tested tissue. For example, calmodulin/calcium dependence or dual cAMP and cGMP specificity would require different assay conditions [16].

(c) In tissues with high concentrations of guanine deaminase, [^{3}H]-xanthine can be produced from the [^{3}H]-guanosine. [^{3}H]-xanthine, in contrast to [^{3}H]-guanosine, will bind to most ion-exchange columns at neutral pH, thus potentially contributing to an underestimation of cGMP PDE activities. However, at pH 6.8 [^{3}H]-xanthine does not bind to the DEAE columns described below.

(d) The columns for PDE assay can be regenerated many times by washing two times with High-salt buffer and then two times with Low-salt buffer.

(e) The procedures for PDE assay can be adjusted to a 125 μL total volume assay by cutting all volumes in half.

2. Notes for the ITC PDE assay (*see* Subheading 3.2).

(a) The substrate concentration of cAMP, cGMP, or cyclic nucleotide analogs in the injection syringe is adjusted to be about 1,000–2,000 times the K_m of the PDE and the volumes can be between 3–20 μL.

(b) Purified recombinant PDEs are diluted in 40 mM MOPS pH 7.5 (concentrations: ~0.1–10 nM). Injecting 10 μL of cyclic nucleotides or their analogs into the reaction cell (1.42 mL) gives a final concentration of between 14 and 28 μM; about 7–14 times above the K_m for PDE2. Volumes and concentrations are adjusted appropriately for other PDEs.

(c) The heat generation is measured for about 15–20 min. The analysis takes about 5 min, so the total time to run one complete experiment is about 45 min. In practical terms about one experiment per hour is typical. In a day one can easily determine in triplicate the K_m, K_{cat}, and K_i for one analog.

(d) Application of ITC for analysis of hydrolysis of cyclic nucleotide analogs by PDEs.

Previously we reported kinetic parameters for several PDEs, obtained by the ITC method in comparison with published kinetic data for the same PDEs, measured by rational PDE assays [11]. That data showed that many of the cyclic nucleotide analogs could act as PDE inhibitors and several were substrates, depending on the types of PDEs expressed in a particular cell or tissue. Here we

Table 1
Kinetic parameters of selected PDEs measured by the ITC

Enzyme	cGMP		cAMP	
	K_m	V_{max}	K_m	V_{max}
PDE1A	8.2 ± 1.0	20 ± 1.0	93 ± 12	41 ± 4.0
PDE1B	5.4 ± 2.7	2.6 ± 0.7	33 ± 3.8	1.5 ± 0.5
PDE1C	4.6 ± 0.1	1.4 ± 1.0	3.2 ± 0.1	16 ± 0.3
PDE2A	31 ± 3.2	176 ± 31	112 ± 33	215 ± 37
PDE4D	n.h.	n.h.	5.5 ± 0.4	63 ± 8.0
PDE5	2.0 ± 0.5	6 ± 1.0	201 ± 17	20 ± 2.9
PDE6	10 ± 1.0	1.0 ± 1.1	823 ± 54	1.6 ± 0.1
PDE8A	n.h.	n.h.	0.6 ± 0.2	43 ± 13
PDE9A	0.2 ± 0.02	1.0 ± 1.1	230 ± 35	21 ± 10
PDE10	1.1 ± 0.03	1.6 ± 0.11	0.2 ± 0.04	0.67 ± 0.01

Data for PDEs1–5 and PDE10 taken from Poppe et al. [11] (K_m, μM; V_{max}, μM/min/mg; n.h. = not hydrolyzed)
For PDE6 and PDE9, V_{max} data, measured with cGMP as substrates, are normalized to 1

provide a somewhat expanded version of those tables by adding new data for PDE8 and PDE9 (Tables 1 and 2). Please note, however, that K_m values for PDE8 appeared to be higher—0.63 μM than was published by other methods. Although it is not clear why this apparent discrepancy exists, it is possible that at the higher enzyme and substrate concentrations, needed for the ITC measurement, a lower affinity form of PDE8 could be formed. Also this method is inherently more accurate for higher K_m PDEs.

Table 2 shows the selectivity and cross-reactivity of cyclic nucleotide analogs determined by the ITC method. Interestingly, 8-Br-cGMP can be a substrate for PDE8A with a K_m as low as 0.55 μM. For PDE9, 8-Br-cAMP is a reasonable substrate (K_m = 5.99 μM), while 8-Br-cGMP and 8-pCPT-cGMP are potentially effective inhibitors in the cell (K_i = 0.77 μM and 1.29 μM respectively).

Therefore, the data obtained by the ITC, underline the importance of careful analysis of all cyclic nucleotide analogs used as probes of cAMP or cGMP signaling pathways, since these analogs could have multiple targets in different types of cells. For example the Epac selective activator, Sp-8-pCPT-2-0-Me- cAMPS, while not a substrate for any PDE is quite a good inhibitor of several of them. Similarly the slightly less effective Epac inhibitor, 8-pCPT-2-0-Me cAMP is a rather good substrate for PDE5.

Table 2
Properties of cyclic nucleotide analogs as PDE substrates and inhibitors by isothermal microcalorimetry

Analog	PDE1A	PDE1B	PDE1C	PDE2	PDE4	PDE5	PDE6	PDE8A	PDE9	PDE10
	K_m/K_i	K_m/K_i	K_m/K_i	K_m/K_i	K_m/K_i	K_m/K_i	K_m/K_i	K_m/K_i	K_m/K_i	K_m/K_i
cAMP	93.1	33.0	3.19	112	5.52	201	823	0.63	230	0.24
8Br-cAMP	32.8	63.6	18.0	38.7	54.4	23.4	68.8	n.e.	5.99	4.83
cGMP	8.23	5.35	4.64	31.0	24.5	2.01	10	n.e.	0.20	23.7
8-Br-cGMP	47.2	4.48	62.8	90.9	30.1	79.2	33	0.55	0.77	23.7
Sp-5,6-DCl-cBMPS	89.6	2.25	55.8	17.7	18.8	15.1	24.5	26.7	n.e.	4.75
6-Bnz-cAMP	n.e.	9.86	323	240	49	68.9	64.6	n.e.	2.27	45.6
8-pCPT-2-0-Me cAMP	50.5	8.57	44.2	14.5	895	3.12	3.51	4.01	n.d.	5.84
Sp-8-pCPT-2-0-Me- cAMPS	0.38	2.02	17.2	n.d.	0.82	0.4	1.05	3.08	n.d.	2.42
Rp-8-Br-cAMPS	n.e.	5.57	37.7	23.3	28.7	94.7	106	n.e.	n.d.	28.1
8-pCPT-cGMP	16.7	8.62	47.1	40	53.4	27.6	41.3	77.7	1.29	6.04
8-Br-Pet-cGMP	2.4	2.13	73.1	98.6	8.84	5.29	11.2	n.e.	n.d.	34.4
Rp-pCPT-cGMPS	42.4	2.1	32.6	52.9	137	49.7	26	7.64	n.d.	38.6
Rp-8-Br-Pet-cGMPS	n.e.	2.5	55.6	0.76	8.1	4.09	13.8	n.e.	2.28	4.98

K_m data for cyclic nucleotides analogs, that can be PDE substrates, are in shaded cells, while ones, that can inhibit PDEs are in non-shaded cells and presented as K_i

Acknowledgments

This work was supported by grants GM083926 and AR056221.

References

1. Beavo JA, Brunton LL (2002) Cyclic nucleotide research–still expanding after half a century. Nat Rev Mol Cell Biol 3:710–718
2. Hofmann F, Bernhard D, Lukowski R, Weinmeister P (2009) cGMP regulated protein kinases (cGK). Handb Exp Pharmacol 191:137–162
3. Bender AT, Beavo JA (2006) Cyclic nucleotide phosphodiesterases: molecular regulation to clinical use. Pharmacol Rev 58:488–520
4. Gillespie PG, Beavo JA (1989) Inhibition and stimulation of photoreceptor phosphodiesterases by dipyridamole and M&B 22,948. Mol Pharmacol 36:773–781
5. Beavo JA, Hardman JG, Sutherland EW (1970) Hydrolysis of cyclic guanosine and adenosine 3′,5′-monophosphates by rat and bovine tissues. J Biol Chem 245:5649–5655
6. Thompson WJ, Appleman MM (1971) Characterization of cyclic nucleotide phosphodiesterases of rat tissues. J Biol Chem 246:3145–3150
7. Yee R, Liebman PA (1978) Light-activated phosphodiesterase of the rod outer segment. Kinetics and parameters of activation and deactivation. J Biol Chem 253:8902–8909
8. Promega (2011) PDE-Glo™ Phosphodiesterase Assay. Techn Bull p10680
9. PerkinElmer (2010) Phosphodiesterase assays for high throughput screening. PerkinElmer Life Sciences Publication, Part# TB353
10. Greengard P, Rudolph SA, Sturtevant JM (1969) Enthalpy of hydrolysis of the 3′ bond of adenosine 3′,5′-monophosphate and guanosine 3′,5′-monophosphate. J Biol Chem 244:4798–4800
11. Poppe H, Rybalkin SD, Rehmann H, Hinds TR, Tang XB, Christensen AE, Schwede F, Genieser HG, Bos JL, Doskeland SO, Beavo JA, Butt E (2008) Cyclic nucleotide analogs as probes of signaling pathways. Nat Methods 5:277–278
12. Beavo JA, Hardman JG, Sutherland EW (1971) Stimulation of adenosine 3′,5′-monophosphate hydrolysis by guanosine 3′,5′-monophosphate. J Biol Chem 246:3841–3846
13. Martins TJ, Mumby MC, Beavo JA (1982) Purification and characterization of a cyclic GMP-stimulated cyclic nucleotide phosphodiesterase from bovine tissues. J Biol Chem 257:1973–1979
14. Todd MJ, Gomez J (2001) Enzyme kinetics determined using calorimetry: a general assay for enzyme activity? Anal Biochem 296:179–187
15. Segel IH (1975) Enzyme kinetics: behavior and analysis of rapid equilibrium and steady-state enzyme systems. New York: Wiley
16. Sonnenburg WK, Rybalkin SD, Bornfeldt KE, Kwak KS, Rybalkina IG, Beavo JA (1998) Identification, quantitation, and cellular localization of PDE1 calmodulin-stimulated cyclic nucleotide phosphodiesterases. Methods 14:3–19

Chapter 4

Radioimmunoassay for the Quantification of cGMP Levels in Cells and Tissues

Ronald Jäger, Dieter Groneberg, Barbara Lies, Noomen Bettaga, Michaela Kümmel, and Andreas Friebe

Abstract

Radioimmunoassay is an established method to determine the amount of a specific substance in a given cell or tissue sample. Commercially available RIA or Elisa are very cost intensive. Here, we describe the generation of radioactive cGMP tracer and the quantification of cGMP. Although working with radioactive material requires experience and care, this method is very sensitive and rather cheap, once it is established.

Key words Nitric oxide, Signal transduction, Guanylyl cyclase, Isotope, Cyclic nucleotide radioimmunoassay

1 Introduction

The second messenger cGMP is produced by many cell types in mammalian and nonmammalian systems. Two different cGMP-producing enzymes are known, peptide- and NO-stimulated guanylyl cyclases. The determination of cGMP levels in a given cell or tissue may provide valuable information on the state of the cGMP signaling cascade; therefore, determination of low cGMP concentrations is desirable. Competition between radioactive cGMP tracer and nonradioactive cGMP for binding to a cGMP-specific antibody is used to determine exact cGMP tissue/cell levels. The radiation of the sample is compared with that of several cGMP standard concentrations to calculate the amount of cGMP. Possible samples are any tissues which can be homogenized after the experiment as well as cells from blood or primary or secondary cell culture. The lowest levels of cGMP to be reproducibly detected range between 2 and 4 fmol. Thereby, this RIA is a sensitive method to analyze endogenously occurring cGMP levels. The protocol we describe here is based on the methods of Gilman [1], Steiner et al. [2], and

Thomas Krieg and Robert Lukowski (eds.), *Guanylate Cyclase and Cyclic GMP: Methods and Protocols*, Methods in Molecular Biology, vol. 1020, DOI 10.1007/978-1-62703-459-3_4, © Springer Science+Business Media, LLC 2013

Brooker et al. [3]. Many steps, however, have been modified to make the method easier and faster. This chapter is divided into two different methodological parts: First, we describe the iodination reaction in which a radioactive cGMP tracer is generated. Due to the long half-life of ^{125}I (60 days) the tracer synthesis has to be repeated only twice a year. Second, the general procedure for the cGMP-RIA is described.

2 Materials

2.1 Tracer Synthesis

1. Sepharose buffer: 50 mM ammonium formate, pH 6.0. Prepare the day before the synthesis, required volume: 200 ml (*see* **Note 1**). Adjust with formic acid to pH 6.0 (*see* **Note 2**).
2. Weigh 1.5 g QAE-A25 Sepharose (GE Healthcare) and let it soak overnight at room temperature in the freshly prepared 50 mM ammonium formate buffer.
3. Tyrosylmethylester (TME)-cGMP stock solution: 2.5 μg/5 μl 2′-*O*-monosuccinylguanosine-3′,5′-cyclic monophosphate, TME-cGMP. Solubilize TME-cGMP (Biolog) in ice-cold 50 mM phosphate buffer, pH 7.4. Prepare 5 μl aliquots and store at −80 °C (*see* **Note 3**).
4. Column elution buffer: 250 mM ammonium formate, pH 6.0. Required volume: 500 ml. Adjust with formic acid to pH 6.0 (*see* **Note 4**).
5. Column equilibration buffer: 50 mM ammonium formate, pH 6.0. Dilute the 250 mM ammonium buffer 1:5 (required volume: 250 ml).
6. Fill a PD-10 column (GE Healthcare) with the soaked QAE-A25 sepharose and equilibrate with ten column volumes freshly prepared equilibration buffer (*see* **Notes 5–7**).
7. 500 mM phosphate buffer: Required volume: 100 ml. Adjust to pH 7.4 with 85 % phosphoric acid.
8. 50 mM phosphate buffer: Dilute the 500 mM phosphate buffer 1:10 with water to a final volume of 100 ml.
9. $Na^{125}I$, 37 MBq (629 GBq/mg), pH 8–11, reductant free (PerkinElmer).
10. Charcoal powder, activated.
11. 25 plastic tubes for elution fractions (min. volume 10 ml).
12. 1-propanol.
13. Chloramine T: 2 mg in 5 ml 50 mM phosphate buffer (*see* **Note 8**).
14. Sodium metabisulfite ($Na_2S_2O_5$): 5 mg in 10 ml 50 mM phosphate buffer (*see* **Note 8**).

15. 250 mM potassium iodide (KI solution): 83 mg in 2 ml water.
16. Thaw one 5 µl aliquot of the TME-cGMP stock solution on ice; dilute in 45 µl of 50 mM phosphate buffer (*see* **Note 3**).

2.2 Radioimmunoassay

1. RIA buffer: 100 mM sodium acetate buffer, pH 6.0. Required volume: 1.5 l; adjust to pH 6.0 with acetic acid (*see* **Note 9**).
2. 1 µM cGMP stock solution: Prepare 10 ml of a 1 µM cGMP solution in water; be very exact when weighing.
3. Determine the exact cGMP concentration using a photometer (molar extinction coefficient for cGMP $\varepsilon_{252} = 13{,}700$) and correct by dilution if necessary.
4. Prepare nine standard cGMP concentrations by dilution of the 1 µM stock solution with RIA buffer. Start with 5.12 nmol/l and dilute this solution 1:2 with RIA buffer eight times; the lowest cGMP standard concentration is 20 pmol/l. Each standard concentration should have a final volume of 5 ml (*see* **Note 10**).
5. Use RIA buffer as the last "standard cGMP concentration" (0 pmol/l).
6. Store each concentration in 100 µl aliquots at −20 °C.
7. 8 mg/ml IgG solution (IgG-*0.8%*): 4 g of porcine gamma globulin in 500 ml RIA buffer. Store 20 ml aliquots at −20 °C (*see* **Note 11**).
8. 0.5 mg/ml IgG solution (IgG-*0.05%*): Dilute 30 ml of the 8 mg/ml IgG solution with 450 ml RIA buffer. Store 20 ml aliquots at −20 °C (*see* **Note 11**).
9. Precipitating solution: 16 % PEG-6000 in 10 mM Tris–HCl, pH 7.4. 1.57 g Tris–hydrochloride in 600 ml water. Adjust to pH 7.4 with sodium hydroxide; add 160 g of polyethylene glycol 6000 (PEG-6000). Finally, add water up to 1 l volume.
10. cGMP-specific antibody (generated in-house, final dilution 1:200,000).
11. Triethylamine (p.a.).
12. Acetic anhydride (99 %, p.a.).
13. RIA tubes: 4.5 ml polystyrene tubes.

3 Methods

3.1 Iodination Reaction

The following steps should be performed in a fume hood, preferably in an isotope lab. Use filter tips whenever pipetting radioactive liquids.

1. Place a 1.5 ml reaction tube under a fume hood (use lead shielding).

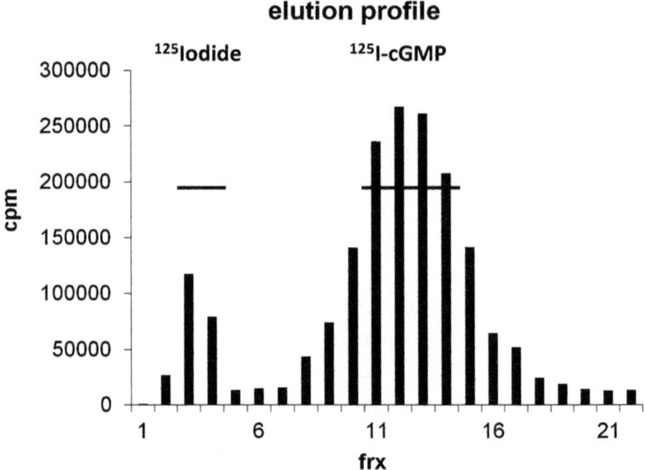

Fig. 1 Elution profile of a PD-10 column. The first peak contains unreacted ^{125}iodide, the second peak contains the iodinated cGMP tracer

2. Mix 30 μl of the 500 mM phosphate buffer with 10 μl of the diluted TME-cGMP.
3. Add 10 μl $Na^{125}I$ (37 MBq; *see* **Note 12**). Use filter tips for mixing in order to avoid pipette contamination.
4. *First charcoal check*! (*See* Subheading 3.2.)
5. Prepare chloramine T and sodium metabisulfite solutions directly prior to starting the reaction.
6. Start the reaction by adding 50 μl chloramine T to the reaction mix; use filter tip and mix by pipetting up and down once. Close the fume hood as radioactive iodine may be generated! (*See* **Note 13**.)
7. Stop the reaction after exactly 50 s by adding 100 μl of the sodium metabisulfite solution and close hood again for few minutes! (*See* **Note 14**.)
8. *Second charcoal check*! (*See* Subheading 3.2.)
9. Dilute the reaction mixture with 100 μl water to reduce ionic strength.
10. Apply reaction volume onto the PD10 column.
11. Start elution by adding 5 ml aliquots of 250 mM ammonium formate, pH 6.0. Collect 25 fractions.
12. Pipet 5 μl of each fraction into RIA tubes and count in a gamma-counter (for a typical elution profile see Fig. 1).
13. Pool 4–5 fractions with the highest radiation and dilute with the same amount of 1-propanol to increase the stability of the tracer.
14. Store the tracer in aliquots at −20 °C.

3.2 Charcoal Check

To determine the iodination efficiency, free ^{125}iodide should be measured before and after the reaction. These steps have to be performed in parallel to the iodination reaction.

1. Prepare a 1.5 ml sample tube with 500 μl of the 250 mM KI solution.
2. *First charcoal check*: Before starting the iodination reaction with chloramine T, wet a pipette tip in the reaction tube containing the $Na^{125}I$ (*see* Subheading 3.1, **step 3**) and stir it in the KI solution. Do not pipette up and down the radioactive solution as the amount of radioactivity sticking to the outside of the tip is enough for analysis.
3. Count 10 μl from the KI solution.
4. Add a pea-sized amount of charcoal powder to the solution, vortex, and centrifuge for 2 min at 13,000 × *g*.
5. Count 10 μl of the supernatant.
6. Proceed the same way with the second charcoal check.

Calculation:

Check 1: Both values (before and after centrifugation) should be very high and almost identical as free iodide does not bind to charcoal.

Check 2: The first value represents the amount of all radioactive substances in the tube (100 % of radioactivity), and the second is the amount of free iodide. In contrast to the TME-cGMP, free iodide does not bind to activated charcoal and is therefore not pelleted.

The efficiency of the iodination reaction should amount to 80–90 %.

3.3 Immunoassay Procedure

Before starting the radioimmunoassay, cGMP from a given sample has to be extracted (*see* **Note 15**). **Steps 1–7** may be performed on the lab bench. For completion of the RIA (**steps 8–14**) we recommend using an isotope lab although shielding is not necessary because of relatively low radioactivity used.

1. Add 100 μl RIA buffer to the sample when working with dried cGMP samples (e.g., from platelets, non-adherent cells, or homogenized tissue). Shake for 30 min at 37 °C to solubilize the cGMP. Occasional vortexing helps to speed up the solubilization process.
2. If you do not know the approximate concentration of cGMP in the sample, use 1:2, 1:10, and 1:100 dilutions in a total volume of 100 μl to be sure of getting a cGMP value within the range of the cGMP standard curve (2–512 fmol).
3. Acetylate samples *and* cGMP standard concentrations: Mix 100 μl of acetic anhydride and 200 μl of triethylamine (prepare

Table 1
Pipetting scheme

Total	Nonspecific binding (NSB)	cGMP standards	Samples
40 μl tracer	50 μl RIA buffer 10 μl RIA buffer 100 μl IgG-*0.05%* 40 μl tracer mix	50 μl RIA buffer 10 μl standard 100 μl antibody mix 40 μl tracer mix	50 μl RIA buffer 10 μl sample 100 μl antibody mix 40 μl tracer mix

Remember to double or triple the amounts of all mixes when performing the RIA in duplicate or triplicate! The tubes for the total activity are not processed any further but only counted on the next day. All samples have to be prepared in duplicate or triplicate!

mixture freshly every time), add 3 μl of this mixture to a sample (100 μl sample/cGMP standard), and vortex immediately! Proceed in a similar fashion with every single sample (*see* **Notes 16** and **17**).

Perform the following steps on ice:

4. Before pipetting the RIA, remember to use duplicates or triplicates for each sample/standard to obtain reliable results. Besides the cGMP samples and the 10 cGMP standard concentrations (0–512 fmol), the assay contains a sample for nonspecific binding (NSB) and a sample for measuring the total activity.
5. It is advisable to label the tubes to retain the order.

   Add 50 μl of RIA buffer to all tubes except for the totals. Then add 10 μl of RIA buffer to the NSB tubes, 10 μl of cGMP standard to the standard tubes and 10 μl of the sample to the sample tubes (*see* Table 1).
6. Add 100 μl antibody mix to all tubes except for the totals and NSB tubes (*see* **Note 18**): Antibody mix consists of IgG-*0.05%* solution + cGMP-specific antibody. The concentration of the antibody has to be determined empirically. Choose an antibody titer that is sufficient to bind 30–50 % of the radioactive tracer (*see* **Note 19**). Add 100 μl IgG-*0.05%* solution to NSB tubes.

   Up to this point, the immunoassay can be prepared on the lab bench. For the following steps we recommend working in an isotope lab.
7. Prepare tracer mix which consists of RIA buffer plus tracer. The activity should be 5,000–10,000 cpm per assay tube. Determine the volume of tracer you need for the number of all tubes (do not forget NSB and total activity!), i.e. × μl tracer + y μl RIA buffer to a final volume of 40 μl per tube. Add 40 μl of the mix to each tube including the totals (*see* **Note 20**).

8. Centrifuge RIA tubes shortly to collect all liquid at the bottom.
9. Incubate samples at 4 °C for at least 4 h, preferably overnight, to reach an antibody–cGMP binding equilibrium.
10. Add 50 μl of IgG-*0.8%* to each sample. Addition of a high amount of IgG makes the protein precipitate visible after centrifugation.
11. Add 3 ml precipitation buffer (4 °C) to each sample and let the antibody–cGMP complex and the added IgG precipitate for 30 min at 4 °C.
12. Centrifuge at 6,000 × *g* for 30 min at 4 °C.
13. Aspirate the supernatant and determine the activity of the pellet with a gamma-counter. Collect the supernatant for appropriate waste disposal.

3.4 Calculation of the cGMP Amount

1. Calculate the average cpm (*Bx*) for each set of duplicate/triplicates.
2. Calculate percent binding (*%Bx/B0*) for every standard concentration *x* and every sample using the following equation: $\%Bx\ /\ B0 = Bx\text{-}NSB(\text{cpm})\ /\ B0\text{-}NSB(\text{cpm}) * 100$

   (*B*0: measured cpm for 0 fmol cGMP, and *Bx*: measured cpm for *x* fmol cGMP.)

   An exemplary Excel sheet is shown in Fig. 2 (*see* **Note 21**).

cGMP radioimmunoassay

	cpm1	cpm2	**cpm**
total	6594	6756	**6675**
NSB	22	44	**33**

standards (fmol cGMP)	cpm1	cpm2	**Bx (cpm)**	**%Bx/Bo**
0	3204	3188	3196	**100**
2	3098	3100	3099	**97**
4	2926	2938	2932	**92**
8	2718	2726	2722	**85**
16	2462	2512	2487	**78**
32	2032	2040	2036	**63**
64	1580	1634	1607	**50**
128	1172	1132	1152	**35**
256	750	796	773	**23**
512	452	486	469	**14**
sample 1	2145	2188	2167	**67**
sample 2	966	980	973	**30**

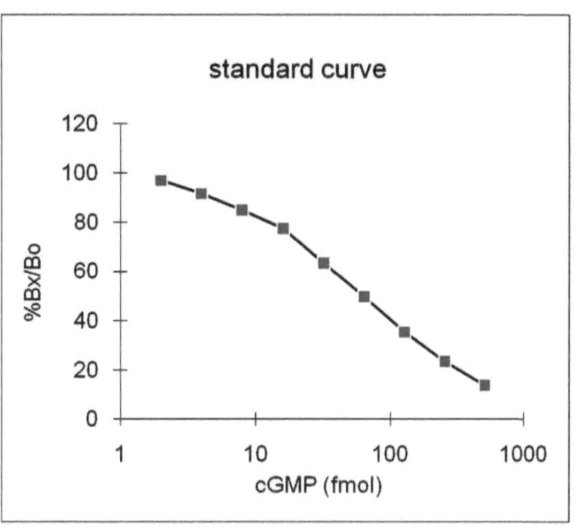

Fig. 2 Exemplary Excel sheet for the calculation of cGMP values

3. Create a sigmoidal standard curve of the calculated (% *Bx*/*B*0) values versus the cGMP standard concentrations using an appropriate software (e.g., GraphPad Prism or Excel).
4. Interpolate the cGMP amounts for every sample.
5. In case you diluted the original samples do not forget to include the dilution factor.
6. cGMP amounts may be normalized to the cell number or the protein concentration of the sample.

4 Notes

1. We routinely use demineralized water for the preparation of all buffers.
2. If you are using a concentrated formic acid (98 %) to adjust the ammonium formate buffer (50 mM, 200 ml), use a 1:10 dilution and add only a few microliters.
3. Keep TME-cGMP solution frozen as long as possible while preparing the reaction.
4. For adjusting the 250 mM ammonium formate buffer to pH 6.0 you will need only approximately 15 μl of 98 % formic acid.
5. To prepare the PD-10 column cut off the sealed end at the notch and push a frit with the smooth side down into the column. Load the column by gently pipetting the sepharose using a pipette tip with a wide opening. Let the buffer run out (sepharose settles by gravity flow), but do not let it run dry! Load sepharose to the upper ring mark of the column and insert a second frit carefully below the ring (smooth side down). Avoid compressing the sepharose and avoid a gap between sepharose and the upper frit. Column has to be able to hold 5 ml buffer above the upper frit (size of the elution fractions).
6. In our experience it is beneficial for the column flow to let the frits soak in buffer for a couple of minutes.
7. A commercially available buffer reservoir may be used to facilitate the equilibration of the column.
8. Dissolve chemicals right before starting the reaction. It is advisable to weigh the chemicals in advance and to calculate the amount of buffer needed but to keep chemicals and buffer separate until the start of the synthesis.
9. RIA buffer is stable for 6 weeks at 4 °C. Aliquots (e.g., 50 ml) can be stored at −20 °C for over 1 year.
10. We routinely control exact dilution with an analytical balance.

11. The IgG-*0.05%* solution is used to prevent the binding of the cGMP antibody to the walls of the plastic tubes. The IgG-*0.8%* solution is used to precipitate the cGMP antibody. Formation of a visible pellet by the high IgG concentration is practical when aspirating the supernatant.
12. Please use utmost caution when pipetting radioactive material!
13. During the reaction, radioactive NaI is oxidized to I_2 by chloramine T. I_2 cannot be expected to react quantitatively with the TME-cGMP. Therefore protection against evaporating radioactive I_2 is mandatory. Take care that the hood is working properly.
14. A reaction time of 50 s has proven to be optimal under our conditions. If the charcoal check indicates insufficient reaction, one may reduce or increase the reaction time to reach the optimal efficiency of cGMP labeling. Be aware that a longer reaction time may lead to the formation of macromolecular complexes which may reduce the total cGMP concentration.
15. There are numerous ways to extract cGMP. When using platelets we stop the reaction by adding ice-cold ethanol (final concentration 66 %) and dry the samples in a speed-vac [4]. Adherent cells can be incubated in a HEPES-based buffer (10 mM HEPES, 154 mM NaCl, 5.6 mM KCl, 2 mM $CaCl_2$, 1 mM $MgCl_2$, 3.6 mM $NaHCO_3$, pH 7.4, 5.6 mM glucose) and the incubation is stopped by adding HCl (final concentration 0.01 mM) to release cGMP [5]. The supernatant can then be directly used in the acetylation reaction. cGMP from tissue can be isolated by snap freezing and subsequent homogenization in a HEPES- or phosphate-based buffer which includes phosphatase inhibitors using a glass–glass homogenizer. After centrifugation of the homogenate, the supernatant contains the cGMP (to be dried in a speed-vac) and the pellet can be used for the determination of the protein concentration [6]. Samples in the dried state are storable for months.
16. Acetylation should be performed in a fume hood to prevent inhalation of triethylamine or acetic anhydride vapors. cGMP acetylation raises the affinity in case the antibody has been generated against 2′-*O*-succinyl-cGMP and leads to an up to 40-fold increase in assay sensitivity. Acetic anhydride and triethylamine should be used immediately after mixing. If the solution turns yellow/brown (after 15 min), do not use it any more.
17. Acetylated samples should be used within 7 days as acetylation is reversible. Keep samples at 4 °C.
18. Prepare a little more antibody mix than you need for all samples and mix well to evenly distribute the antibody. Remember to include the amount for duplicates or triplicates.

19. We determine the optimal antibody concentration after each tracer synthesis. Titers routinely tested are 1:10K, 1:20K, 1:50K, 1:100K, 1:200K and 1:500K diluted in IgG-*0.05%*. We use our in-house-generated antibody against 2′-*O*-succinyl-cGMP (first described in Friebe et al. [7]) at a titer of 1:200,000. When using a commercially available antibody, the optimal titer has to be determined.
20. The totals only contain 40 μl of tracer mix. These tubes are not treated as further described in the protocol but instead are counted as they are in order to determine the absolute amount of radiation in each RIA tube.
21. Be aware that increasing cold cGMP reduces antibody binding of radioactive ^{125}I-cGMP and, therefore, leads to a decrease in bound radiation (*Bx*).

References

1. Gilman AG (1972) A protein binding assay for adenosine 3':5'-cyclic monophosphate. Proc Natl Acad Sci USA 67:305–312
2. Steiner AL, Wehmann RE, Parker CW, Kipnis DM (1972) Radioimmunoassay for the measurement of cyclic nucleotides. Adv Cyclic Nucleotide Res 2:51–61
3. Brooker G, Harper JF, Terasaki WL, Moylan RD (1979) Radioimmunoassay of cyclic AMP and cyclic GMP. Adv Cyclic Nucleotide Res 10:1–33
4. Friebe A, Müllershausen F, Smolenski A, Walter U, Schultz G, Koesling D (1998) YC-1 potentiates NO- and CO-induced cGMP effects in human platelets. Mol Pharmacol 54:962–967
5. Müllershausen F, Russwurm M, Koesling D, Friebe A (2004) In vivo reconstitution of the negative feedback in nitric oxide/cGMP signaling: role of phosphodiesterase type 5 phosphorylation. Mol Biol Cell 15: 4023–4030
6. Müllershausen F, Lange A, Mergia E, Friebe A, Koesling D (2006) Desensitization of NO/cGMP signaling in smooth muscle: blood vessels versus airways. Mol Pharmacol 69:1969–1974
7. Friebe A, Mergia E, Dangel O, Lange A, Koesling D (2007) Fatal gastrointestinal obstruction and hypertension in mice lacking nitric oxide-sensitive guanylyl cyclase. Proc Natl Acad Sci USA 104(18):7699–7704

Chapter 5

Hyperspectral Imaging of FRET-Based cGMP Probes

Thomas C. Rich, Andrea L. Britain, Tiffany Stedman, and Silas J. Leavesley

Abstract

In recent years a variety of fluorescent probes for measurement of cGMP signals have been developed (Nikolaev et al., Nat. Methods 3:23–25, 2006; Honda et al., Proc Natl Acad Sci USA 98:2437–42, 2001; Nausch et al., Proc Natl Acad Sci USA 105:365–70, 2008). The probes are comprised of known cGMP binding sites—e.g., from phosphodiesterase type 5 (PDE5) or protein kinase G (PKG)—attached to fluorescent proteins. Binding of cGMP triggers conformational changes that alter the emitted fluorescence. In the case of Förster resonance energy transfer (FRET)-based probes, binding of cGMP alters the distance between the donor and acceptor fluorophores and thus alters FRET. However, FRET-based probes inherently have low signal-to-noise ratios, limiting the utility of these probes. Here we describe the use of hyperspectral imaging and analysis approaches to increase the signal-to-noise ratio of FRET-based cGMP measurements. These approaches are appropriate for monitoring changes in cGMP signals either in cell populations using a spectrofluorimeter or in single cells using spectral microscope systems with appropriate spectral filtering capabilities.

Key words Phosphodiesterase, Guanylyl cyclase, Natriuretic peptide receptor, cGMP signals, Spectroscopy, Microscopy

1 Introduction

Techniques for real-time measurement of intracellular signals have dramatically increased our understanding of second messenger signaling processes [1–5]. Optical approaches, including Förster resonance energy transfer (FRET)-based measurements, offer technically straightforward techniques for monitoring agonist-induced changes in second messenger levels. However, many of the optical approaches for measuring second messenger signals suffer from low dynamic range and high background—especially in autofluorescent tissues. Low dynamic range and high background signals are particularly problematic for FRET-based cGMP probes. Here we describe the use of hyperspectral imaging and analysis

Thomas Krieg and Robert Lukowski (eds.), *Guanylate Cyclase and Cyclic GMP: Methods and Protocols*, Methods in Molecular Biology, vol. 1020, DOI 10.1007/978-1-62703-459-3_5, © Springer Science+Business Media, LLC 2013

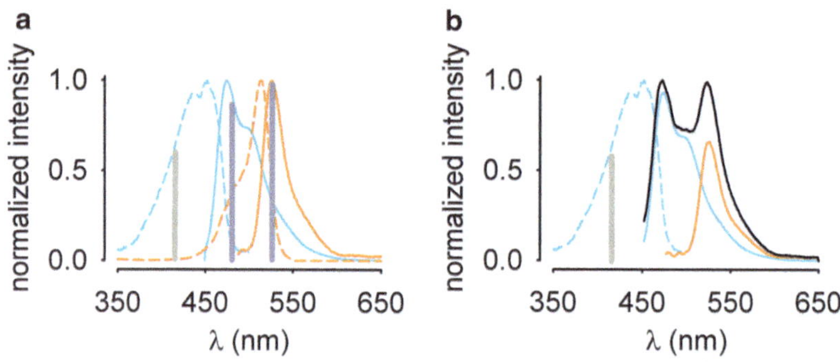

Fig. 1 Schematic depicting traditional and hyperspectral measurement of FRET signals. *Dashed lines* indicate the excitation spectra for CFP (*blue*) and YFP (*orange*). *Solid lines* depict the emission spectra. (**a**) Standard approaches for measurement of FRET. The *light grey line* depicts the excitation wavelength, 415 ± 10 nm, near the peak excitation for CFP. The *dark grey lines* depict the wavelengths at which the emission is detected, 480 ± 10 and 520 ± 10 nm, near the emission peaks for CFP and YFP, respectively. (**b**) Hyperspectral approaches for measurement of FRET. The *light grey line* again depicts the excitation wavelength for CFP. The *black line* depicts the measured emission spectrum, typically ≥ 32 wavelengths. The *solid blue* and *orange lines* depict the abundance of CFP and YFP in the signal. In other words, the *blue* and *orange lines* depict the contribution of CFP and YFP, respectively, to the measured fluorescence signal (*black line*). Under ideal conditions, FRET can be accurately assessed using either approach described above. However, biology seldom provides ideal conditions. Hyperspectral assessment of FRET signals offers increased signal-to-noise characteristics and increased dynamic range compared to standard measurement approaches. As such, hyperspectral approaches are often better suited for measurement of FRET signals in biological preparations

approaches for the measurement of FRET in cell populations using a spectrofluorimeter or in single cells using spectral epifluorescence microscope systems. The cGMP sensor used is comprised of a cGMP binding site from phosphodiesterase type 5 (PDE5) sandwiched between the cyan fluorescent protein (CFP) and yellow fluorescent protein (YFP), as described in [1, 6].

Hyperspectral imaging and analysis approaches were developed by the National Aeronautics and Space Administration (NASA) to solve remote sensing problems such as feature identification in satellite images. These technologies offer tremendous potential for the study of biological systems [7]. The schematic in Fig. 1 depicts the basic theory for application of hyperspectral imaging approaches to the assessment of FRET signals. The dashed lines indicate the excitation spectra of cyan fluorescent protein (CFP, blue lines) and yellow fluorescent protein (YFP, orange lines). The solid lines depict the emission spectra. In traditional microscopy, a single filter is used for excitation of CFP (440 ± 10 nm, light grey band), and two filters are used for emission, near the peaks for CFP and YFP (480 ± 10 nm and 520 ± 10 nm, dark grey bands). For fluorescent signals with adequate signal-to-noise characteristics, this approach is sufficient. However, FRET signals are often small, with low signal-to-noise ratios and limited dynamic range. Hyperspectral imaging and analysis approaches allow for FRET signals to be measured with increased signal-to-noise ratios, often at the cost of longer acquisition times. In hyperspectral imaging, samples are typically excited at a specific wavelength and emission intensity is measured at many wavelengths, often ≥32 wavelengths (Fig. 1b, black line). The range and number of wavelengths are chosen to ensure that sufficient spectral information is available for subsequent quantification of the CFP and YFP components. The amplitude of CFP and YFP components in the fluorescence emission is assessed and FRET is estimated by taking the ratio of the areas under the CFP and YFP signals. The fluorescence emission spectrum is acquired for each pixel; thus FRET is estimated on a pixel-by-pixel basis. This approach increases both the signal-to-noise ratio and dynamic range of FRET measurements [8]. This overall approach is also capable of detecting fluorescent proteins in highly autofluorescent backgrounds [9]. Here we describe hyperspectral assessment of FRET signals in cell populations using a spectrofluorimeter, and in single cells using an acousto-optical tunable filter (AOTF)-based microscope system or spectral confocal microscope system (*see* **Note 1**).

2 Materials

FRET responses are highly dependent on both pH and temperature. As such, extracellular solutions should be buffered (≥10 mM HEPES) and allowed to equilibrate within the experimental chamber (5–10 min, depending on chamber volume and temperature difference).

2.1 Components for Hyperspectral Measurements Using a Spectrofluorimeter

1. Spectrofluorimeter. A stirred cuvette spectrofluorimeter such as the PTI QuantaMaster 40.
2. Disposable cuvettes. Standard 10 × 10 × 48 mm plastic cuvettes are appropriate for spectral FRET measurements (*see* **Note 2**). Smaller volume cuvettes may also be employed.
3. Extracellular buffer solution. We typically use a buffer solution containing 145 mM NaCl, 4 mM KCl, 20 mM HEPES, 10 mM D-Glucose, 1 mM $MgCl_2$, and 1 mM $CaCl_2$, pH 7.3 (*see* **Note 3**).
4. Cells expressing cDNA constructs encoding either CFP-alone, YFP-alone, or the CFP-YFP cGMP FRET probe. Cell culture and lipid-reagent and viral transfection protocols for FRET-based sensors are described in [1, 8, 10]. The ratio of cDNA to transfection reagent or the multiplicity of infection (MOI) for viral transfection is cell type dependent. *See* **Note 4** which outlines fluorescent protein transfection protocols used for HEK293 cells.

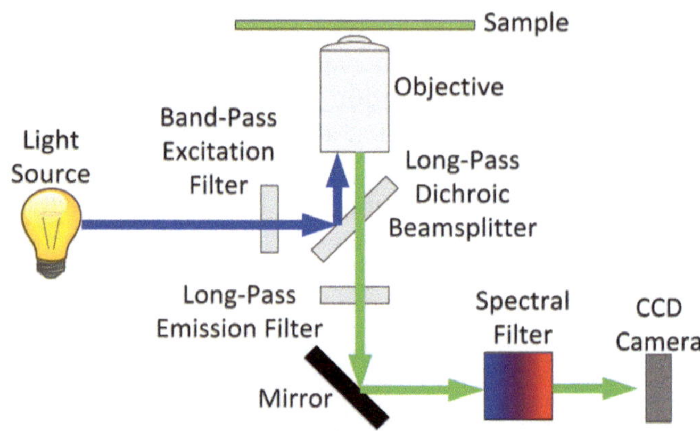

Fig. 2 Schematic of epifluorescence widefield microscope for use in hyperspectral imaging applications. Excitation light from a xenon arc lamp travels through a light guide, band pass filter, a long-pass dichroic beam splitter, and the objective in order to illuminate the sample. Emitted fluorescence travels through the long-pass beam splitter, is reflected off of a mirror, travels through the spectral emission filter, and is then measured by the CCD camera

5. Reagents: Stock solutions of the following reagents are stored in single-use aliquots at −20 °C (typical stock solution concentrations given parenthetically): atrial natriuretic peptide (ANP, 10 μM), sodium nitroprusside (SNP, 100 mM), 3-isobutyl-1-methylxanthine (IBMX, 500 mM), sildenafil (100 μM), cGMP (10 mM). The vehicle for stock solutions is either extracellular buffer or dimethylsulfoxide (DMSO).

2.2 Components for Hyperspectral Measurements Using an Epifluorescence Widefield Microscope

1. Light path. A standard xenon arc lamp and appropriate excitation and dichroic filters. For CFP and YFP FRET pairs, a 415 ± 10 nm excitation filter and a 450 nm long-pass dichroic beam splitter are appropriate (*see* Fig. 2 and **Note 5**).
2. Spectral emission filter. An acousto-optical filter (AOTF) is the gold standard for emission-based hyperspectral approaches. We have used a variable-bandwidth AOTF with a spectral range of 450–800 nm (HSi-300, Chromodynamics, Inc.).
3. Multi-ion discharge lamp. A multi-ion discharge lamp (Lightform, Inc.) is used to verify correct calibration of the AOTF.
4. NIST-traceable light source. A National Institute of Standards and Technology (NIST)-traceable light source is required for calibration of wavelength-dependent nonlinearities along the light path (*see* **Note 6**). We have typically used a NIST-traceable Tungsten-Halogen lamp (LS-1-CAL, Ocean Optics).
5. cGMP sensor-expressing cells plated on glass coverslips. *See* Subheading 2.1, **item 4**, and **Note 4**.

2.3 Data Analysis Software

1. Software for analysis of spectrofluorimeter measurements. Analysis of emission intensities measured using a spectrofluorimeter is typically performed using custom scripts coded in the MATLAB (MathWorks) programming environment, described in [8]. Sample MATLAB code has been provided. This code can be modified to fit more complex or batch analysis needs (*see* **Note 7**).
2. Software for analysis of hyperspectral images. Analysis of hyperspectral image stacks is more computationally involved than analysis of spectrofluorimetric data. Hyperspectral image stacks are three-dimensional sets of images in which two dimensions represent the field of view and the third dimension represents the emission wavelength. The strength of this approach is the ability to analyze the spectra for each pixel at each time point. For this analysis we typically employ ENVI (Exelis Visual Information Solutions) software in conjunction with custom scripts coded in Matlab (*see* **Note 8**).

3 Methods

Procedures are typically carried out at room temperature.

3.1 Hyperspectral Measurements Using a Spectrofluorimeter

1. Adjust spectrofluorimeter settings (for CFP-YFP FRET pair). Several settings in the spectrofluorimeter acquisition software must be set prior to data acquisition:
   - (i) The excitation wavelength should be set to 415 nm.
   - (ii) The instrument should be set to scan the emission spectrum from 450 to 650 nm in 1 nm increments. We have typically used a dwell time of 0.05 s for each wavelength and averaged 20 emission scans (*see* **Note 9**).
   - (iii) Excitation and emission slit widths should be adjusted to minimize cross talk.
   - (iv) The speed of the stir bar should be adjusted to maintain cells in suspension (*see* **Note 10**), without causing cell lysis. The same stir bar speed is appropriate for lysed cell preparations. A well-stirred solution is required to maintain the density of cells (or of the FRET probe in lysed cell preparations) in the light path and for adequate mixing of reagents (e.g., following addition of ANP).
2. Measure the background/blank signal using a cuvette containing 4 mL buffer and stir bar only. If the background signal contains significant noise characteristics, it is desirable to average several background scans.
3. Estimate the cGMP sensitivity of the FRET probe. In order to estimate the sensitivity of the FRET probe to changes in cGMP

one must measure the FRET response at known cGMP levels. Since it is at best difficult to quantitatively control intracellular cGMP levels, these measurements are typically conducted in lysed cell preparations. PDE activity is inhibited such that cGMP concentrations remain constant throughout the measurement. cGMP concentration is altered by adding known amounts of cGMP to the cuvette. A standard range of cGMP concentrations used to assess the sensitivity of cGMP sensors is 0.05–50 μM cGMP. Several steps are required to estimate the sensitivity of FRET-based or other fluorescent cGMP probes.

(i) Cells expressing the cGMP FRET probe (*see* **Note 4**) are detached and resuspended in buffer solution (1×10^6 cells/mL of buffer). Cells are lysed with 20 strokes of a dounce. Other lysis techniques should provide similar results. 4 mL of lysed cells and appropriate-sized stir bar are added to a disposable cuvette. The cuvette is placed in the spectrofluorimeter and allow adequate time for the temperature to equilibrate.

(ii) A baseline emission scan is conducted using the settings described in **step 1**.

(iii) 500 μM IBMX (PDE inhibitor) and the baseline emission scan is repeated.

(iv) Starting with the lowest cGMP concentration (e.g., 0.05 μM), cGMP is added to reach desired concentration and the emission scan is repeated. This is repeated each cGMP concentration to construct desired cGMP response curve (e.g., 0.05–50 μM cGMP).

(v) The FRET response at each cGMP concentration is analyzed as described in Subheading 3.1, **steps 6–8**. The resultant FRET cGMP concentration response curve is fit using the Hill equation to estimate the cGMP concentration that elicits a half maximal response ($K_{1/2}$) and the Hill coefficient (the predicted Hill coefficient is 1 for one binding site cGMP probes).

4. Perform intact cell measurements. Intact cell measurements are made using each of three different constructs: (1) cells expressing the FRET-based cGMP sensor, (2) cells expressing CFP alone, and (3) cells expressing YFP alone. The measurements made using the CFP and YFP alone allow to control for photobleaching as well as potential quenching of fluorophores by NO donors or other guanylyl cyclase activators (*see* **Note 11**). The steps required to make spectrofluorimeter-based measurements in cell populations are as follows:

   (i) Add 4 mL of buffer containing intact cells ($1–10^6$ cells/mL) and stir bar to a cuvette, place the cuvette in the spectrofluorimeter, and allow temperature to equilibrate.

(ii) Perform at least two baseline scans.

(iii) Add agonists/antagonists to trigger changes in intracellular cGMP levels and conduct emission scans. Agonists of particulate and soluble guanylyl cyclase include 0.1–100 nM ANP and 100 μM SNP; antagonists of PDEs include 100–500 μM IBMX, a broadband PDE inhibitor, and 100 nM sildenafil, a PDE5-specific inhibitor. Experiments should be conducted in a stirred cuvette to ensure adequate distribution of agonists/antagonists.

(iv) Emission scans should be repeated throughout the time course of the experiment—until the response reaches steady state.

(v) Emission scans should *not* be automatically averaged by acquisition software when measuring the time course of cGMP signals. Software packages often only save the averaged dataset and not the individual sweeps (resulting in a potentially slow sampling rate). A more appropriate method for filtering the responses is to apply a sliding window filter after analysis of the FRET response at each time point (Subheading 3.1, **steps 6–8**). A simple sliding window filter can be implemented by making use of the function "slidefun" in Matlab.

5. Estimate maximal cGMP response. If calibrated cGMP measurements (with respect to lysed cell cGMP $K_{1/2}$) are required, the FRET response at minimum and maximum cGMP levels must be estimated in intact cells (*see* **Note 12**) using the following steps.

(i) Basal cGMP is maintained at low levels in most cell types; thus the basal measurements described in Subheading 3.1, **step 4(ii)** are a reasonable approximation for the minimum cGMP response.

(ii) At the end of an experiment, the maximal cGMP response is estimated by addition of saturating PDE inhibitor (e.g., 500 μM IBMX) and saturating concentrations of guanylyl cyclase agonist (e.g., 100 nM ANP or 100 μM SNP). It may be difficult to demonstrate that this is indeed the maximal cGMP response due to potential compartmentalization of cGMP and the FRET-based cGMP sensor (*see* **Note 12**).

(iii) Calibrate the cGMP response as follows. Estimate the FRET response at each time point using the collected spectra as described in Subheading 3.1, **steps 6–8**. Normalize the FRET responses based upon the minimal and maximal cGMP responses. Solve the Hill equation for the cGMP concentration at each time point using the estimated $K_{1/2}$ and Hill coefficient of the FRET probe (Subheading 3.1, **step 3**).

6. Correct for background fluorescence by subtracting the background/blank signal (Subheading 3.1, **step 2**) from all experimental trials.
7. Analyze fluorescence emission spectra using linear unmixing. We have used the "lsqnonneg" algorithm in Matlab, which estimates the solution to the equation $C \times x = d$, where C represents the reference library, x represents the amounts (abundances) of donor and acceptor, and d represents the measured spectrum (*see* **Note 13**). The CFP-alone and YFP-alone spectra must be used as components of the reference library. If significant photobleaching (or other changes) in CFP or YFP emission intensities was observed during control experiments, then the time dependence of these changes should also be estimated and incorporated into calculation of CFP and YFP levels. A sample Matlab script used to unmix fluorescent spectra is included in **Note 7**.
8. Calculate the FRET level. In order to calculate the FRET levels, two methods may be used: (1) the ratio of the YFP emission intensity (at the peak YFP emission wavelength) divided by the CFP emission intensity (at the peak CFP emission wavelength) may be calculated; or (2) the ratio of the YFP abundance divided by the CFP abundance, as provided by linear unmixing, as described above. In our previous studies, method 2 has better signal-to-noise characteristics. In specific, because data from many wavelengths are used, the measurements are more resistant to noise at any one wavelength. Thus, method 2 allows for better discrimination between changes in FRET levels.

3.2 Hyperspectral Measurements Using an Epifluorescence Widefield Microscope

1. A schematic of a generic epifluorescence widefield microscope setup for use in hyperspectral imaging is depicted in Fig. 2. Excitation light is transmitted through a light guide and band pass excitation filter. Light is then reflected off of a long-pass dichroic beam splitter, passes through the objective, and illuminates the sample. Emitted fluorescence travels through the long-pass beam splitter, is reflected off of a mirror, and travels through a spectral emission filter to the CCD camera. Light attenuation occurs throughout the light path in a wavelength-dependent manner necessitating flat-field correction of the acquired image stack (*see* Subheading 3.2, **step 5**).
2. Adjust epifluorescence widefield microscope system settings.
   (i) Choose appropriate excitation filters for the CFP-YFP pair, such as a 415 nm filter with a 20 nm bandwidth (full width at half maximum) and a 450 nm long-pass dichroic beam splitter (*see* **Note 5**).

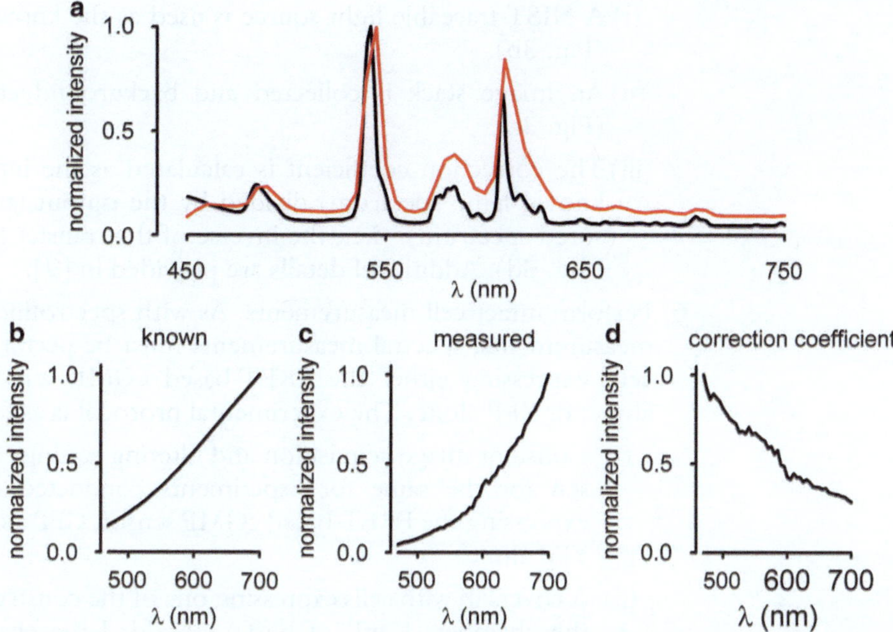

Fig. 3 Calibration of a spectral microscope system. (**a**) The multi-ion discharge lamp spectrum measured with a spectral epifluorescent microscope system (*red line*) aligned with the spectrum measured with the spectrometer (*black line*), confirming that the wavelength calibration of the HSi-300 AOTF (Chromodynamics, Inc.) was accurate. (**b**) Known spectrum of the NIST-traceable light source (input). (**c**) Spectrum of the NIST-traceable light source measured using a hyperspectral epifluorescent microscope system (output). (**d**) The correction coefficient used to achieve a flat spectral response is calculated by dividing the known spectrum by the measured spectrum (input/output)

(ii) Set the acquisition software associate with the AOTF to collect fluorescence emission from 450 to 700 nm, with a 5 nm spacing.

(iii) Adjust camera settings to maximize the dynamic range of the image at the wavelengths of peak intensity (*see* **Note 14**).

3. Confirm the calibration of the AOTF. In order to confirm the factory calibration of the AOTF (or other spectral filter system), a multi-ion discharge lamp is placed on the microscope stage and a spectral image stack is obtained. The spectrum of the image stack is then compared to the spectrum of the multi-ion lamp, as measured using a spectrometer or other appropriate NIST-calibrated device (Fig. 3a). Details are provided in [9, 11].
4. Measure the background/blank signal using a blank coverslip mounted in a chamber containing 1 mL buffer.
5. Correct the wavelength response of the spectral microscope system. Epifluorescence microscope systems do not provide a flat spectral response. Thus, it is necessary to correct the spectral information in each image stack acquired to compensate for wavelength-dependent attenuation (*see* **Note 15**).

(i) A NIST-traceable light source is used as the known input (Fig. 3b).

(ii) An image stack is collected and background corrected (Fig. 3c).

(iii) The correction coefficient is calculated as the input (the known lamp spectrum) divided by the output (the measured spectrum)—i.e., the inverse of the transfer function (Fig. 3d). Additional details are provided in [9].

6. Perform intact cell measurements. As with spectrofluorimeter measurements, spectral measurements must be performed on cells expressing either the FRET-based cGMP sensor, CFP alone, or YFP alone. The experimental protocol is as follows.

   (i) Constant image acquisition and filtering settings are chosen for the same for experiments conducted on cells expressing the FRET-based cGMP sensor, CFP alone, and YFP alone.

   (ii) A coverslip with cells expressing one of the constructs into the chamber, 1 mL of buffer, is added the chamber is placed in the microscope system, and temperature is allowed to equilibrate.

   (iii) A field of view with cells expressing the fluorescent protein(s) of interest is chosen. Multiple fields of view may be chosen when using microscopes with automated stages; however, acquiring images from multiple fields of view will markedly reduce the sampling rate.

   (iv) At least two baseline image stacks for each field of view are collected.

   (v) Agonists/antagonists to trigger changes in intracellular cGMP levels are added (*see* **Note 16**). The acquisition time required to record agonist-induced cGMP responses will depend upon both the agonist and the cell type. For example, in vascular smooth muscle cell cultures, a typical ANP- or SNP-induced cGMP response would reach steady state within 10 min [3, 12] whereas in pulmonary microvascular endothelial cells, typical agonist-induced responses reach steady state within 1 h [13] (*see* **Note 17**).

7. Estimate minimal and maximal cGMP responses. As with measurements using the spectrofluorimeter, calibrated cGMP measurements require estimation of minimal and maximal cGMP-mediated FRET responses.

   (i) Basal cGMP is maintained at low levels in most cell types; thus the basal measurements described in Subheading 3.2, **step 3** are a reasonable approximation for the minimal FRET response.

   (ii) The maximal FRET response may be estimated by addition of saturating PDE inhibitor (e.g., 500 μM IBMX) and saturating concentrations of guanylyl cyclase agonist (e.g., 100 nM ANP or 100 μM SNP) and acquiring spectral image stacks (*see* **Note 12**).

(iii) Calibrate the cGMP response (*see* Subheading 3.1, **step 5**).

8. Correct for background fluorescence by subtracting the background/blank signal from all image stacks.
9. Apply the correction coefficient to achieve a flat spectral response. Note that the correction coefficient should be applied to both the FRET hyperspectral images as well as all controls (donor and acceptor).
10. Analyze spectral image stacks. Analysis is conceptually similar to spectrofluorimetry analysis. However, due to the large amount of data contained in a spectral image, analysis requires efficient analysis algorithms and software. The steps required to analyze spectral image stacks are outlined below.

    (i) As in spectrofluorimetry analysis (*see* Subheading 3.1, **step 7**), create a library containing the spectrum of CFP and YFP using the spectra acquired from cells expressing either CFP alone or YFP alone.

    (ii) Analyze hyperspectral images using linear unmixing in ENVI, or similar spectral analysis software (*see* **Note 8**).

    (iii) Calculate the ratio of unmixed YFP to unmixed CFP by dividing the unmixed YFP image by the unmixed CFP image.

    (iv) If there is significant change in either the YFP-alone or CFP-alone controls (e.g., due to photobleaching), the YFP and CFP unmixed images should be corrected before calculating their ratio.

    (v) Quantify dynamics of cGMP in individual cells or regions by selecting appropriate regions of interest (ROI) and calculating the average (mean) FRET signal for the region. In addition, the standard deviation may provide some estimate of the variability of cGMP within the ROI.

    (vi) If analyzing a time series or multiple images, perform **steps (ii)**–(**v**) for every image in the series. The resultant FRET images—from **step (v)**—can then be combined into a single time series image stack.

4 Notes

1. Spectral confocal microscope systems are typically customized to order. It is important to specify a spectral resolution appropriate for separating the donor and acceptor emission spectra.

Confocal microscopes with gratings often allow 2.5–10 nm wavelength spacing, depending on the confocal system and gratings selected.

2. For the spectral measurements described here most plastic cuvettes will suffice. However, if UV illumination is required, light attenuation through most plastic cuvettes impedes adequate illumination. Should this be the case, quartz cuvettes may be employed. Alternatively, we have found that SARSTEDT disposable cuvettes (No. D-51588) minimally attenuate UV illumination.
3. Typical extracellular solutions will suffice provided that they are adequately buffered at an appropriate pH and allowed to equilibrate to the desired temperature in the experimental chamber—FRET measurements are exquisitely sensitive to both pH and temperature [14].
4. *Fugene 6 reagent-based transfection.* HEK293 cells are plated at ~60 % confluence either in 35 mm dishes with coverslips (microscopy) or in 100 mm dishes (spectroscopy). Cells are transfected with constructs encoding either CFP, YFP, or the FRET-based cGMP sensor using the Fugene 6 reagent (Promega, Madison, WI) with 1 μg cDNA and 3 μL Fugene 6 reagent per 35 mm dish or 6 μg cDNA and 18 μL Fugene 6 reagent per 100 mm dish. Cells are assayed 48–56 h post-transfection. Typically 85–95 % of HEK293 cells express the transfected fluorescent protein construct.

Adenovirus-mediated transfection. HEK293 cells are plated at~60 % in 35 mm dishes with coverslips or in 100 mm dishes. Cells are infected with adenovirus encoding CFP, YFP, or the FRET-based cGMP sensor using an MOI of 10 PFU/cell. Two hours post infection, 1 mM hydroxyurea is added to the media to inhibit viral replication. Cells are assayed 48–56 h post infection. Typically, 95–100 % of HEK293 cells express the fluorescent protein construct. A similar procedure is used for adenovirus-mediated transfection of endothelial and smooth muscle cells, with the exceptions that the MOI required for expression is ~100 PFU/cell and the addition of hydroxyurea is not required. Under these conditions 60–80 % of endothelial and vascular smooth muscle cells express fluorescent proteins using this protocol.

5. Custom dichroic mirrors are required to minimize wavelength-dependent attenuation of emitted light on spectral widefield fluorescence microscopes. For the CFP-YFP FRET pair, we have previously used a 458 nm long-pass dichroic beam splitter (FF458-Di02, Semrock, Inc.). A 450 nm long-pass dichroic beam splitter would allow better estimates of the CFP abundance.
6. For calibrating spectral microscopy systems, the NIST-traceable lamp should have a spectral output overlapping the range of

wavelengths that can be detected. We have used a calibrated tungsten-halogen lamp (LS-1-CAL, Ocean Optics, Inc.).

7. Sample MATLAB script used for spectral unmixing. This script was written using MATLAB version 2012a. The open variables are column vectors that list the wavelengths, the normalized emission intensity measured at each wavelength for the donor and acceptor (e.g., CFP and YFP emission spectra), and the emission intensity measured at each wavelength for the FRET pair, as indicated.

```
clear all
% Open Variables
load Wavelength
load Donor_Spectrum
load Acceptor_Spectrum
load FRET_spectrum
% Define spectral library
library=[Donor_Spectrum Acceptor_Spectrum];
% Unmix FRET spectrum
abundance=lsqnonneg(library,FRET_Spectrum);
donor_abundance=abundance(1,1);
acceptor_abundance=abundance(2,1);
scaled_donor = Donor_Spectrum*donor_abun-
dance;
scaled_acceptor=Acceptor_Spectrum*acceptor_
abundance;
% Calculate FRET efficiency
FRET_spectral = acceptor_abundance/donor_
abundance;
figure(1)
plot(Wavelength,FRET_Spectrum,'g',
Wavelength,scaled_donor,'b--',Wavelength,
scaled_acceptor,'r.','LineWidth',2)
xlabel('Wavelength (nm)','FontSize',12)
ylabel('Intensity (counts)','FontSize',12)
legend('FRET Spectrum','Estimated Donor
Spectrum','Estimated Acceptor Spectrum');
hleg=title({'Estimated Contributions of
Donor and Acceptor to FRET';
['FRET_S_P_E_C_T_R_A_L=',num2str(FRET_
spectral)]},'FontSize',12); set(hleg,
'FontSize',12)
```

8. Spectral analysis software is also available as separate toolboxes offered in imaging software associated with most major microscope manufacturers. In our experience, it is not always clear which analysis algorithm is being applied in these software

packages, and, importantly, the analysis can result in a loss of information in spectrally analyzed images. This can be problematic for quantitative analysis of spectral image data.

9. In our experience, averaging 20 emission scans results in spectra with sufficient signal-to-noise characteristics for accurate FRET measurements. However, depending upon expression levels and the quantum efficiency of the donor and acceptor used, it may be possible to average fewer emission scans, resulting in a decreased scan time and increased sample rate. This is important if performing kinetic studies.
10. Intact cell measurements using a spectrofluorimeter are typically conducted using cells in suspension in a stirred cuvette. This configuration allows for adequate mixing time of reagents and a consistent fluorescence signal in the light path. It is also possible to use cells plated on glass coverslips mounted in a specially designed cuvette such that the cells on the coverslip are in the light path. This setup would be beneficial if the experimental design required real-time measurement of cGMP signals in a configuration with cells attached to a substrate while allowing subsequent biochemical analysis of the same cells—e.g., measurement of cGMP levels using enzyme immunoassay (*see* ref. 12).
11. Recent studies in our lab clearly demonstrate that small changes in pH, temperature, and the concentration of reactive oxygen species can have markedly different effects on the fluorescence emission of CFP and YFP [8, 14, 15]. These differential effects can lead to apparent changes in FRET (that are not triggered by changes in cyclic nucleotide concentration). Thus, it is critical to measure and subsequently correct for effects that changes in the intracellular environment have on individual fluorophores.
12. It is important to realize that in some cell types, it is difficult to elicit uniform cGMP levels throughout the cell. Thus, calibration of FRET response in intact cells may not accurately reflect intracellular cGMP concentrations. In these cases, only qualitative changes of cGMP levels can be inferred.
13. The amounts (abundances) of CFP and YFP estimated by the "lsqnonneg" algorithm represent the relative amount of CFP or YFP signal with respect to the spectra in the reference library. This is analogous to the area under the curve for each spectrum in the reference library.
14. On our system (Cascade 512B, Photometrics), typical settings for fluorescent protein measurement are an EM gain of 3,800 and an acquisition time of 1,000 ms. In spectral microscopy, it is important not to oversaturate images at any wavelength, as

this would result in spectral artifacts, which could be incorrectly interpreted.

15. Although spectral imaging studies have been performed without flat-field spectral correction, we have found that this correction greatly increases the accuracy of the linear unmixing analysis. The increased accuracy is achieved in two ways. First, the background (or dark) spectrum is subtracted from the data of interest. This ensures that when the background-subtracted spectral data are analyzed, the analysis routine is not trying to fit background signals (such as stray room light) with the known spectra in the spectral library (the donor and acceptor spectra). Second, the background-subtracted data are multiplied by a factor that corrects for the spectral transfer function of the microscope. This ensures that data will be comparable across multiple systems (for example, between a microscope and a spectrofluorimeter). This step also removes wavelength-dependent artifacts from the spectra. In particular, we have noticed that when this method corrects for short-bandwidth spectral artifacts ("bumps" or "dips" in the spectra), the resulting data are able to be fit to the library with a much lower least-square error (as much as a 50 % reduction in the least-square error).
16. If the microscope system used to measure FRET responses is not equipped with a dedicated perfusion system, reagents can be pipetted directly into the buffer solution in the imaging chamber. Prior to addition to the chamber, reagents should be diluted in extracellular buffer solution. In order to minimize mixing time, the volume of the solution should be *at least* 10 % of the volume of solution in the imaging chamber.
17. The time course and amplitude of cGMP responses can vary substantially from cell to cell, even within a single field of view. To adequately sample cell-to-cell variability of agonist-induced responses, experiments should be conducted on *at least* three fields of view (requiring cells plated on three different coverslips) per experimental condition (e.g., addition of 10 nM ANP) for *at least* three different transfections (e.g., on *at least* three different days). When using a 40× objective, each field of view typically contains 3–5 cells suitable for cGMP measurements. For experiments on primary cultures, cells from at least three different isolations should be assessed.

Acknowledgments

This work was supported by NIH grants T32HL076125 and P01HL066299 and the Center for Lung Biology, University of South Alabama, College of Medicine.

References

1. Nikolaev VO, Gambaryan S, Lohse MJ (2006) Fluorescent sensors for rapid monitoring of intracellular cGMP. Nat Methods 3:23–25
2. Honda A, Adams SR, Sawyer CL, Lev-Ram V, Tsien RY, Dostmann WR (2001) Spatiotemporal dynamics of guanosine 3′,5′-cyclic monophosphate revealed by a genetically encoded, fluorescent indicator. Proc Natl Acad Sci USA 98:2437–2442
3. Nausch LW, Ledoux J, Bonev AD, Nelson MT, Dostmann WR (2008) Differential patterning of cGMP in vascular smooth muscle cells revealed by single GFP-linked biosensors. Proc Natl Acad Sci USA 105:365–370
4. Tsien RY, Miyawaki A (1998) Seeing the machinery of live cells. Science 280:1954–1955
5. Hill SJ, Williams C, May LT (2010) Insights into GPCR pharmacology from the measurement of changes in intracellular cyclic AMP; advantages and pitfalls of differing methodologies. Br J Pharmacol 161:1266–1275
6. Nikolaev VO, Lohse MJ (2009) Novel techniques for real-time monitoring of cGMP in living cells. Handb Exp Pharmacol 191:229–243
7. Harris AT (2006) Spectral mapping tools from the earth sciences applied to spectral microscopy data. Cytometry 69A:872–879
8. Rich TC, Britain AL, Byrne MA, Alvarez D, and Leavesley SJ (2013) Hyperspectral imaging approaches applied to FRET-based measurements of localized cAMP signals in pulmonary endothelial cells. Am J Respir Crit Care Med 187:A1724
9. Leavesley S, Wang X, Rich TC (2010) Assessing FRET response of fluorescent proteins in varying cellular microenvironments and equipment configurations. XXV Congress Int Soc Adv Cytometry A92
10. Leavesley SJ, Annamdevula N, Boni J, Stocker S, Grant K, Troyanovsky B, Rich TC, Alvarez DF (2012) Hyperspectral imaging microscopy for identification and quantitative analysis of fluorescently-labeled cells in highly autofluorescent tissue. J Biophotonics 5: 67–84
11. Blackman BE, Heimann J, Horner K, Wang D, Richter W, Rich TC, Conti M (2011) PDE4D and PDE4B function in distinct subcellular compartments in mouse embryonic fibroblasts. J Biol Chem 286:12590–12601
12. Lerner J, Zucker R (2004) Calibration and validation of confocal spectral imaging systems. Cytometry A 62:8–34
13. Piggott LA, Hassell KA, Berkova Z, Morris AP, Silberbach M, Rich TC (2006) Natriuretic peptides and nitric oxide stimulate cGMP synthesis in different cellular compartments. J Gen Physiol 128:3–14
14. Zhu B, Strada S, Stevens T (2005) Cyclic GMP-specific phosphodiesterase 5 regulates growth and apoptosis in pulmonary endothelial cells. Am J Physiol 289:L196–L206

Chapter 6

Visualization of cGMP with cGi Biosensors

Martin Thunemann, Natalie Fomin, Christian Krawutschke, Michael Russwurm, and Robert Feil

Abstract

Cyclic guanosine 3′–5′-monophosphate (cGMP) is an important signaling molecule in physiology, pathophysiology, and pharmacological therapy. It has been proposed that the functional outcome of an increase of cGMP in a given cell largely depends on the existence of global versus local cGMP pools. The recent development of genetically encoded fluorescent biosensors for cGMP is a major technical advance in order to monitor the spatiotemporal dynamics and compartmentalization of cGMP signals in living cells. Here we give an overview of the available cGMP sensors and how they can be used to visualize cGMP. The focus is on the fluorescence resonance energy transfer (FRET)-based cGi-type sensors (Russwurm et al., Biochem J 407:69–77, 2007), which are currently among the most useful tools for cGMP imaging in cells, tissues, and living organisms. We present detailed protocols that cover the entire imaging experiment, from the isolation of primary cells from cGi-transgenic mice and adenoviral expression of cGi sensors to the description of the setup required to record FRET changes in single cells and tissues. In-cell calibration of sensors and data evaluation is also described in detail and the limitations and common pitfalls of cGMP imaging are discussed. Specifically, we outline the use of FRET microscopy to visualize cGMP in murine smooth muscle cells (from aorta, bladder, and colon) and cerebellar granule neurons expressing cGi sensors. Most of the protocols can be easily adapted to other cell types and cGMP indicators and can be used as general guidelines for cGMP imaging in living cells, tissues and, eventually, whole organisms.

Key words Nitric oxide, Natriuretic peptide, Guanylyl cyclase, Protein kinase G, Phosphodiesterase, Transgenic mice, Adenovirus, VSMC, Cerebellum, Green fluorescent protein, Microscopy, Cell imaging, Intracellular signaling

1 Introduction

The cyclic nucleotide, cyclic guanosine 3′–5′-monophosphate (cGMP), is a ubiquitous second messenger that controls many cellular functions, from contraction to growth and survival [1]. In mammals, cGMP can be generated from GTP either by cytosolic/soluble guanylyl cyclases (sGCs) or by membrane-bound/particulate guanylyl cyclases (pGCs), which are stimulated by NO and natriuretic peptides (ANP, BNP, CNP), respectively [2, 3]. Cyclic GMP exerts its actions through at least three types of cGMP

Thomas Krieg and Robert Lukowski (eds.), *Guanylate Cyclase and Cyclic GMP: Methods and Protocols*, Methods in Molecular Biology, vol. 1020, DOI 10.1007/978-1-62703-459-3_6,

receptors, cyclic nucleotide-gated (CNG) cation channels [4], cGMP-dependent protein kinases (cGKs) [5], and cGMP-regulated phosphodiesterases (PDEs), which degrade cAMP and/or cGMP [6].

In addition to "classical" short-term effects like smooth muscle relaxation and platelet inhibition, cGMP is also involved in learning and memory [7] and regulates long-term processes like cell growth, differentiation, and survival [8]. Dysfunctions of the cGMP signaling cascade have been linked to a number of psychiatric and neurodegenerative disorders [9] and cardiovascular diseases such as metabolic syndrome [10] and arterial hypertension [11]. Drugs that increase the intracellular cGMP concentration are successfully used in patients, for instance, NO-releasing organic nitrates for the treatment of angina pectoris or inhibitors of the cGMP-specific PDE5, such as sildenafil (Viagra), for erectile dysfunction and pulmonary hypertension. Furthermore, sGC "stimulators" that sensitize the enzyme towards NO (riociguat, [12]) are in clinical development for the treatment of pulmonary hypertension [13]. "NO-heme mimetics" that replace the heme group of sGC and activate the enzyme by mimicking the conformation of the NO-heme complex (cinaciguat [14], ataciguat [15]) are tested in clinical trials for the treatment of decompensated heart failure, neuropathic pain, and peripheral arterial disease [16–18]. Recent preclinical studies indicate that cGMP-elevating drugs might also be effective in conditions that are associated with abnormal cell growth and degeneration such as the development of tumors [19, 20], Alzheimer's disease [21], retinal degeneration [22], and noise-induced hearing loss [23].

The molecular and cellular mechanisms that underlie cGMP's multiple roles in (patho-)physiology and therapy are, however, not well understood. It is assumed that the spatiotemporal dynamics of changes in the cGMP concentration is important for signaling specificity [24]. Distinct cGMP signaling compartments may be established by the interplay between cGMP generation and degradation via specific guanylyl cyclases and PDEs, respectively, combined with physical barriers that limit cGMP diffusion. Such structural barriers could be present in zones, where the cytoskeleton or the endoplasmic reticulum approaches the plasma membrane. For example, NO-sensitive sGC may generate global cytoplasmic cGMP signals, whereas the natriuretic peptide-activated pGCs may be involved in the formation of local cGMP microdomains near the plasma membrane [25, 26]. It is a major challenge in cGMP research to improve our knowledge of intracellular cGMP compartments and their functional relevance. Moreover, it is largely unknown when, where and how much cGMP is produced in a given tissue of a living mammalian organism under normal and pathological conditions or during pharmacotherapy with cGMP-elevating drugs.

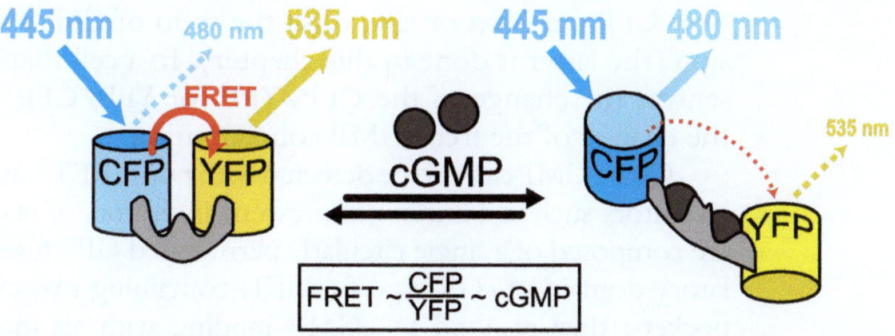

Fig. 1 Working principle of FRET-based cGi-type cGMP biosensors. In cGi biosensors, the tandem cGMP-binding sites of the bovine cGKI (*grey*) are flanked by CFP and YFP; sensor variants with different sensitivities for cGMP have been engineered (cGi-500, cGi-3000, cGi-6000, where the number denotes the apparent EC_{50} value in nM). In the absence of cGMP, FRET occurs from excited CFP to YFP leading to light emission from YFP. Upon cGMP binding, the cGi biosensor undergoes a conformational change that causes a decrease in FRET efficiency. Thus, light emission from YFP at 535 nm is reduced, while emission from CFP at 480 nm is increased. The FRET efficiency of cGi-type biosensors is depicted as the CFP/YFP emission ratio, which increases upon cGMP binding to the biosensor

Genetically encoded fluorescent biosensors based on variants of the green fluorescent protein (GFP) are powerful tools for real-time imaging of cGMP in native cells [27]. Most of the currently available sensor proteins detect cGMP by the principle of fluorescence resonance energy transfer (FRET). FRET relies on the radiationless transfer of energy from an excited donor fluorophore (e.g., the cyan fluorescent protein, CFP) to an acceptor fluorophore (e.g., the yellow fluorescent protein, YFP) via dipole–dipole coupling [28]. For FRET to occur, the donor and acceptor must be in close proximity (typically 2–6 nm) and appropriately oriented to each other, and there should be an overlap of at least 30 % between the donor's emission and acceptor's excitation spectrum [29]. Most FRET-based cGMP sensors consist of one or two cGMP-binding sites (derived from PDEs or cGKs) sandwiched between CFP and YFP, which serve as FRET donor and acceptor, respectively (Fig. 1). The binding of cGMP to the sensor results in a conformational rearrangement that causes a change in the distance and/or orientation of the fluorophores to each other. Thereby, cGMP binding can induce either an increase or a decrease of the FRET efficiency, depending on the particular sensor. The change in FRET efficiency can be monitored by conventional epifluorescence microscopy by measuring the emitted light of both donor (CFP) and acceptor (YFP) upon excitation of the donor (CFP). Note that a change in FRET efficiency from CFP to YFP results in a coincident change of the YFP and CFP emission intensities in opposite directions. If, for example, cGMP binding induces a decrease in FRET efficiency, this results in a decrease in YFP emission and an increase in CFP emission (Fig. 1). A change in FRET efficiency is depicted by calculating either the ratio of YFP over CFP emission or vice versa the ratio of CFP over YFP emission (the latter is done in this chapter). In a cell that expresses the sensor, the change of the CFP/YFP (or YFP/CFP) ratio reflects the change of the free cGMP concentration.

Cyclic GMP can also be detected with non-FRET-based fluorescent indicators such as FlincGs (fluorescent indicators of cGMP). FlincGs are composed of a single circularly permutated GFP fused to the regulatory domain of cGK type I (cGKI) containing two cGMP-binding pockets; they respond to cGMP binding with an increase in GFP fluorescence intensity [30]. Furthermore, the ion current generated by olfactory CNGA2 channels in the presence of cGMP at the plasma membrane has been used to monitor cGMP [25, 26].

In Table 1 we list a selection of available fluorescence-based cGMP sensors and their properties. To be useful for real-time monitoring of the spatial and temporal pattern of cGMP signals in intact cells, a sensor should fulfill the following criteria: (1) the structural change upon cGMP binding should be fast and reversible; (2) the cGMP-induced change of fluorescence should be strong; (3) the sensitivity and selectivity for cGMP should allow the detection of physiological concentrations of cGMP without cross-activation by cAMP; in this regard, cGMP sensors with EC_{50} values for cGMP of ~1 μM and for cAMP of >50 μM should perform well in most applications; (4) expression of the sensor should not have unwanted side effects such as buffering of the intracellular cGMP or general toxicity; (5) it should be assured that the localization of the sensor overlaps with the cellular compartment to be studied, for instance, by using targeted sensor variants.

In a systematic approach, Russwurm and colleagues designed and tested more than 50 FRET-based indicator constructs containing different regions of single or tandem cGMP-binding domains derived from PDE5 or cGKI [31]. Three constructs that contain the tandem cGMP-binding sites of cGKI, named cGi-500, cGi-3000, and cGi-6000 (for "cGMP indicator" with an apparent EC_{50} of 500 nM, 3,000 nM, and 6,000 nM, respectively) appeared particularly useful for cGMP imaging with respect to speed, amplitude, and reversibility of the cGMP-induced FRET change as well as sensitivity to cGMP and cross-activation by cAMP (Table 1 and [31]). The cGi sensors cover a range of cGMP affinities, with EC_{50} values between 500 nM and 6 μM, corresponding well to the affinities of known cGMP effector proteins such as cGKs. Thus, although it is difficult to predict the actual basal and stimulated cGMP concentrations in a given cell in vivo, cGi sensors should be capable of monitoring physiologically relevant cGMP levels. Note that the binding of cGMP to cGi-type sensors induces a *decrease* of the FRET efficiency. The FRET changes in cGi sensors are usually presented as donor-to-acceptor emission ratio (CFP/YFP), so that an *increase* in the cGMP concentration is reflected by an *increase* in the emission ratio making data presentation more intuitive (Fig. 1).

Table 1
Selected fluorescence-based cGMP sensors and some of their properties

Name	cGMP-binding domain	Fluorophores	Detection mode[a]	Signal change (%)[b]		EC_{50} cGMP (μM)	EC_{50} cAMP (μM)	Selectivity EC_{50}(cAMP)/EC_{50}(cGMP)	Reference
				In vitro	Intracellular[c]				
CGY-Del1	cGKI cGMP-BD A and B[d]	CFP/YFP	FRET↑	n.d.	24[(Y/C)]	1	100	100	[46]
cygnet-2	cGKI cGMP-BD A and B[e]	CFP/YFP	FRET↓	38[(C/Y)]	24[(C/Y)]	1.9	185	100	[47]
cGi-500	cGKI cGMP-BD A and B	CFP/YFP	FRET↓	77[(C/Y)]	38[(C/Y)]	0.5	>100	>200	[31]
cGi-3000	cGKI cGMP-BD A and B	CFP/YFP	FRET↓	72[(C/Y)]	37[(C/Y)]	3	>100	>30	[31]
cGi-6000	cGKI cGMP-BD A and B	CFP/YFP	FRET↓	58[(C/Y)]	30[(C/Y)]	6	>1,000	>166	[31]
δ-FlincG[f]	cGKI cGMP-BD A and B	cpEGFP	Fluorescence↑[g]	75	75	0.17	48	280	[30]
			Ratiometric↑[h]	250[h]	n.d.	0.49	48	100	
cGES-GKIB	cGKI cGMP-BD B	CFP/YFP	FRET↓	n.d.	30[(C/Y)]	5	485	100	[48]
cGES-DE5	PDE5 GAF A	CFP/YFP	FRET↑	16[(Y/C)]	30[(Y/C)]	1.5	630	420	[48]
Red cGES-DE5[i]	PDE5 GAF A	Sapphire/RFP	FRET↑	10[(R/S)]	15[(R/S)]	0.04	>100	>1,000	[32]
Cygnus[j]	PDE5 GAF A	mTagBFP/sREACh	Fluorescence↓[j]	10	20	1	400	400	[33]

[a]Arrows indicate signal increase (↑) or decrease (↓) of the sensor upon cGMP binding
[b]In case of FRET-based sensors, signal changes were determined via emission ratios of the fluorophores as indicated in brackets; C, CFP; Y, YFP; R, RFP; S, Sapphire
[c]Apparent intracellular FRET changes may vary depending on the setup used for measurements (e.g., filter sets)
[d]Including functional kinase domain

(continued)

Table 1
(continued)

[e]Including silenced kinase domain

[f]This sensor is not based on FRET, but on the fluorescence change of a circularly permuted GFP (cpEGFP)

[g]Determined using 480 nm excitation and 510 nm emission

[h]Determined using ratiometric measurement with excitation at 410 and 480 nm and emission at 510 nm (given as 480/410 nm ratio)

[i]This is a "red" FRET-based cGMP indicator; it can be used for ratiometric measurement with excitation of GFP-derived Sapphire and emissions of Sapphire and red fluorescent protein (RFP) dimer2

[j]This is a "blue" cGMP sensor based on FRET between monomeric Tag blue fluorescent protein (mTagBFP) as FRET donor and super resonance energy accepting chromoprotein (sREACh) as FRET acceptor; sREACh is a so-called "dark YFP quenching FRET acceptor" that does not emit light upon excitation; cGMP binding causes an increase of the FRET efficiency and, therefore, a decrease of mTagBFP emission; single-channel measurement of mTagBFP with 405 nm excitation and 440 nm emission is performed

Abbrevations: cGKI cGMP-BD, cGMP-binding domain of cGMP-dependent protein kinase type I; n.d., not determined; PDE5 GAF, cGMP-binding GAF domain of phosphodiesterase 5 (GAF: domain first identified in c*G*MP-specific phosphodiesterases, ***Anabaena*** *A*denylyl cyclases and *E. coli F*hlA)

This chapter focuses on the use of FRET-based cGi-type biosensors to visualize cGMP in individual smooth muscle and neuronal cells of mice. However, the protocols can also be used as general guidelines for cGMP imaging of cultured cells, isolated tissues and, eventually, living mice or other intact organisms. Details about cGMP detection with alternative cGMP indicators such as cygnets, cGES-type and FlincG-type sensors can be found in Table 1 and other chapters of this book (*see* Chapters 5, 7, and 8). To image cGMP and other signaling molecules such as cAMP or Ca^{2+} in the same cell, it is important to use fluorescent probes that are spectrally compatible. Indeed, cGMP indicators have been developed that consist of fluorescent proteins other than CFP and YFP (e.g., the "red" sensor, red cGES-DE5, and the "blue" sensor, cygnus); these sensors can be combined with "conventional" CFP/YFP-based biosensors (Table 1 and [32, 33]). The feasibility to visualize cGMP during signal transduction in a living cell or organism will be widely useful for basic research on cyclic nucleotides as well as for drug development and target validation. Moreover, the use of genetically encoded biosensors will greatly advance our general understanding of information processing in health and disease. Having these powerful tools, we envision a future where we can watch brains think [34].

2 Materials

General requirements are access to an animal facility for mice and resources for the genotyping of transgenic mice (for detailed protocols, *see* ref. 35) as well as a standard cell culture laboratory with tissue culture hood (class II biological safety cabinet), water bath, and CO_2 incubator. All solutions should be made with autoclaved, deionized water (≥18 MΩ). To prepare solutions of compounds used for cell isolation and culture (e.g., enzyme solutions) the original container should be opened only inside the tissue culture hood.

2.1 Isolation and Culture of Primary Smooth Muscle Cells

1. Dissection tools: regular and fine scissors and forceps (Dumont #5, Dumont #5/45).
2. Stereomicroscope for dissection of mice (magnification 6.5–50×).
3. 3.5 and 10 cm petri dishes (bacterial-grade).
4. 12-well plates (cell culture-grade) equipped with autoclaved round glass coverslips (20 mm diameter).
5. Netwell meshes (24 mm diameter, 74 μm mesh size).
6. PBS (pH 7.4): Phosphate-buffered saline (PBS) with 135 mM NaCl (7.89 g/L), 3 mM KCl (0.22 g/L), 8 mM Na_2HPO_4

(1.42 g/L $Na_2HPO_4 \times 2\ H_2O$), 2 mM KH_2PO_4 (0.27 g/L), adjust pH to 7.4 with HCl or NaOH, autoclave, store at room temperature.

7. Ca^{2+}-free medium (pH 7.4): 85 mM Na L-glutamate (15.91 g/L Na L-glutamate × H_2O), 60 mM NaCl (3.51 g/L), 10 mM HEPES (2.38 g/L), 5.6 mM KCl (0.42 g/L), 1 mM $MgCl_2$ (0.20 g/L $MgCl_2 \times 6\ H_2O$), adjust pH to 7.4 with HCl, autoclave, store at 4 °C.
8. 100 mg/mL BSA: Dissolve 0.5 g bovine serum albumin (BSA) in 5 mL Ca^{2+}-free medium, sterilize by filtration, store in 0.5 mL aliquots at −20 °C.
9. 100 mg/mL DTT: Dissolve 0.5 g dithiothreitol (DTT) in 5 mL Ca^{2+}-free medium, sterilize by filtration, store in 0.5 mL aliquots at −20 °C.
10. 7 mg/mL Papain: Dissolve 100 mg Papain in 14.29 mL Ca^{2+}-free medium, store in 0.5 mL aliquots at −20 °C.
11. 10 mg/mL Collagenase: Dissolve 100 mg Collagenase in 10 mL Ca^{2+}-free medium, store in 0.5 mL aliquots at −20 °C.
12. 10 mg/mL Hyaluronidase: Dissolve 100 mg Hyaluronidase in 10 mL Ca^{2+}-free medium, store in 0.5 mL aliquots at −20 °C.
13. Enzyme solution A (5 mL, prepare shortly before use): To 4.4 mL Ca^{2+}-free medium, add 500 μL Papain (final concentration: 0.7 mg/mL), 50 μL DTT (final concentration: 1 mg/mL), 50 μL BSA (final concentration: 1 mg/mL).
14. Enzyme solution B (5 mL, prepare shortly before use): To 3.95 mL Ca^{2+}-free medium, add 500 μL Collagenase (final concentration: 1 mg/mL), 500 μL Hyaluronidase (final concentration: 1 mg/mL), 50 μL BSA (final concentration: 1 mg/mL).
15. DMEM: Dulbecco's Modified Eagle Medium with 4.5 g/L glucose, sodium pyruvate, and stable glutamine.
16. FBS: heat-inactivated fetal bovine serum (FBS), store in 50 mL aliquots at −20 °C.
17. 100× Pen/Strep: Penicillin (10,000 U/mL)/Streptomycin (10,000 μg/mL), store in 5 mL aliquots at −20 °C.
18. Culture medium with 5 % or 10 % FBS (for culture of adenovirally infected or transgenic cells, respectively): Add 25 mL or 50 mL FBS and 5 mL Pen/Strep to 500 mL DMEM, store at 4 °C for up to 2 months.
19. Serum-free culture medium: Add 5 mL Pen/Strep to 500 mL DMEM, store at 4 °C for up to 2 months.
20. Recombinant, replication-deficient adenoviruses for expression of cGi-500, cGi-3000, and cGi-6000 (*see* **Note 1**).

2.2 Isolation and Culture of Primary Cerebellar Granule Neurons

1. Dissection tools: regular and fine scissors and forceps (Dumont #5, Dumont #5/45), razor blades.
2. Stereomicroscope for dissection of mice (magnification 6.5–50×).
3. 3.5 and 10 cm petri dishes (bacterial-grade).
4. 12-well plates (cell culture-grade) equipped with autoclaved and PDL-coated round glass coverslips (20 mm diameter).
5. Netwell meshes (24 mm diameter, 74 μm mesh size).
6. Autoclaved Pasteur pipettes.
7. 10 μg/mL PDL: Dissolve 1 mg Poly-D-lysine hydrobromide in 1 mL H_2O by shaking at 37 °C for 2 h and then at 4 °C overnight (*see* **Note 2**). Dilute PDL to 100 mL with H_2O to a final concentration of 10 μg/mL, store at 4 °C for up to 2 months.
8. PDL-coated coverslips: 1 day before isolation of cells, coat autoclaved 20 mm glass coverslips in 12-well plates with 1 mL 10 μg/mL PDL. Incubate overnight inside the tissue culture hood at room temperature (*see* **Note 3**). On the next day, collect PDL for reuse (it can be reused twice) and wash glass coverslips three times with H_2O. Dry open multi-well plates for 2 h inside the tissue culture hood before plating cells.
9. PBS (pH 7.4): Phosphate-buffered saline (PBS) with 135 mM NaCl (7.89 g/L), 3 mM KCl (0.22 g/L), 8 mM Na_2HPO_4 (1.42 g/L $Na_2HPO_4 \times 2\ H_2O$), 2 mM KH_2PO_4 (0.27 g/L), adjust pH to 7.4 with HCl or NaOH, autoclave, store at room temperature.
10. 10× Krebs buffer (pH 7.4) without D-Glucose: 1.24 M NaCl (72.5 g/L), 54 mM KCl (4 g/L), 5 mM NaH_2PO_4 (0.7 g/L $NaH_2PO_4 \times H_2O$). Adjust pH to 7.4 with NaOH, autoclave, store at 4 °C.
11. 1 M D-Glucose (198.17 g/L D-Glucose × H_2O), sterilize by filtration, store at 4 °C.
12. 3.82 % $MgSO_4$: Dissolve 3.82 g $MgSO_4 \times 7\ H_2O$ in 100 mL H_2O, sterilize by filtration, store at 4 °C.
13. 1.2 % $CaCl_2$: Dissolve 1.2 g $CaCl_2 \times 2\ H_2O$ in 100 mL H_2O, sterilize by filtration, store at 4 °C.
14. 0.3 % BSA solution (prepare always fresh on the day of preparation): Weigh 450 mg of bovine serum albumin (BSA) into a beaker glass, add 132.84 mL H_2O, 15 mL 10× Krebs buffer, 2.16 mL 1 M D-Glucose and 1.2 mL 3.82 % $MgSO_4$. Adjust the pH to 7.4 with NaOH, sterilize by filtration, store at 4 °C.
15. 2.5 % Trypsin: Dissolve 1 g Trypsin in 40 mL H_2O, store in 5 mL aliquots at −80 °C.

16. 1 % DNase: Dissolve 100 mg DNase in 10 mL H_2O, store in 0.3 mL aliquots at −20 °C.
17. Trypsin inhibitor.
18. MEM: Minimum essential medium (MEM) without glutamine.
19. 50 mg/mL Gentamicin.
20. 100× L-glutamine (200 mM), store in 1 mL aliquots at −20 °C.
21. 50× B27 supplement (*see* **Note 4**), store in 2 mL aliquots at −20 °C.
22. FBS: heat-inactivated fetal bovine serum (FBS), store in 50 mL aliquots at −20 °C.
23. High K^+ (HK) medium: Weigh 825 mg KCl into an autoclaved beaker glass, add 50 mL from a medium bottle with 500 mL MEM. After the KCl has dissolved, sterilize the solution by filtration and transfer it back into the original medium bottle. Add 1 mL of Gentamicin (50 mg/mL), store at 4 °C for up to 2 months.
24. CGN medium: To 88 mL HK medium, add 9 mL FBS, 0.9 mL 100× L-glutamine, and 2 mL 50×B27 supplement, store at 4 °C for up to 1 month.
25. 100 mM Ara-C: Dissolve 100 mg Cytosine-*β-D*-arabinofuranoside hydrochloride (Ara-C) in 3.58 mL H_2O, store in 1 mL aliquots at −20 °C.

2.3 FRET-Based cGMP Imaging

A setup to analyze cGMP via ratiometric FRET epifluorescence microscopy can be assembled in many different ways, starting from virtually every fluorescence microscope [36]. In the following section, we describe our setup that is based on an inverted microscope in an air-conditioned (~21 °C) darkroom.

1. cGi-expressing cells isolated from cGi-transgenic mice or infected with cGi-encoding adenovirus, and grown on glass coverslips (Subheadings 3.1 and 3.2).
2. Inverted microscope (Axiovert 200 with 1.0/1.6× Optovar lens, Carl Zeiss).
3. Fluorescence-grade objectives with 10× and 40× magnification (Plan NeoFluar 10×/0.30; EC Plan NeoFluar 40×/1.30 Oil, Carl Zeiss).
4. Computer-controlled light source with electronic shutter, e.g., a rapid filter switching device (Oligochrome, TILL Photonics, *see* **Note 5**).
5. FRET filter set: 445/20 nm CFP excitation filter and 470 nm dichroic mirror (AHF or Chroma Technology).

6. Beam splitter (Micro-Imager DUAL-View, Photometrics) with 05-EM insert (516 nm dichroic mirror, 480/50 nm CFP and 535/40 nm YFP emission filters, *see* **Note 6**) placed between microscope and camera.
7. YFP filter set: 497/16 nm excitation filter, 516 nm dichroic mirror, 535/22 nm emission filter (AHF or Chroma Technology, *see* **Note 7**).
8. Cooled electron-multiplying charged-coupled device (EM-CCD) camera (Retiga 2000R, QImaging).
9. Superfusion system: FPLC pump (Pharmacia P-500, GE Healthcare), FPLC injection valves (Pharmacia V-7, GE Healthcare), superfusion chamber (self-made or commercially available), vacuum pump with adjustable vacuum, sample loops (e.g., 2, 5, 20 mL), tubing, connectors, hypodermic needles, and syringes (*see* Fig. 2 and **Note 8**).
10. Silicon vacuum grease: For convenient application, fill grease into 1 mL syringe and use without needle.
11. Image acquisition and online analysis software (Live Acquisition, TILL Photonics, *see* **Note 9**), image analysis software for offline analysis (Offline Analysis, TILL Photonics; ImageJ [37]), data analysis software (Microsoft Excel, Microsoft; Origin, OriginLab Corp.).
12. 1 M D-Glucose (198.17 g/L D-Glucose × H_2O), sterilize by filtration, store at 4 °C.
13. Imaging buffer (pH 7.4): 140 mM NaCl (8.18 g/L), 5 mM KCl (0.373 g/L), 1.2 mM $MgCl_2$ (0.296 g/L $MgSO_4$ × 7 H_2O), 2.0 mM $CaCl_2$ (0.222 g/L), 5 mM HEPES (1.19 g/L). Adjust to pH 7.4 with NaOH, autoclave, store at room temperature. Add 10 mL 1 M D-Glucose to 1,000 mL imaging buffer before use (final concentration ~10 mM D-Glucose).
14. 20 % denatured ethanol.
15. 100 mM DEA/NO: Dissolve 50 mg 2-(*N,N*-diethylamino)-diazenolate-2-oxide diethylammonium salt (DEA/NO, NO-releasing drug that stimulates sGC) in 2.42 mL ice-cold 10 mM NaOH (*see* **Note 10**), store in 50 μL aliquots at −20 °C.
16. 10 mM GSNO: Dissolve 1 mg *S*-Nitrosoglutathione (GSNO, stimulates sGC) in 0.297 mL ice-cold H_2O. Prepare freshly every day, store on ice until use.
17. 20 mM ODQ: Dissolve 10 mg 1*H*-[1,2,4]oxadiazolo[4,3-a] quinoxalin-1-one (ODQ, sGC inhibitor) in 2.67 mL DMSO, store in 100 μL aliquots at −20 °C.
18. 100 μM ANP: Dissolve 0.1 mg atrial natriuretic peptide (ANP, 1-28, rat, stimulates the particulate guanylyl cyclase, GC-A) in 0.327 mL H_2O, store in 50 μL aliquots at −20 °C.

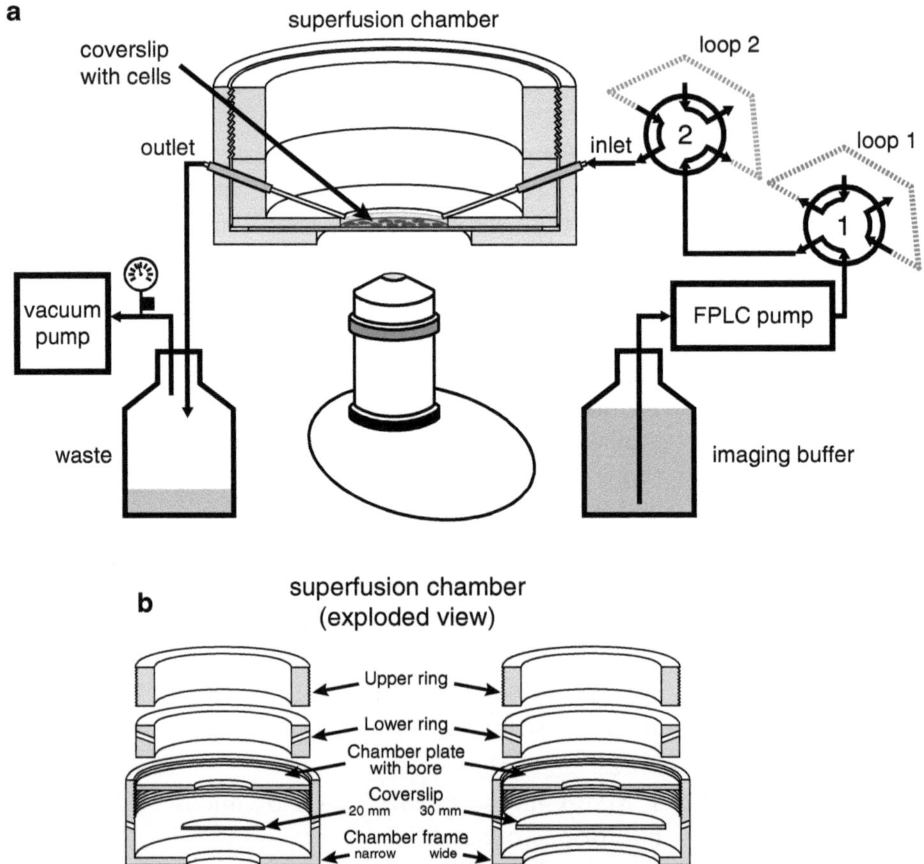

Fig. 2 Superfusion system for cGMP imaging. (**a**) The imaging buffer is continuously supplied (1 mL/min) by a FPLC pump through two injection valves (1, 2) onto the cells, which attach to a glass coverslip that serves as bottom of the superfusion chamber. The "chamber plate" is placed on top of the coverslip and defines the geometry of the superfusion chamber (for details on chamber assembly, see panel **b**). The buffer is continuously removed from the chamber via suction by an adjustable vacuum pump. Drug solutions are loaded into sample loops (loop 1, loop 2) that are attached to the injection valves. By changing the valve position, the drug solution can be delivered to the cells. Two valves connected in series can be used to apply different drugs simultaneously (e.g., a PDE inhibitor from loop 1 followed by DEA/NO together with the same PDE inhibitor from loop 2). The valve settings shown in the figure would lead to drug application from loop 2, but not from loop 1. (**b**). Exploded view on a cross section of the superfusion chamber. Variants with a narrow (*left*) or wide (*right*) opening of the chamber frame are shown, which accommodate either 20 mm or 30 mm coverslips. The coverslips serve as bottom of the chamber. The "chamber plate" with its central oval-shaped bore is placed on top of the coverslip. It defines the superfusion chamber and its maximal volume. Each type of chamber plate can be used with both 20 and 30 mm coverslips. The upper ring is screwed into the metal frame and tightens the chamber assembly. The inlet and outlet needles are directed through bores in the lower ring and the chamber frame until they approach the borders of the chamber plate openings (*see* also panel **a**); fine-adjustment of the outlet needle can be used to regulate the liquid level during superfusion

19. 100 μM CNP: Dissolve 0.5 mg C-type natriuretic peptide (CNP, stimulates the particulate guanylyl cyclase, GC-B) in 2.275 mL H_2O, store in 50 μL aliquots at −20 °C.

20. 3 mM Sildenafil: Dissolve 50 mg Sildenafil citrate (inhibits cGMP-specific PDE5) in 25 mL H_2O, store in 1.5 mL aliquots at −20 °C.
21. 500 mM IBMX: Dissolve 1.0 g 3-Isobutyl-1-methylxanthine (IBMX, unspecific PDE inhibitor) in 9.0 mL DMSO, store in 1.5 mL aliquots at −20 °C.

2.4 In-Cell Calibration of cGMP FRET Sensors

1. cGi-expressing cells (e.g., from cGi-transgenic mice) grown on glass coverslips (Subheading 3.1).
2. ICM (pH 7.3): Intracellular-like medium (ICM) with 125 mM KCl (9.32 g/L), 19 mM NaCl (1.11 g/L), 1 mM EGTA (0.38 g/L), 10 mM HEPES (2.38 g/L), 0.33 mM $CaCl_2$ (37 mg/L). Adjust the pH to 7.3 with KOH, autoclave, store at room temperature.
3. 50 mM β-Escin: Dissolve 55 mg β-Escin in 1.0 mL H_2O. Incubate at 37 °C for complete dissolution, store in 100 μL aliquots at −20 °C in the dark.
4. 100 mM cGMP: Dissolve 100 μmol cGMP in 1.0 mL H_2O, store in 100 μL aliquots at −20 °C.

3 Methods

3.1 Isolation and Culture of Primary Smooth Muscle Cells

Because the cGMP signaling system is of outstanding importance in smooth muscle cells (SMCs), we provide a detailed protocol for the isolation and culture of primary SMCs from different origins, namely vascular smooth muscle cells (VSMCs) from aorta, bladder smooth muscle cells (BSMCs), and colon smooth muscle cells (CSMCs). The smooth muscle tissues are obtained from cGi-transgenic mice that express the cGi-500 sensor under the control of the smooth muscle-specific SM22α promoter (MT, MR, and RF, unpublished data). Alternatively, cells can be isolated from wild-type mice and infected with an adenovirus encoding one of the cGi sensors.

1. Use 3–5 mice with an age of 1–3 months (cGi-transgenic mice or wild-type mice for infection with cGi-encoding adenovirus, *see* **Note 11**).
2. Prepare 3 × 3.5 cm and several 10 cm petri dishes with PBS.
3. Sacrifice animals by CO_2 inhalation. Do not perform cervical dislocation, as the aorta might get disrupted, making its isolation more difficult.
4. If transgenic mice are used, collect tail tips for re-genotyping.
5. Wet fur of the animals with 70 % ethanol.
6. Open abdominal and thoracic cavity.

7. Collect bladder and colon in a 10 cm petri dish with PBS.
8. In order to isolate the aorta, remove diaphragm, remaining intestines and esophagus as well as liver and spleen. Align the body with the tail on the left and the head on the right (for right-handed persons). The alignment helps to keep orientation during aorta excision, especially when remaining blood, fat, and/or muscle may prevent a clear view on the aorta. Cut ascending vessels, trachea, and esophagus below the pharynx, grab the heart with a forceps and lift it carefully to set the aorta under some tension. Cut along the spine towards the tail. Avoid rupture of the aorta during dissection; it is very difficult to find its loose end in the body. Transfer the excised aorta (still attached to heart and lung) into a 10 cm petri dish with PBS.
9. Under a stereomicroscope, isolate the smooth muscle tissue from bladder, colon and aorta. Remove as much fat as possible, because it interferes with the subsequent isolation steps.
   - To isolate bladder SMCs, remove surrounding fat and cut the bladder open. Grab the urothelium and peel it off the smooth muscle layer.
   - To isolate colon SMCs, remove remaining feces from the colon by washing with PBS using a syringe with bended needle, and transfer it into a new 10 cm petri dish with PBS. Remove remaining mesenteries, fat and blood vessels. Make a cut at the proximal part of the colon. Starting at this cut, peel off the smooth muscle layer from the enteric tissue. Once a small piece of the smooth muscle layer has been separated from the enteric tissue, it can easily be peeled off the whole colon.
   - To isolate vascular SMCs from aorta, carefully remove the heart, thymus, lung, and remainders of the airways and the esophagus (*see* **Note 12**). Clean the vessel carefully from blood, fat, and connective tissue by using two fine forceps.
10. Place the dissected smooth muscle tissues into 3.5 cm petri dishes with PBS. Up to this point the preparation should not take longer than 45 min.
11. Transfer the petri dishes into a tissue culture hood and cut the tissues into ~5 mm pieces using scissors. Transfer the pieces into 15 mL tubes with enzyme solution A (1.5 mL for 3–5 aortae or bladders, 2 mL for 3–5 colons). Incubate for 45 min in a water bath at 37 °C; invert every 15 min.
12. Centrifuge at 200 × *g* for 2 min. Discard the supernatant and suspend the tissue fragments in enzyme solution B by gentle shaking; use the same volume as for solution A (*see* previous **step 11**).
13. Incubate for 10–15 min in a water bath at 37 °C. After ~7 min in enzyme solution B, resuspend the digestion mixture with a 1,000 μL pipette. If the tissue pieces occlude the opening of

the pipette tip, incubate for another ~3 min. Resuspend ten times and continue the incubation. After 5 min, resuspend again with a 1,000 μL pipette until the majority of tissue pieces disappears and the solution becomes turbid. Even if some tissue pieces remain, stop the digestion after max. 15 min by adding culture medium to a final volume of 10 mL and resuspend (*see* **Note 13**).

14. Centrifuge at 200 × *g* for 7 min; discard the supernatant and resuspend the cell pellet containing the SMCs (VSMCs, BSMCs, or CSMCs) in 1.0 mL culture medium.
15. Mix 18 μL of the cell suspension with 2 μL trypan blue and count viable and dead (trypan blue-positive) cells in a cytometer (*see* **Note 14**). Calculate the titer of viable cells; the viability should be ≥90 % with a yield of ~2×10^5 cells per aorta, bladder, or colon.
16. Adjust the number of cells in culture medium to 6×10^4 VSMCs/mL, 4×10^4 BSMCs/mL, 3×10^4 CSMCs/mL. Plate 1.0 mL of each cell suspension per well into 12-well plates equipped with 20 mm coverslips. This corresponds to a plating density of 1.8×10^4 VSMCs/cm^2, 1.1×10^4 BSMCs/cm^2, and 0.9×10^4 CSMCs/cm^2. From 3–5 mice, the yield is ~10 wells with VSMCs, ~20 wells with BSMCs, and ~30 wells with CSMCs (*see* **Note 15**).
17. Grow the cells in culture medium at 37 °C and 6 % CO_2. Use medium with 10 % FBS for cGi-transgenic VSMCs, BSMCs, and CSMCs; use medium with 5 % FBS for wild-type VSMCs that are subsequently infected with cGi-encoding adenovirus. Change the medium 3 days after plating the primary cells.
18. For cGMP imaging of cGi-transgenic SMCs, change the culture medium to serum-free culture medium when the cells are ~70 % confluent (after 4–7 days); keep the cells for another 24 h in culture before cGMP imaging (*see* **Note 16**). Results with cGi-transgenic VSMCs are shown in Fig. 3a–e. Similar results are obtained with cGi-transgenic BSMCs and CSMCs.
19. For adenoviral infection, grow primary wild-type VSMCs for 5 days in culture medium with 5 % FBS; then change medium to serum-free culture medium and add ~1/10 of the medium volume of adenoviral supernatant; after 24 h add FBS to 5 %. Keep the cells for another 24 h in culture before cGMP imaging (*see* **Note 16**). Results with cGi adenovirus-infected VSMCs are shown in Fig. 4.

3.2 Isolation and Culture of Primary Cerebellar Granule Neurons

Besides in smooth muscle, the cGMP signaling system plays a role in many other cell types and organs, for instance, in the development and function of the nervous system. As an example of cGMP FRET imaging in neural cells, we present a protocol

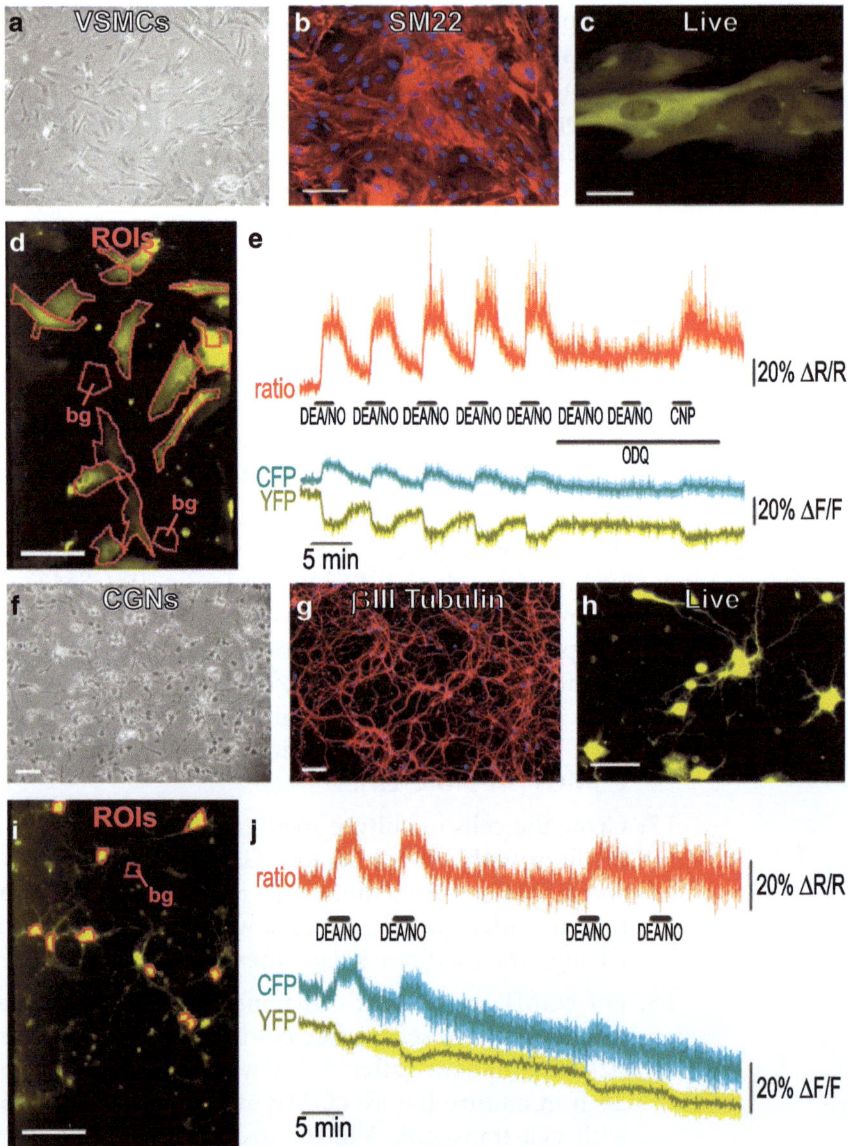

Fig. 3 FRET measurements with cGi-500 VSMCs and cGi-6000 CGNs isolated from transgenic mice. (**a**) Phase contrast image of VSMCs after 4 days in primary culture. (**b**) Immunofluorescence staining of VSMCs with an antiserum against the smooth muscle marker SM22α (*red*); nuclear counterstain with Hoechst 33258 (*blue*). (**c**) Live YFP fluorescence of cGi-500-transgenic VSMCs shows the expression and cytosolic localization of the sensor. (**d**) cGi-500-transgenic VSMCs after 7 days in primary culture (the last 24 h in serum-free medium) used for the imaging experiment shown in panel **e**. ROIs and two background regions (bg) are indicated in *red*; the two background regions were used to account for uneven illumination of the microscopic field of view at this magnification. (**e**) Imaging experiment with VSMCs showing percent changes of CFP and YFP fluorescence ($\Delta F/F$) and of the CFP/YFP ratio ($\Delta R/R$). Data represent means ± SEMs of the ROIs shown in panel **d**. Cells were repeatedly stimulated with 75 nM DEA/NO. In the presence of 10 μM ODQ, the response to DEA/NO was abolished, while 100 nM CNP was still able to increase cGMP levels. (**f**) Phase contrast image of CGNs after 2 days in primary culture. (**g**) Immunofluorescence staining with a monoclonal antibody against the neuronal marker βIII-tubulin (*red*); nuclear counterstain with Hoechst

for the isolation and imaging of primary cerebellar granule neurons (CGNs). The CGNs are obtained from postnatal day 7 (P7) cGi-transgenic mice that express the cGi-6000 sensor under the control of a cytomegalovirus (CMV) promoter (MT, NF, MR, and RF, unpublished data). Since CGNs are the most abundant neuronal cells in the cerebellum, this protocol yields a relatively high number of cells and a largely homogeneous culture. Note that primary cell cultures can also be obtained from other cells of the brain such as astrocytes or hippocampal neurons. These cell types can also be isolated from CMV-cGi-6000 mice and analyzed by cGMP FRET imaging, but isolation protocols require embryonic or newborn mice and different culture conditions (e.g., [38]).

1. Set up mouse breeding to obtain cGi-transgenic progeny. Separate pregnant females and check for pup delivery (postnatal day 0). On postnatal day 5, pups can be labeled with a waterproof marker pen and tail tips can be collected for genotyping to confirm the presence of the cGi transgene in the pups (*see* **Note 17**). For one preparation, 3–5 transgenic pups are required.
2. On the day of preparation (postnatal day 7), prepare 5 × 50 mL centrifuge tubes (tubes 1–5) in the tissue culture hood:
   - Tube 1: Add 30 mL 0.3 % BSA solution.
   - Tube 2: Add 30 mL 0.3 % BSA solution and, shortly before use, 300 μL 2.5 % Trypsin.
   - Tube 3: Dissolve 7.8 mg Trypsin inhibitor in 15 mL 0.3 % BSA solution (thorough mixing and warming to 37 °C facilitates its dissolution), add 150 μL 3.82 % $MgSO_4$ and, shortly before use, 150 μL 1 % DNase.
   - Tube 4: Add 17 mL 0.3 % BSA solution and 8 mL from tube 3; discard 10 mL, so that 15 mL remain in the tube.
   - Tube 5: Add 12.5 mL 0.3 % BSA solution, 100 μL 3.82 % $MgSO_4$, and 15 μL 1.2 % $CaCl_2$.

Fig 3 (continued) 33258 (*blue*). (**h**) Live YFP fluorescence of cGi-6000-transgenic CGNs shows sensor expression and its localization in cell bodies and neurites. (**i**) cGi-6000-transgenic CGNs after 5 days in primary culture used for the imaging experiment shown in panel **j**. ROIs defining the cell bodies and a background region (bg) are indicated in *red*. (**j**) Imaging experiment with CGNs showing percent changes of CFP and YFP fluorescence ($\Delta F/F$) and of the CFP/YFP ratio ($\Delta R/R$). Data represent means ± SEMs of the ROIs shown in panel **i**. Upon multiple stimulations with 100 nM DEA/NO, cells responded repeatedly with cGMP increases. Increased rundown of CFP and YFP fluorescence in comparison to VSMCs (panel **e**) might result from stronger bleaching due to illumination at higher magnification; if necessary, imaging conditions can be optimized according to **Note 22**. Scale bars in panels **a**, **b**, **d**, **i**: 100 μm; **c**: 25 μm; **f–h**: 50 μm

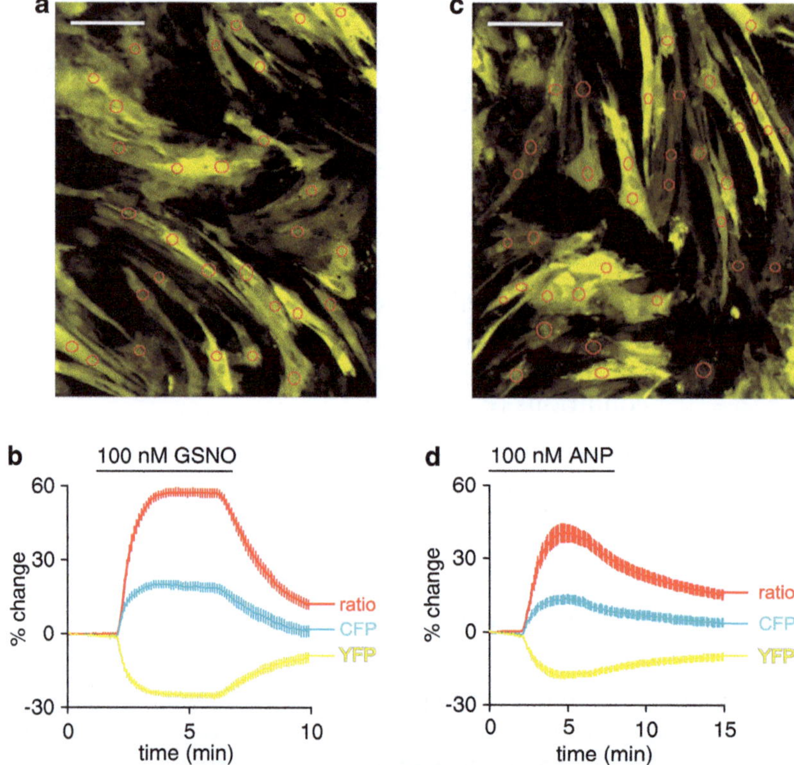

Fig. 4 FRET measurements in adenovirally infected VSMCs. (**a**) and (**c**) Live YFP fluorescence of the cGi-6000 indicator delivered to VSMCs by adenoviral infection. ROIs used for the measurements shown in panel **b** and **d** are indicated in *red*; background regions are not shown. (**b**) and (**d**) Percent changes of CFP fluorescence, YFP fluorescence and CFP/YFP ratio values elicited by 100 nM GSNO (panel **b**) or 100 nM ANP (panel **d**). Data represent means ± SEM of the ROIs shown in panels **a** and **c**. Scale bars in panels **a** and **c**, 100 μm

3. Distribute part of the remaining 0.3 % BSA solution into 10 × 3.5 cm petri dishes (3 mL/dish); keep them on ice for preparation of the brains.
4. Wash dissection instruments thoroughly. Keep them in ethanol and rinse in autoclaved H_2O before use.
5. Use 3–5 mouse pups of postnatal day 7. Dip the pup's head for 1 s into a beaker with 70 % ethanol and then cut the neck and let the head fall into a 50 mL tube with ice-cold PBS.
6. Collect the tail tip for re-genotyping.
7. Under a stereomicroscope, dissect the brains in a petri dish with cold 0.3 % BSA solution. Take the head with forceps and remove the skin. To open the skull, hold it at the nose and cut the skull with fine scissors above the brain from caudal to rostral. Pull the skull apart to each side with bend forceps. Disconnect the brain from the skull by cutting at the olfactory bulb.
8. Store the brains in another petri dish with 0.3 % BSA solution on ice while dissecting the remaining brains.

9. After all brains have been dissected, remove the cerebella from them. Under high magnification (40×), remove meninges and blood vessels from the cerebella. This is critical because fibroblasts from these tissues would overgrow the neurons in culture. Transfer the cerebella into a new petri dish with 0.3 % BSA solution on ice. Up to this point the preparation should not take longer than 1 h.
10. In the tissue culture hood, mince the cerebella in a 3.5 cm petri dish with 0.3 % BSA solution using a razorblade until the suspension becomes turbid (~5 min). A thorough disintegration is critical for a high cell yield. Transfer the minced tissue into tube 1 using a Pasteur pipette.
11. Centrifuge for 5 min at 240 × *g*. Remove the supernatant and resuspend the pellet in 30 mL solution from tube 2 (with Trypsin). Incubate for 15 min in a 37 °C water bath. Invert tube after 5 and 10 min.
12. Using a 25 mL pipette, resuspend the cells and transfer the suspension into tube 4 (with diluted Trypsin inhibitor and DNase). Centrifuge for 5 min at 240 × *g*. Remove the supernatant and resuspend the cell pellet with a 25 mL pipette in 7 mL solution from tube 3 (with Trypsin inhibitor and DNase). Resuspend 20 times with a 10 mL pipette and then ten times with a Pasteur pipette. Then add 12 mL solution from tube 5.
13. Pass the cell suspension through a netwell mesh into a new 50 mL tube.
14. Centrifuge for 5 min at 240 × *g*. Remove the supernatant and resuspend the cell pellet in 5 mL CGN medium. Mix 18 μL of the cell suspension with 2 μL trypan blue and count viable and dead (trypan blue-positive) cells in a cytometer. Calculate the titer of viable cells. The viability should be ≥90 % and the yield should be ~4×10^6 cells per cerebellum. Plate 4×10^5 cells in 1.0 mL CGN medium per 12-well equipped with a PDL-coated coverslip (1.1×10^5 cells/cm^2, *see* **Note 18**). From 3–5 pups, ~30–50 wells with cells can be obtained.
15. Grow the cells at 37 °C and 6 % CO_2. Imaging experiments can be performed starting from day 2 after plating. Results with cGi-transgenic CGNs are shown in Fig. 3f–j. We do not recommend to starve cells (e.g., by withdrawal of serum or supplement) prior to imaging, because CGNs appear to be sensitive to changes in culture conditions.
16. If cells shall be cultured for more than 4 days, add Ara-C to a final concentration of 5 μM to the medium 24 h after plating to suppress proliferation of fibroblasts and glial cells (*see* **Note 19**).
17. Change medium every 2–3 days according to the efficiency of cell attachment, the amount of cell debris and microglial cells (*see* **Note 20**).

3.3 FRET-Based cGMP Imaging

Here we provide a protocol to perform FRET-based cGMP measurements by conventional ratiometric epifluorescence microscopy of cells expressing cGi-type cGMP biosensors. To measure cGMP-induced FRET changes, the CFP FRET donor is excited at 445 nm and the CFP and YFP emissions are simultaneously recorded at 480 nm and 535 nm, respectively, using a beam splitter. In a given cell, the ratio of CFP/YFP emission depends on the concentration of free cGMP (Subheading 1 and Fig. 1). Basic components of a FRET imaging setup are the microscope platform (in our case an inverted epifluorescence microscope), a superfusion system with a chamber that accommodates the cells, a computer-controlled light source, and a beam splitter to separate CFP and YFP emission signals and to record them simultaneously with a single CCD camera (Subheading 2.3). The way of drug application appears to be important for successful imaging experiments. To perfuse the cells with imaging buffer and to apply cGMP-elevating or other drugs, we use a custom-built superfusion system. It consists of a FPLC pump, injection valves for drug application, and a self-made superfusion chamber (*see* Fig. 2 and **Note 8**). This setup provides stable and precise superfusion of the cells with imaging buffer and test compounds. It also circumvents the need for manipulations during image acquisition, such as the manual application of drugs, which might cause movement artifacts or focus drifts (discussed in ref. 36). The protocol describes FRET-based cGMP imaging in cultured SMCs or CGNs expressing cGi-type biosensors in their cytosol. Drugs are applied by bulk superfusion of the cells. However, the same or a slightly modified setup can also be used for:

- Other cell types
- Other FRET-based cGMP indicators
- Sensors targeted to subcellular compartments
- Local drug application with micromanipulator-driven pipettes or drug-coated beads (e.g., [39])
- Simultaneous cGMP imaging and detection of other second messengers (e.g., Fura-2-based Ca^{2+} imaging, [40])
- cGMP imaging of isolated mouse tissues *ex vivo* and, eventually, anesthetized mice (e.g., in the cremaster muscle, [41])

Typical results of cGMP imaging experiments with cGi sensors in transgenic VSMCs and CGNs are shown in Fig. 3. Comparable results are obtained with VSMCs that were infected with a cGi-encoding adenovirus (Fig. 4).

1. cGMP imaging is performed in an air-conditioned darkroom at ~21 °C.
2. Check the beam splitter for alignment of CFP and YFP channels every time before starting a FRET imaging session.

3. Install sample loops of appropriate volumes at the injection valves, e.g., a 2 mL loop for 2 min superfusion and a 20 mL loop for 20 min superfusion at 1 mL/min. Connect the imaging buffer reservoir (1 L) to the FPLC pump and flush the superfusion system including the sample loops with imaging buffer for 10 min at 5 mL/min (Fig. 2a).
4. Place a coverslip with cGi-expressing cells (Subheadings 3.1 and 3.2) in a 3.5 cm petri dish filled with imaging buffer.
5. Assemble the superfusion chamber. Place silicon grease on the chamber frame and the chamber plate to seal it (Fig. 2b). Mount the coverslip with the cells facing to the inside of the chamber. Place the chamber plate on top of the coverslip, add 200 μL imaging buffer to cover the cells, and finish the chamber assembly (Fig. 2b). Take care that the chamber is not leaky. Clean the glass coverslip on the outside and fix the chamber on the microscope stage.
6. Place inlet and outlet needles into the superfusion chamber and start the superfusion with imaging buffer at 1 mL/min. Adjust the vacuum and the level of the outlet needle so that steady superfusion of the cells is obtained (Fig. 2a).
7. By using the YFP filter set, identify a region with fluorescent cells of appropriate brightness. Dim cells with signal-to-background ratios ≤2.5 should not be used as well as extremely bright cells, in which strong sensor expression could interfere with cell functions. Check also for cell morphology and sensor localization; the sensor should be homogeneously distributed in the cytosol without localized depositions ("bright spots"). Readjust the inlet and outlet needles, if their positions have changed during the setup of the system.
8. Acquire a still image of the fluorescent cells. Select regions of interests (ROIs) to be imaged as well as a background region without fluorescent cells (*see* **Note 21**).
9. Adjust the camera settings (pixel binning, gain), exposure time and acquisition cycle interval (the time from the beginning of one acquisition to the beginning of the next acquisition, *see* **Note 22**). Ensure that the hardware settings are correct (FRET filter cube with CFP excitation filter, 470 nm dichroic mirror, light path through the beam splitter to the camera) and start the experiment.
10. Record images during superfusion with imaging buffer at 1 mL/min until a stable baseline is obtained (*see* **Note 23**).
11. Dilute test compounds in imaging buffer to their final concentrations. Typical final concentrations are: DEA/NO and GSNO, 20–500 nM; ANP and CNP, 10–250 nM; IBMX, 100–500 μM; Sildenafil, 10–30 μM; ODQ, 1–10 μM. Note

that cGi sensors do not show responses when the commonly used concentrations of membrane-permeable cGMP analogues are applied to intact cells (*see* **Note 24**). To account for dead volumes of the superfusion system (tubing, syringes), the volume of a drug solution should be ~25 % larger than the volume of the sample loop. With syringes, load the compounds via the injection valves into the respective sample loops (*see* Fig. 2a and **Note 25**).

12. Apply the test compounds at 1 mL/min via valve switching (Fig. 2a) and note down the time when the drugs were applied. To terminate the superfusion of a compound before the sample loop has been completely flushed, switch the valve back to the loading position and note down the time. Flush the sample loops with imaging buffer before loading the next drug solution.
13. Upon drug application, follow CFP/YFP ratio changes as well as changes in the individual CFP and YFP channels to recognize potential artificial CFP/YFP ratio changes (*see* **Notes 9** and **25**).
14. After the imaging session, flush the complete superfusion system including injection valves, sample loops and all connective tubing with H_2O and then with 20 % ethanol. Store the system in 20 % ethanol. If necessary, clean the outlet needle from aspirated silicon grease.
15. Perform offline analysis of the acquired images. Redefine ROIs (*see* **Note 26**) and determine CFP and YFP emission intensities of the ROIs and at least one non-fluorescent background region.
16. Process raw FRET data by performing background correction, CFP/YFP ratio calculation and baseline normalization as detailed in Table 2. To make data interpretation more intuitive, use the CFP/YFP ratio, which increases when the cGMP concentration rises and the cGi sensor binds cGMP.

3.4 In-Cell Calibration of cGMP FRET Sensors

Absolute cGMP concentrations can be estimated from CFP/YFP ratio changes measured with cGMP FRET sensors, if the sensor has been calibrated with defined cGMP concentrations. The calibration can be done in a fluorescence spectrometer using the purified sensor protein or cytosolic extracts from cells expressing the sensor. Alternatively, the sensor can be calibrated in permeabilized cells under conditions, which are similar to FRET studies with intact cells. Here we describe in-cell calibration of cGi biosensors in transgenic VSMCs by using the same imaging setup as for the FRET measurements in native cells (Subheadings 2.3 and 3.3). In order to clamp defined intracellular cGMP concentrations, the cell membrane is permeabilized by superfusion with β-Escin in intracellular-

Table 2
Calculation of the background-corrected and baseline-normalized CFP/YFP ratio ΔR/R

Measured parameters		
	CFP channel	YFP channel
ROI	$\mathrm{CFP_{ROI}}(t)$	$\mathrm{YFP_{ROI}}(t)$
Background	$\mathrm{CFP_{bg}}(t)$	$\mathrm{YFP_{bg}}(t)$
Calculated parameters		
	CFP/YFP ratio	
1. Background correction	$R(t) = \dfrac{\mathrm{CFP_{ROI}}(t) - \mathrm{CFP_{bg}}(t)}{\mathrm{YFP_{ROI}}(t) - \mathrm{YFP_{bg}}(t)}$	
2. Baseline normalization	$R_0 = \dfrac{1}{n}\sum_{t_0}^{t_1} R(t)$	
3. Signal change relative to baseline	$\dfrac{\Delta R}{R} = \dfrac{R(t) - R_0}{R_0} \times 100\%$	

From time-lapse recordings of CFP and YFP emission intensities in a fluorescent ROI and a non-fluorescent background region (bg), the background-corrected CFP/YFP ratio "$R(t)$" is calculated. The index "(t)" indicates the time dependency of parameters. For baseline normalization, n ratio values obtained from t_0 to t_1 during the baseline period (e.g., from the start of the experiment to the first drug application) are averaged, leading to R_0. With R_0, ratio changes can be normalized to the baseline and denoted as $\Delta R/R$. Background correction and baseline-normalization can also be applied directly to CFP and YFP emission intensities leading to $\Delta F/F$ values (not shown)

like medium (ICM), which mimics the cytosolic milieu (high $[K^+]$, low $[Na^+]$ and $[Ca^{2+}]$). The conditions for permeabilization must be optimized, so that cGMP can freely pass through the cell membrane, while the cells remain viable and the sensor protein is retained inside them. Upon superfusion with a range of known cGMP concentrations a calibration curve can be established. With this curve unknown cGMP concentrations can be estimated from $\Delta R/R$ values measured in intact cells. Figure 5 shows the in-cell calibration of transgenic VSMCs expressing the cGi-500 sensor.

1. cGMP imaging is performed in an air-conditioned darkroom at ~21 °C.
2. Dilute 5 μL of the β-Escin stock solution (50 mM) in 5 mL ICM to obtain a 50 μM working solution.
3. Prepare a dilution series of cGMP in ICM (10 mL for each concentration) that covers a range of 10 nM to 100 μM cGMP (e.g., 10, 50, 100, 500, 1,000, 10,000, 100,000 nM). Set up the microscope and superfusion system as described in Subheading 3.3. Connect 5 mL and 2 mL loops to the injection valves 1 and 2, respectively (Fig. 2a). Place a coverslip with cGi-expressing VSMCs (Subheading 3.1) in the superfusion chamber and cover the cells with 300 μL imaging buffer. Select

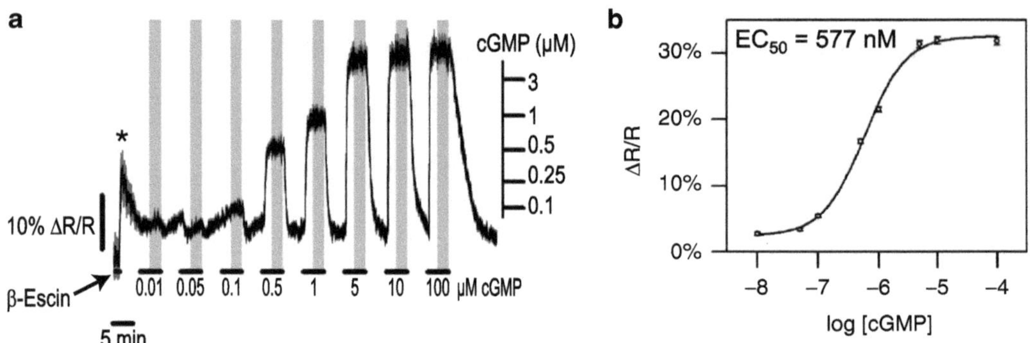

Fig. 5 In-cell calibration of cGi-500 in permeabilized VSMCs. (**a**) After permeabilization with β-Escin, cells were superfused with increasing cGMP concentrations (*black bars*) followed by a washout after each cGMP application. Note that the first peak (*asterisk*) is most likely caused by morphological changes resulting from permeabilization. For each cell, the $\Delta R/R$ values for each cGMP concentration were averaged during 160 s of the plateau (indicated by *grey bars*). These values were then used for calibration; data shown are mean ± SEM of 22 cells. The cGMP scale bar at the *right* was generated using the calibration curve shown in panel **b**. (**b**) To establish the cGMP calibration curve, $\Delta R/R$ values (mean ± SEM) were pooled from 88 cells measured in three independent experiments (including the experiment shown in panel **a**). The data were used to fit a dose–response curve with an EC_{50} value of 577 nM as described in Subheading 3.4

an appropriate field of view and define ROIs and background regions. Start superfusion with ICM at 1 mL/min and begin image acquisition.

4. For cell permeabilization, superfuse the cells for 80 s with 50 μM β-Escin via the 2 mL sample loop (*see* **Note 27**). Then wash with ICM for ~5 min until a stable baseline is obtained.
5. Superfuse cells with each cGMP dilution for 5 min via the 5 mL sample loop; start with the lowest cGMP concentration. After each cGMP application, wash the cells with ICM until the CFP/YFP ratio has returned to baseline (usually within 5 min, *see* **Note 28**).
6. For data evaluation, redefine ROIs, perform background subtraction, and calculate the normalized CFP/YFP ratio changes, $\Delta R/R$, as described in Subheading 3.3 and Table 2. For baseline normalization, use images acquired after permeabilization and before the first cGMP application. Estimate maximal $\Delta R/R$ values from the plateau of each cGMP concentration.
7. Draw a calibration curve with log[cGMP] versus $\Delta R/R$ and perform a nonlinear regression analysis using the dose–response function

$$\left(\frac{\Delta R}{R}\right)=\left(\frac{\Delta R}{R}\right)_{\text{min}}+\frac{\left(\frac{\Delta R}{R}\right)_{\text{max}}-\left(\frac{\Delta R}{R}\right)_{\text{min}}}{1+10^{(\log EC_{50}-\log[\text{cGMP}])h}}$$

to estimate $\log(EC_{50})$, $(\Delta R/R)_{min}$, $(\Delta R/R)_{max}$, and the Hill constant h.

8. To estimate unknown cGMP concentrations from $\Delta R/R$ values measured in native cells, solve the dose–response function for [cGMP] using EC_{50}, $(\Delta R/R)_{min}$, $(\Delta R/R)_{max}$, and h obtained from the calibration:

$$[\mathrm{cGMP}] = \mathrm{EC}_{50}\left(\frac{\left(\frac{\Delta R}{R}\right)_{max} - \left(\frac{\Delta R}{R}\right)_{min}}{\left(\frac{\Delta R}{R}\right) - \left(\frac{\Delta R}{R}\right)_{min}} - 1\right)^{-\frac{1}{h}}$$

The cGMP concentration can be estimated from the "linear" part of the calibration curve, e.g., between ~100 nM and ~3 μM cGMP for cGi-500 (Fig. 5).

4 Notes

1. Adenoviruses are pathogenic for humans (biosafety level 2); consult your biosafety officer about the safety requirements. Adenoviruses are created as described [42]. A detailed protocol has been published [43]; follow this protocol until **step 23**, then repeat **steps 21** and **22** until a sufficient titer is reached (as described in **step 25**); do not perform CsCl purification, but use the viral supernatants for infection; store the virus supernatants at −20 °C.
2. The dissolution of high molecular weight PDL is a slow process; ensure complete dissolution by incubation as indicated.
3. Homogeneous coating of the glass coverslips with PDL is critical for cell attachment. Otherwise, cells will not attach properly, will form large clusters, or will die soon after plating. Besides having a homogeneous PDL solution, take care that coverslips are fully covered with the solution and that the coverslips did not swim up during overnight incubation. In this case, invert the coverslip and use the side which was in contact with the PDL solution.
4. B-27 is a serum-free supplement to support survival of cultured neurons plated without astrocytes as feeder cells. It contains vitamins, essential fatty acids, hormones, and anti-oxidants [44].
5. We recommend a bright light source (e.g., a monochromator or filter switching device with a 150 W Xenon arc bulb) with an electronic shutter that illuminates the cells only when an image is acquired. Continuous illumination leads to bleaching and phototoxicity.

6. Due to the spectral properties of the fluorophores and emission filters, the CFP emission "bleeds through" into the YFP channel. In principle, a correction factor for the microscope setup can be obtained in control experiments with cells expressing only CFP. However, because the stoichiometry of CFP and YFP is fixed in the single-molecule cGi biosensors, the correction for CFP "bleed-through" would lead to slightly higher CFP/YFP ratio changes without altering the signal-to-noise ratio. Therefore, it is generally not required.
7. When using the filter cube for FRET measurements, the emitted light is filtered by emission filters in the beam splitter attached to the camera, but not by an emission filter inside the filter cube. Due to the lack of this emission filter, it is not possible to examine the cells by eye. To find and examine fluorescent cells before FRET measurements, we use another filter cube carrying the YFP filter set including an emission filter.
8. Our superfusion chamber (Fig. 2b) is made of two variants of a chamber frame to accommodate coverslips with 20 mm or 30 mm diameter; the coverslips serve as bottom of the superfusion chamber. On top of the coverslip resides a "chamber plate" with an oval-shaped opening, which defines the volume of the actual superfusion chamber. We use plates with oval bores of 5 × 14 mm or 8 × 20 mm defining a chamber volume of 50 μL or 200 μL, respectively (Fig. 2b). For superfusion, inlet and outlet needles made from hypodermic needles are placed at the borders of the chamber plate openings (Fig. 2a). In combination with a 30 mm coverslip the chamber can be used with low working distance objectives (e.g., oil objectives) without the problem that the objective collides with the chamber frame. The chamber plate with the larger opening (8 × 20 mm) is helpful if only a small fraction of the cells is suitable for imaging and, thus, a larger field of view needs to be examined, or if tissues isolated from cGi-transgenic mice are analyzed.
9. The acquisition software needs to support the beam splitter, so that the emission intensities of CFP and YFP and the CFP/YFP ratio of selected ROIs will be displayed on-line during the experiment. Note that an *increase* in the CFP/YFP ratio that is caused by cGMP binding to the sensor results from an *increase* in CFP emission at 480 nm and a simultaneous *decrease* in YFP emission at 535 nm (*see* Figs. 3 and 4). However, the CFP/YFP ratio will also be altered, if CFP and YFP emissions change into in the same direction but with unequal intensity, for instance, due to unequal bleaching (*see* **Note 22**). Thus, it is important to follow changes of the single CFP and YFP channels to detect artificial CFP/YFP ratio changes. For data analysis it is helpful to generate templates or macros for automated processing of the evaluation steps.

10. At neutral pH and ~21 °C, DEA/NO decays spontaneously and releases NO with a half-life of 16 min. The alkaline pH of 10 mM NaOH prevents the decomposition of DEA/NO. While working with them, DEA/NO stock solutions should be kept on ice.
11. Mice older than 3 months can also be used, but the cell yield as well as the transgene expression might be reduced.
12. Aorta, veins, esophagus and airways can be distinguished by their morphology. Note that after successful dissection, the aorta is long (~2.5 cm) and can easily be followed from the periphery to the heart.
13. The time of digestion in enzyme solution B is critical; over-digestion will lead to a dramatic reduction in the yield of viable cells. Preparations of BSMCs and CSMCs should be incubated shorter (10 min) than VSMCs (15 min). When older mice (>3 months of age) or new enzyme batches are used, the digestion time in enzyme solution B should be optimized in pilot experiments. To remove undigested pieces of tissue after resuspension in culture medium, the suspension can be filtered through a netwell mesh insert (74 μm mesh size). After filtration, wash the netwell mesh insert with another 5 mL of culture medium and add the 5 mL to the cell suspension.
14. If digested under optimal conditions, SMCs have an elongated morphology directly after isolation; this feature distinguishes them from round-shaped non-SMCs such as endothelial cells. The SMCs isolated from different tissues differ in size, in the order VSMCs < BSMCs < CSMCs. Note that SMCs will change their morphology soon after plating to become round-shaped before they attach to the coverslip.
15. Cells can be plated also on 30 mm coverslips in 6-well plates. Plate 3 mL/well of the VSMC, BSMC or CSMC cell suspension.
16. BSMCs and CSMCs grow to confluence within 5–6 days, VSMCs within 6–8 days after plating. Confluent cells can be subcultured in 1:3 ratios for up to three passages. Note that cell passaging, culture confluence and the presence or absence of serum may affect the expression level and/or basal activity of components of the cGMP signaling pathway as well as other cell functions. Therefore, it is important to perform imaging experiments under standardized conditions (e.g., sub-confluent primary SMC cultures, which were serum-starved for ~24 h).
17. Drawing a number on the pup's belly works well in our hands. To ensure that the labeling is not lost until preparation on postnatal day 7, redraw the label on postnatal day 6.
18. The isolated CGNs are post-mitotic; therefore, their density will not increase during cell culture. For certain applications,

for example, if long-term culture (≥7 days) is desired, the amount of plated cells should be increased to 3×10^5 cells/cm^2. Cells can be plated also on 30 mm coverslips in 6-well plates (use 3 mL cell suspension or 1.2×10^6 cells per well).

19. Ara-C interferes with DNA synthesis. Thus, it is toxic for proliferating non-neuronal cells that contaminate the CGN culture, such as fibroblasts and glial cells, but not for the post-mitotic CGNs themselves.

20. The medium should be equilibrated in the CO_2 incubator before it is added to the cells. For medium change, remove half of the old medium and replace it with new medium, so that survival factors secreted by the CGNs won't be completely removed. Contaminating microglia cells attach only weakly to other cells and glass coverslips. If the amount of microglial cells is high, remove them by tapping against the culture dish and a complete medium change.

21. At 10×–16× magnification, up to 25 cells (SMCs, CGNs) can be analyzed in one imaging experiment. Here, a given ROI can define the whole cell or only part of it. However, to analyze single cells at the subcellular level, a higher magnification (40×–63×) is recommended. Here, ROIs can be drawn highlighting different subcellular compartments (e.g., neurites and cell body of a neuron, or perinuclear region and cytosol of a SMC).

22. Typical exposure times at 10×–16× magnification are 100–350 ms (at 4×4 pixel binning) with an acquisition cycle interval of 2 s; at 40×–63× magnification we use exposure times of 20–80 ms (at 1×1 or 2×2 pixel binning) with an acquisition cycle interval of 6 s. If other devices are used (e.g., dimmer light sources or cameras with higher sensitivity), these parameters should be optimized to obtain sufficient signal-to-background ratios (≥2.5), and to keep photobleaching as low as possible. Longer exposure times lead to increased signal-to-background ratios, but the prolonged illumination will also increase bleaching of the sensor and phototoxicity. If the extent of bleaching of CFP and YFP is different ("unequal bleaching"), artificial CFP/YFP ratio changes will be observed (*see* also **Note 9**). To reduce bleaching, the acquisition cycle interval can be increased at constant exposure time, or the exposure time can be reduced while the sensitivity of detection is improved by increasing the camera pixel binning. The first strategy leads to a reduced temporal resolution, while the latter leads to a reduced spatial resolution. All parameters (exposure time, acquisition cycle interval and camera pixel binning) should be optimized to the experimental question (temporal versus spatial resolution, amount of bleaching, signal-to-background ratio).

23. The baseline images are important for subsequent normalization and should be recorded over ≥30 acquisition cycles. If the baseline is unstable, make sure that the liquid level in the superfusion chamber does not fluctuate. If no stable baseline can be obtained, try a different field of view.
24. Albeit sensors bind membrane-permeable cGMP analogues such as 8-pCPT-cGMP or 8-Br-PET-cGMP in vitro, these compounds do not elicit FRET responses when applied extracellularly in typical concentrations to intact cells [31]. This is probably caused by the low intracellular concentrations reached by these analogues in intact cell studies.
25. Before drugs are loaded into the sample loops, flush them with imaging buffer to remove any remaining compounds. When loading the sample loops, avoid air bubbles as they disturb the superfusion. If DEA/NO is used, be aware that as soon as the stock (in 10 mM NaOH) is diluted in imaging buffer (pH 7.4), NO is released with a half-live of 16 min (at ~21 °C) (*see* **Note 10**). Always prepare fresh dilutions of DEA/NO in imaging buffer and standardize the time between dilution and application. Pre-dilutions of DEA/NO (e.g., to 100 μM) can be prepared at alkaline pH (in 10 mM NaOH) and stored on ice for up to 1.5 h. To exclude artificial CFP/YFP ratio changes due to the presence of organic compounds (e.g., DMSO) or acids/bases leading to pH changes [45], vehicle controls should be performed.
26. During offline analysis, the recorded data should be examined for the correct alignment of the acquired CFP and YFP images. Check also whether initially drawn ROIs are correctly defined for the whole experiment or if they need to be redefined because of cell movements or changes in cell morphology.
27. These conditions were successfully used for primary VSMCs from cGi-500-transgenic mice as well as for transfected murine embryonic fibroblasts expressing cGi-500, cGi-3000, or cGi-6000. Permeabilized cells have a somewhat rounded morphology, but stably retain the FRET sensor for at least 90 min. For other cell types it might be necessary to optimize β-Escin concentration and incubation time. To determine the optimal incubation time for permeabilization with a given concentration of β-Escin, cells are imaged during continuous superfusion with a low concentration of cGMP (e.g., 100 nM cGMP for cGi-500). Then β-Escin is applied until the FRET signal begins to change indicating successful permeabilization of the plasma membrane.
28. Returning to baseline conditions between the applications of different cGMP concentrations allows one to detect and normalize for baseline drifts, which might be caused by unequal bleaching of the sensor (*see* **Note 22**).

Acknowledgments

We thank Barbara Birk, Caroline Vollmers, Erika Mannheim, Ursula Krabbe, and Fred Eichhorst for expert technical assistance, Simone Di Giovanni for advice with neuronal cell culture, Gisela Drews and Peter Krippeit-Drews for providing the superfusion chamber, and Kübra Gülmez, Phillip Messer, Annyesha Mohanty, and Christine Wenz for their contributions to cGMP imaging experiments. Special thanks go to Susanne Feil and Lai Wen for reading the manuscript, to Lai Wen for providing transgenic mice for the calibration experiment, to Thomas Ott for microinjection of transgenes into mouse oocytes, and to Lutz Pott, Anke Gallhoff, and Kirsten Bender for introduction to the adenoviral system. We also thank all past and present members of our laboratories for critical discussions and the Deutsche Forschungsgemeinschaft for financial support.

References

1. Beavo JA, Brunton LL (2002) Cyclic nucleotide research—still expanding after half a century. Nat Rev Mol Cell Biol 3(9):710–718. doi:10.1038/nrm911
2. Friebe A, Koesling D (2003) Regulation of nitric oxide-sensitive guanylyl cyclase. Circ Res 93(2):96–105. doi:10.1161/01.RES.0000082524.34487.31
3. Kuhn M (2003) Structure, regulation, and function of mammalian membrane guanylyl cyclase receptors, with a focus on guanylyl cyclase-A. Circ Res 93(8):700–709. doi:10.1161/01.RES.0000094745.28948.4D
4. Biel M, Michalakis S (2009) Cyclic nucleotide-gated channels. Handb Exp Pharmacol 191:111–136. doi:10.1007/978-3-540-68964-5_7
5. Hofmann F, Feil R, Kleppisch T et al (2006) Function of cGMP-dependent protein kinases as revealed by gene deletion. Physiol Rev 86(1):1–23. doi:10.1152/physrev.00015.2005
6. Francis SH, Blount MA, Corbin JD (2011) Mammalian cyclic nucleotide phosphodiesterases: molecular mechanisms and physiological functions. Physiol Rev 91(2):651–690. doi:10.1152/physrev.00030.2010
7. Kleppisch T, Feil R (2009) cGMP signalling in the mammalian brain: role in synaptic plasticity and behaviour. Handb Exp Pharmacol 191:549–579. doi:10.1007/978-3-540-68964-5_24
8. Kemp-Harper B, Feil R (2008) Meeting report: cGMP matters. Sci Signal 1(9):pe12. doi:10.1126/stke.19pe12
9. Menniti FS, Faraci WS, Schmidt CJ (2006) Phosphodiesterases in the CNS: targets for drug development. Nat Rev Drug Discov 5(8):660–670. doi:10.1038/nrd2058
10. Martel G, Hamet P, Tremblay J (2010) Central role of guanylyl cyclase in natriuretic peptide signaling in hypertension and metabolic syndrome. Mol Cell Biochem 334(1–2):53–65. doi:10.1007/s11010-009-0326-8
11. Ehret GB, Munroe PB, Rice KM et al (2011) Genetic variants in novel pathways influence blood pressure and cardiovascular disease risk. Nature 478(7367):103–109. doi:10.1038/nature10405
12. Stasch JP, Becker EM, Alonso-Alija C et al (2001) NO-independent regulatory site on soluble guanylate cyclase. Nature 410(6825):212–215. doi:10.1038/35065611
13. Stasch JP, Hobbs AJ (2009) NO-independent, haem-dependent soluble guanylate cyclase stimulators. Handb Exp Pharmacol 191:277–308. doi:10.1007/978-3-540-68964-5_13
14. Stasch JP, Schmidt PM, Nedvetsky PI et al (2006) Targeting the heme-oxidized nitric oxide receptor for selective vasodilatation of diseased blood vessels. J Clin Invest 116(9):2552–2561. doi:10.1172/JCI28371
15. Schindler U, Strobel H, Schonafinger K et al (2006) Biochemistry and pharmacology of novel anthranilic acid derivatives activating heme-oxidized soluble guanylyl cyclase. Mol Pharmacol 69(4):1260–1268. doi:10.1124/mol.105.018747

16. Schmidt HH, Schmidt PM, Stasch JP (2009) NO- and haem-independent soluble guanylate cyclase activators. Handb Exp Pharmacol 191:309–339. doi:10.1007/978-3-540-68964-5_14
17. Sanofi-Aventis (2008–2009) Efficacy and safety study of Ataciguat versus placebo in patients with neuropathic pain (SERENEATI). In: ClinicalTrials.gov [Internet]. National Library of Medicine (US). http://clinicaltrials.gov/ct2/show/NCT00799656 NLM Identifier: NCT00799656. Accessed 31 May 2012
18. Sanofi-Aventis (2007–2008) Efficacy and safety of HMR1766 in patients with fontaine stage II peripheral arterial disease (ACCELA). In: ClinicalTrials.gov [Internet]. National Library of Medicine (US). http://clinicaltrials.gov/ct2/show/NCT00443287 NLM Identifier: NCT00443287. Accessed 31 Mar 2012
19. Fukumura D, Kashiwagi S, Jain RK (2006) The role of nitric oxide in tumour progression. Nat Rev Cancer 6(7):521–534. doi:10.1038/nrc1910
20. Kashiwagi S, Tsukada K, Xu L et al (2008) Perivascular nitric oxide gradients normalize tumor vasculature. Nat Med 14(3):255–257. doi:10.1038/nm1730
21. Calabrese V, Mancuso C, Calvani M et al (2007) Nitric oxide in the central nervous system: neuroprotection versus neurotoxicity. Nat Rev Neurosci 8(10):766–775. doi:10.1038/nrn2214
22. Olshevskaya EV, Ermilov AN, Dizhoor AM (2002) Factors that affect regulation of cGMP synthesis in vertebrate photoreceptors and their genetic link to human retinal degeneration. Mol Cell Biochem 230(1–2):139–147. doi:10.1023/A:1014248208584
23. Jaumann M, Dettling J, Gubelt M et al (2012) cGMP-Prkg1 signaling and Pde5 inhibition shelter cochlear hair cells and hearing function. Nat Med 18(2):252–259. doi:10.1038/nm.2634
24. Fischmeister R, Castro LR, Abi-Gerges A et al (2006) Compartmentation of cyclic nucleotide signaling in the heart: the role of cyclic nucleotide phosphodiesterases. Circ Res 99(8): 816–828. doi:10.1161/01.RES.0000246118.98832.04
25. Castro LR, Verde I, Cooper DM et al (2006) Cyclic guanosine monophosphate compartmentation in rat cardiac myocytes. Circulation 113(18):2221–2228. doi:10.1161/CIRCULATIONAHA.105.599241
26. Piggott LA, Hassell KA, Berkova Z et al (2006) Natriuretic peptides and nitric oxide stimulate cGMP synthesis in different cellular compartments. J Gen Physiol 128(1):3–14. doi:10.1085/jgp.200509403
27. Nikolaev VO, Lohse MJ (2009) Novel techniques for real-time monitoring of cGMP in living cells. Handb Exp Pharmacol 191:229–243. doi:10.1007/978-3-540-68964-5_11
28. Förster T (1946) Energiewanderung und Fluoreszenz. Naturwissenschaften 33(6):166–175. doi:10.1007/bf00585226
29. Zaccolo M (2004) Use of chimeric fluorescent proteins and fluorescence resonance energy transfer to monitor cellular responses. Circ Res 94(7):866–873. doi:10.1161/01.RES.0000123825.83803.CD
30. Nausch LW, Ledoux J, Bonev AD et al (2008) Differential patterning of cGMP in vascular smooth muscle cells revealed by single GFP-linked biosensors. Proc Natl Acad Sci USA 105(1):365–370. doi:10.1073/pnas.0710387105
31. Russwurm M, Mullershausen F, Friebe A et al (2007) Design of fluorescence resonance energy transfer (FRET)-based cGMP indicators: a systematic approach. Biochem J 407(1):69–77. doi:10.1042/BJ20070348
32. Niino Y, Hotta K, Oka K (2009) Simultaneous live cell imaging using dual FRET sensors with a single excitation light. PLoS One 4(6):e6036. doi:10.1371/journal.pone.0006036
33. Niino Y, Hotta K, Oka K (2010) Blue fluorescent cGMP sensor for multiparameter fluorescence imaging. PLoS One 5(2):e9164. doi:10.1371/journal.pone.0009164
34. Lemke EA, Schultz C (2011) Principles for designing fluorescent sensors and reporters. Nat Chem Biol 7(8):480–483. doi:10.1038/nchembio.620
35. Feil S, Valtcheva N, Feil R (2009) Inducible Cre mice. Methods Mol Biol 530:343–363. doi:10.1007/978-1-59745-471-1_18
36. Gesellchen F, Stangherlin A, Surdo N et al (2011) Measuring spatiotemporal dynamics of cyclic AMP signaling in real-time using FRET-based biosensors. Methods Mol Biol 746:297–316. doi:10.1007/978-1-61779-126-0_16
37. Rasband WS (1997–2012) ImageJ. U. S. National Institutes of Health, Bethesda, MD, USA. http://imagej.nih.gov/ij/
38. Kaech S, Banker G (2006) Culturing hippocampal neurons. Nat Protoc 1(5):2406–2415. doi:10.1038/nprot.2006.356
39. Shelly M, Lim BK, Cancedda L et al (2010) Local and long-range reciprocal regulation of cAMP and cGMP in axon/dendrite formation. Science 327(5965):547–552. doi:10.1126/science.1179735
40. von Hayn K, Werthmann RC, Nikolaev VO et al (2010) Gq-mediated Ca2+ signals inhibit adenylyl cyclases 5/6 in vascular smooth muscle cells.

Am J Physiol Cell Physiol 298(2):C324–C332. doi:10.1152/ajpcell.00197.2009

41. Zhang J, Chen L, Raina H et al (2010) In vivo assessment of artery smooth muscle [Ca2+]i and MLCK activation in FRET-based biosensor mice. Am J Physiol Heart Circ Physiol 299(3):H946–H956. doi:10.1152/ajpheart.00359.2010
42. He TC, Zhou S, da Costa LT et al (1998) A simplified system for generating recombinant adenoviruses. Proc Natl Acad Sci USA 95(5):2509–2514
43. Luo J, Deng ZL, Luo X et al (2007) A protocol for rapid generation of recombinant adenoviruses using the AdEasy system. Nat Protoc 2(5):1236–1247. doi:10.1038/nprot.2007.135
44. Brewer GJ, Torricelli JR, Evege EK et al (1993) Optimized survival of hippocampal neurons in B27-supplemented neurobasal, a new serum-free medium combination. J Neurosci Res 35(5):567–576. doi:10.1002/jnr.490350513
45. Griesbeck O, Baird GS, Campbell RE et al (2001) Reducing the environmental sensitivity of yellow fluorescent protein. Mechanism and applications. J Biol Chem 276(31):29188–29194. doi:10.1074/jbc.M102815200
46. Sato M, Hida N, Ozawa T et al (2000) Fluorescent indicators for cyclic GMP based on cyclic GMP-dependent protein kinase Ialpha and green fluorescent proteins. Anal Chem 72(24):5918–5924. doi:10.1021/ac0006167
47. Honda A, Adams SR, Sawyer CL et al (2001) Spatiotemporal dynamics of guanosine 3′,5′-cyclic monophosphate revealed by a genetically encoded, fluorescent indicator. Proc Natl Acad Sci USA 98(5):2437–2442. doi:10.1073/pnas.051631298
48. Nikolaev VO, Gambaryan S, Lohse MJ (2006) Fluorescent sensors for rapid monitoring of intracellular cGMP. Nat Methods 3(1):23–25. doi:10.1038/nmeth816

Chapter 7

Advances and Techniques to Measure cGMP in Intact Cardiomyocytes

Konrad R. Götz and Viacheslav O. Nikolaev

Abstract

Förster resonance energy transfer (FRET)-based biosensors are powerful tools for real-time monitoring of signaling events in intact cells using fluorescence microscopy. Here, we describe a highly sensitive method which allows FRET-based measurements of the second messenger cGMP in adult mouse ventricular myocytes. Such measurements have been challenging before, primarily due to relatively low cGMP concentrations in cardiomyocytes and limited sensitivity of the available biosensors. With our new technique, one can reliably measure dynamic changes in cGMP upon stimulation of myocytes with natriuretic peptides and other physiological and pharmacological ligands.

Key words cGMP, FRET, Imaging, Fluorescence, Biosensor

1 Introduction

Reliable measurements of cGMP in cells and tissues, especially in cardiomyocytes, have been challenging. Standard biochemical assays, such as antibody-based radioimmunoassays or enzyme-linked immunosorbent assays, involve thousands of cells and have no spatial resolution at the subcellular level. In the adult heart, cGMP can be found in much lower concentrations than cAMP and acts in a compartmentalized fashion [1, 2], so that robust novel techniques to measure this second messenger with high sensitivity, temporal and spatial resolution are highly desirable.

In the last decade, several optical and non-optical methods to measure cGMP in single intact cells have been developed. Among these techniques, electrophysiological recordings using ectopically expressed cyclic nucleotide-gated channels as the sensors for subsarcolemmal cGMP deserve particular attention since they uncovered differential contributions of various phosphodiesterase (PDE) families in compartmentation of cGMP in adult rat ventricular myocytes. In particular, PDE2 has been shown to

Thomas Krieg and Robert Lukowski (eds.), *Guanylate Cyclase and Cyclic GMP: Methods and Protocols*, Methods in Molecular Biology, vol. 1020, DOI 10.1007/978-1-62703-459-3_7,

exclusively control cGMP signals stimulated by natriuretic peptide receptors, while both PDE2 and PDE5 have been associated with cGMP pools produced by the soluble guanylyl cyclase [3]. However, certain limitations of this technique such as the restriction of the sensor to the subsarcolemmal compartment and low cGMP/cAMP selectivity of these channels prevent its further broad application.

In parallel, several groups have pioneered fluorescent cGMP biosensors which allow visualization of intracellular cGMP with high temporal and spatial resolution [4]. Most of these sensors are based on a partially truncated cGMP-dependent protein kinase molecule which changes its conformation upon binding of cGMP and which can be visualized by Förster resonance energy transfer (FRET) between cyan and yellow fluorescent proteins fused to the kinase backbone [5–7] or by the change in fluorescence of circularly permuted green fluorescent protein attached to its C-terminus [8]. These sensors represent a major advance in cGMP imaging but have certain drawbacks, either relatively low sensitivity (~1 μM affinity for cGMP, which is about 5–10 times less than the affinity of the kinase) or again very low cGMP/cAMP selectivity. The same is true for the sensors based on single regulatory GAF domains from PDE2 and PDE5 which were initially developed in our laboratory [9]. This was the reason for the fact that we and others could not achieve robust and reliable recordings of cGMP in adult cardiomyocytes using such biosensors. In contrast, they proved extremely useful in neonatal myocytes which produce more cGMP and show compartmentalized signaling by this second messenger [10, 11].

To expand this exciting field of research onto adult cardiomyocytes, one ideally requires a biosensor with both high affinity for cGMP and good cGMP/cAMP selectivity. Very recently, one of the older PDE5-based biosensors initially developed in our laboratory using cyan and yellow fluorescent proteins as donor and acceptor fluorophores has been modified by exchanging them to green (T-Sapphire) and red (Dimer2) fluorescent proteins. Unexpectedly, this modification led to an ~50-fold increase in the affinity for cGMP (30 nM vs. 1.5 μM) together with retained low affinity for cAMP (>1 mM) which made this biosensor (called red cGES-DE5) ideal for cGMP measurements in adult cardiomyocytes [12]. Here, we describe the method which can be used to isolate adult ventricular cardiomyocytes from mice transgenically expressing this sensor in heart muscle cells under the α-myosin heavy chain promoter (manuscript under consideration in Circulation) and to perform FRET measurements of intracellular cGMP in these cells.

2 Materials

All materials for cell isolation and cell culture should be made up with ultrapure water and sterilized via a 0.2 μm filter prior to use.

2.1 Cardiomyocyte Isolation

1. 10× stock perfusion buffer: 1,130 mM NaCl, 47 mM KCl, 6 mM KH_2PO_4, 6 mM Na_2HPO_4, 12 mM $MgSO_4$, 120 mM $NaHCO_3$, 100 mM $KHCO_3$, 100 mM HEPES, 300 mM taurine, 0.32 mM phenol red dissolved in water. Weigh 66.0 g NaCl, 3.5 g KCl, 816.6 mg KH_2PO_4, 1.07 g $Na_2HPO_4 \cdot 2H_2O$, 2.96 g $MgSO_4 \cdot 7H_2O$, 10.1 g $NaHCO_3$, 10.1 g $KHCO_3$, 23.8 g HEPES, 37.5 g taurine (120 mg phenol red). Add water up to a volume of 1,000 mL. Filter via a 0.2 μm sterile filter and store at 4 °C.
2. 2,3-butanedione monoxime (BDM) stock solution: 500 mM BDM in water. Weigh 2.52 g BDM (B0753, Sigma). Add water up to a volume of 50 mL. Filter via a 0.2 μm sterile filter, make 2 mL aliquots, and store at −20 °C. Warm up before use to dissolve particles.
3. Bovine serum albumin (BSA) stock solution: 10 % BSA in water. Weigh 5 g BSA (A8806, Sigma). Add water up to a volume of 50 mL. Filter via a 0.2 μm sterile filter, make 800 μL aliquots, and store at −20 °C. Warm up before use.
4. 100 mM calcium chloride solution. Weigh 1.47 g $CaCl_2 \cdot 2H_2O$. Add water up to 100 mL. Filter via a 0.2 μm sterile filter and store at 4 °C.
5. 10 mM calcium chloride solution. Dilute 100 mM calcium chloride solution 1–10 with water. Filter via a 0.2 μm sterile filter and store at 4 °C.
6. Liberase DH solution: 4.2 mg/mL of Liberases in water. Dissolve 50 mg of Liberase DH (Roche, 05401054001) in 12 mL of sterile water, reconstitute on ice for 20 min, aliquot under sterile conditions into 150 μL aliquots, keep at −20 °C.
7. Trypsin solution: 2.5 % of trypsin in water. Order the ready-to-use 2.5 % trypsin solution from Invitrogen (15090), aliquot to 300 μL, and store at −20 °C.
8. Phosphate-buffered saline: Dulbecco's calcium and magnesium-free PBS cell-culture grade, any manufacturer (e.g., Sigma D8537). Store at 4 °C.
9. Cell-culture medium: Supplement the minimum essential medium (MEM, Invitrogen, 51200-046) with 0.1 % BSA (1:100 from the stock solution; see Subheading 2.1, **item 3**), 2 mM L-glutamine, 10 mM BDM, antibiotics (100 U/mL penicillin, 100 μg/mL

streptomycin), and insulin-transferrin-selenium supplement A (10 mg/L insulin, 5.5 mg/L transferrin, 6.6 μg/L sodium selenite, 110 mg/L sodium pyruvate final concentrations; use the 100× stock from Invitrogen, 51300-044). This medium can be stored at −20 °C and warmed up on the day of cardiomyocyte isolation.

10. 1 mg/mL laminin solution (Sigma, L2020).
11. 1× perfusion buffer. Weigh 100 mg glucose, transfer to a cylinder, add water up to 80 mL, mix well to dissolve the powder. Add 10 mL of 10× stock perfusion buffer and 2 mL of BDM stock solution. Make up to 100 mL and filter sterilize.
12. Digestion buffer. Prepare 20 mL of digestion buffer, containing 19.6 mL of the 1× perfusion buffer, 2.5 μL of 100 mM calcium chloride solution, 300 μL of liberase solution, and 300 μL of trypsin solution. Add enzymes straight before use.
13. Stopping buffer 1: 2.25 mL of the 1× perfusion buffer, 250 μL BSA solution, and 1.25 μL of 100 mM calcium chloride solution.
14. Stopping buffer 2: 9.5 mL of the 1× perfusion buffer, 500 μL BSA solution, and 3.75 μL of 100 mM calcium chloride solution.
15. Buffer for FRET imaging (buffer A): NaCl 144 mM, KCl 5.4 mM, $MgCl_2$ 1 mM, $CaCl_2$ 1 mM, HEPES 10 mM in water. Adjust the pH to 7.3, store at room temperature.

2.2 Heart Perfusion System

A simple perfusion system for cardiomyocyte isolation using a Langendorff preparation can be purchased or assembled in a customized fashion. It should include the following components:

1. Peristaltic pump to allow flow rates up to 3 mL/min.
2. Water bath with adjustable temperature in the range of 37–45 °C to provide a constant temperature of 37 °C for solutions arriving at the heart.
3. Plastic tubing with adapters and mounting for standard syringe needles.
4. Metal or plastic cannula, e.g., made from syringe needles (*see* **Note 1**).
5. Stereomicroscope used to cannulate the heart.

2.3 FRET Imaging System

Several FRET imaging systems are commercially available or can be built up from commercially available components as previously described. Typically, they include:

1. A light source, for example a standard fluorescent lamp or a monochromator-based light source.

2. Inverted fluorescent microscope equipped with a 40–100× oil-immersion objective and a filter cube contacting BP405/20 excitation filter and DCLP455 dichroic mirror.
3. Beam splitter which splits the emission light into donor and acceptor channels (*see* **Note 2**).
4. CCD camera.
5. Computer with an imaging software (*see* **Note 3**).

3 Methods

3.1 Cardiomyocyte Isolation

1. Rinse and equilibrate the heart perfusion system with the 1 × perfusion buffer. Turn on the heating approx. 30 min prior to isolation.
2. Draw up 2.5 mL of the digestion buffer and transfer into a sterile 10–20 mL beaker.
3. Anesthetize a mouse, rapidly excise the heart, and place it into a Petri dish with a room temperature cell-culture grade phosphate buffer saline.
4. Under a stereomicroscope, cannulate the heart via aorta and tightly fasten it with 1–2 thread loops around the cannula.
5. Connect the cannula to the heart perfusion system and let the perfusion buffer run through for 3 min.
6. Change to the digestion buffer and perfuse for 9 min (*see* **Note 4**).
7. Stop the pump and place the heart into the beaker with 2.5 mL of digestion buffer. Cut off the atria and mince the ventricles for 30 s using small scissors.
8. Add 2.5 mL of the stopping buffer 1 and dissociate the tissue by pulling the suspension up and down using an insulin syringe without a needle for 3 min.
9. Filter the cell suspension via a 150 μm cell-culture mesh and transfer it into a 15 mL Falcon tube. Leave cardiomyocytes to settle down for 10 min.
10. Carefully remove the supernatant using a transfer pipette and resuspend the cells in 10 mL of the stopping buffer 2.
11. Gradually increase the calcium concentration in the cell suspension up to 1 mM by adding the following solutions with 4 min interval: 50 μL of 10 mM calcium chloride solution for 2 times, 100 μL of 10 mM calcium chloride solution, and 30 and 50 μL of 100 mM calcium chloride solution.
12. Place the tube into a cell-culture incubator and let the cells sediment for 10–15 min.

13. Carefully remove the supernatant leaving approx. 1.2 mL of the solution.
14. Resuspend the cells by gentle shaking and plate the suspension dropwise onto glass coverslides, ~50 μL per each slide (*see* **Note 5**).
15. Place the plates into the incubator and let the cells adhere for 30 min.
16. Cover each well with 1–2 mL of the pre-warmed cell-culture medium. Place back into the incubator and use for FRET measurements within 24 h.

3.2 FRET Measurements

1. Place a coverslide with adherent cells into a measuring chamber.
2. Rinse once with 400 μL of the buffer A and cover with 400 μL of the fresh buffer.
3. Place the chamber onto the microscope and find a nice properly attached cardiomyocyte using transmission light (*see* **Note 6**).
4. Start the imaging software and excite the cells with a short pulse of 405 nm light to check the fluorescence in both green and red channels.
5. Adjust the exposure time to achieve good-quality images without much photobleaching.
6. Bring the cell into focus and start a time-lapse recording. Monitor the FRET ratio online and treat the cell with solutions of interest to analyze changes of the ratio which reflect intracellular cGMP concentrations (*see* **Notes 7** and **8**).
7. Finish the experiment, save the data, and begin with a new one.
8. Analyze the data offline to calculate corrected FRET ratios (*see* **Note 9**).

4 Notes

1. For mouse hearts, we typically use 20–21G cannulas and grind off the sharp part of the needle.
2. We typically use the DV2 DualView (Photometrics). It should be equipped with the 565dcxr dichroic mirror along with BP515/30 and BP590/40 emission filters, which are optimal for T-Sapphire and Dimer2, respectively. There are alternative products such as Optosplit or ORCA-D2 CCD camera. The latter already includes a beam splitter for 2-channel imaging.

3. Imaging software is capable of performing time-lapse image acquisition and online rationing. Individual donor and acceptor fluorescence intensities and the calculated FRET ratio are usually visible on the screen during any experiment. Several off-the-shelf FRET imaging systems are available from such companies as Visitron Systems, Leica Microsystems, and TILLPhotonics. They are usually equipped with proprietary imaging software adapted for FRET imaging, e.g., VisiView, Meta Imaging Series, and Live Acquisition, respectively.
4. Use the perfusion rates of 1.90–2 mL/min.
5. We use 24 mm diameter glass coverslides which fit the standard Autofluor chambers. The sterile (autoclaved) coverslides should be placed into 6-well plates and coated with laminin before plating the cells. Apply 100 μL of laminin solution per coverslide, draw up to reuse, and let the glass surface dry.
6. A nice cell should be elongated and striated, the edges should not be blunt; if the cell starts contracting, we search for another one. One can check whether the cell is attached properly just by removing a small amount of buffer and carefully pipetting this solution back into the chamber.
7. For cGMP measurements, we record an image every 5–10 s. Figure 1 shows an example of a FRET recording. In this example, a cell was treated with the C-type natriuretic peptide (CNP) at room temperature (by accurately pipetting the peptide solution into the sample chamber) which increases cGMP levels, as monitored over time as a decrease of FRET ratio. Before adding CNP we wait for at least 50 s to get a stable baseline. To apply ligands during FRET experiments, various self-built or commercially available perfusion systems can be used.
8. FRET measurements can be performed at room temperature or at 37 °C depending on the nature of experiment. We have noticed that only freshly isolated cells of very good quality respond robustly to CNP. Keeping cells in culture for prolonged periods of time (>24 h) might result in a lower performance and less reliable data.
9. To analyze the data offline, refer to our previously published protocol [13]. This analysis includes background subtraction and, optionally, corrections for bleed-through and photobleaching. For unimolecular biosensors such as red cGES-DE5, the bleed-through correction is not critical, since donor and acceptor proteins are expressed at equimolar levels. This correction results in bigger relative amplitude of the signal but does not change the shape of the trace.

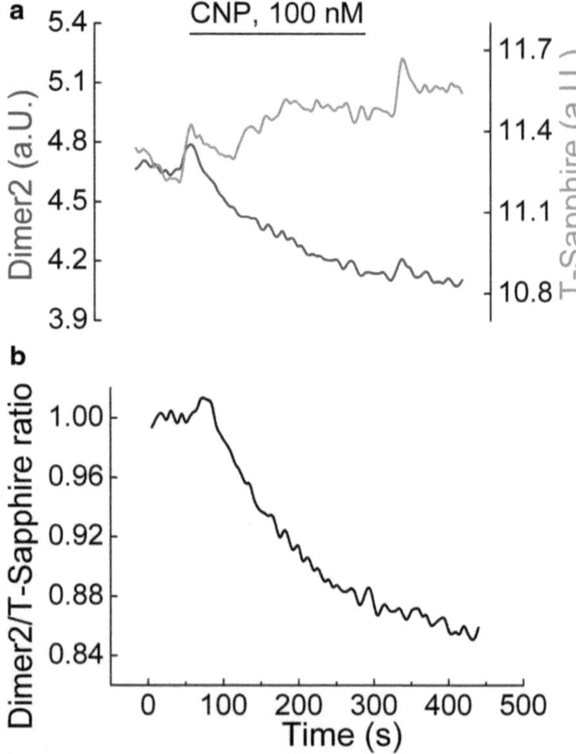

Fig. 1 Representative FRET measurement of cGMP in an adult mouse ventricular myocyte expressing the red cGES-DE5 sensor and stimulated with C-type natriuretic peptide (CNP). Increase in intracellular cGMP causes a decrease in the acceptor (Dimer2) and an increase in the donor (T-Sapphire) fluorescence (**a**). The normalized acceptor/donor ratio reflects the amount of FRET and is inversely proportional to cytosolic cGMP levels (**b**)

Acknowledgments

This work was supported by the Deutsche Forschungsgemeinschaft (grant NI 1301/1-1 and SFB 1002 to V.O.N) and University of Göttingen Medical Center ("pro futura" grant to V.O.N.).

References

1. Castro LR, Schittl J, Fischmeister R (2010) Feedback control through cGMP-dependent protein kinase contributes to differential regulation and compartmentation of cGMP in rat cardiac myocytes. Circ Res 107:1232–1240
2. Fischmeister R, Castro LR, Abi-Gerges A, Rochais F, Jurevicius J, Leroy J, Vandecasteele G (2006) Compartmentation of cyclic nucleotide signaling in the heart: the role of cyclic nucleotide phosphodiesterases. Circ Res 99:816–828
3. Castro LR, Verde I, Cooper DM, Fischmeister R (2006) Cyclic guanosine monophosphate compartmentation in rat cardiac myocytes. Circulation 113:2221–2228
4. Nikolaev VO, Lohse MJ (2009) Novel techniques for real-time monitoring of cGMP in living cells. Handb Exp Pharmacol 191:229–243
5. Honda A, Adams SR, Sawyer CL, Lev-Ram V, Tsien RY, Dostmann WR (2001) Spatiotemporal dynamics of guanosine

3′,5′-cyclic monophosphate revealed by a genetically encoded, fluorescent indicator. Proc Natl Acad Sci USA 98:2437–2442

6. Russwurm M, Mullershausen F, Friebe A, Jager R, Russwurm C, Koesling D (2007) Design of fluorescence resonance energy transfer (FRET)-based cGMP indicators: a systematic approach. Biochem J 407:69–77
7. Sato M, Hida N, Ozawa T, Umezawa Y (2000) Fluorescent indicators for cyclic GMP based on cyclic GMP-dependent protein kinase Ialpha and green fluorescent proteins. Anal Chem 72:5918–5924
8. Nausch LW, Ledoux J, Bonev AD, Nelson MT, Dostmann WR (2008) Differential patterning of cGMP in vascular smooth muscle cells revealed by single GFP-linked biosensors. Proc Natl Acad Sci USA 105:365–370
9. Nikolaev VO, Gambaryan S, Lohse MJ (2006) Fluorescent sensors for rapid monitoring of intracellular cGMP. Nat Methods 3:23–25
10. Stangherlin A, Gesellchen F, Zoccarato A, Terrin A, Fields LA, Berrera M, Surdo NC, Craig MA, Smith G, Hamilton G et al (2011) cGMP signals modulate cAMP levels in a compartment-specific manner to regulate catecholamine-dependent signaling in cardiac myocytes. Circ Res 108:929–939
11. Mongillo M, Tocchetti CG, Terrin A, Lissandron V, Cheung YF, Dostmann WR, Pozzan T, Kass DA, Paolocci N, Houslay MD et al (2006) Compartmentalized phosphodiesterase-2 activity blunts beta-adrenergic cardiac inotropy via an NO/cGMP-dependent pathway. Circ Res 98:226–234
12. Niino Y, Hotta K, Oka K (2009) Simultaneous live cell imaging using dual FRET sensors with a single excitation light. PLoS One 4:e6036
13. Börner S, Schwede F, Schlipp A, Berisha F, Calebiro D, Lohse MJ, Nikolaev VO (2011) FRET measurements of intracellular cAMP concentrations and cAMP analog permeability in intact cells. Nat Protoc 6:427–438

Chapter 8

Real-Time Monitoring the Spatiotemporal Dynamics of Intracellular cGMP in Vascular Smooth Muscle Cells

Kara F. Held and Wolfgang R. Dostmann

Abstract

Real-time and noninvasive imaging of intracellular second messengers in mammalian cells, while preserving their in vivo phenotype, requires biosensors of exquisite constitution. Here we provide the methodology for utilizing the single wavelength cGMP-biosensor δ-FlincG in aortic vascular smooth muscle cells.

Key words cGMP, Biosensors, Nitric oxide, Vascular smooth muscle, Live-cell imaging

1 Introduction

The intracellular second messenger, cyclic guanosine-3′,5′-monophosphate (cGMP), is a critical modulator of vascular smooth muscle (VSM) in the regulation of arterial vasodilation, essential for the maintenance of blood flow. cGMP is synthesized through activation of soluble guanylyl cyclases (sGC) and particulate guanylyl cyclases (pGC) by nitric oxide (NO) and natriuretic peptides (NPs), respectively [1, 2], and degraded by phosphodiesterases (PDEs) [3]. cGMP interacts with three main classes of downstream proteins: cyclic nucleotide-gated cation channels (CNG) whose main functions are in photoreceptors and olfactory neurons, cGMP-dependent kinases (PKG), and PDEs [4, 5]. PKGs represent a small subfamily of the AGC (PKA, PKG, and PKC)-type serine/threonine kinases that are activated specifically by cGMP [6, 7]. PKG's interactome consists of a limited number of targets, including the large conductance calcium-sensitive potassium (BK_{Ca}) channel, G-substrate, PDE type 5, phospholamban, RhoA, Telokin, vasodilator-stimulated phosphoprotein (VASP), vimentin and myosin-binding subunit (MBS), MYPT1, regulator of G-protein 2 (RGS2), and IP_3 receptor type I-associated cGMP kinase substrate (IRAG) [8–13]. In VSM, the actions of PKG activation ultimately lead to dilation of the vessel through many mechanisms mainly

Thomas Krieg and Robert Lukowski (eds.), *Guanylate Cyclase and Cyclic GMP: Methods and Protocols*, Methods in Molecular Biology, vol. 1020, DOI 10.1007/978-1-62703-459-3_8, © Springer Science+Business Media, LLC 2013

involving the reduction of intracellular calcium. Several mechanisms of note are through IRAG and phospholamban on the sarcoplasmic reticulum, induction of membrane hyperpolarization through the BK_{Ca}-induced hyperpolarization, and L-type voltage-dependent Ca^{2+} channel (L-VDCC), and prevention of myosin contraction through myosin light chain phosphatase (MLCP) activation and RhoA/ROCK inhibition to induce increased dephosphorylation of myosin light chain (MLC) [9, 12–17].

Phosphodiesterase types 1, 2, 3, 5, 6, 9, 10, and 11 all have tissue-specific cGMP degradation activities. VSM mainly expresses PDE5, but has also been shown to contain PDE1 [18, 19]. PDE5 is unique in that it can be phosphorylated to enhance and prolong its activity [3, 19–25]. The level of phosphorylation has been shown to be essential for cGMP dynamics in VSM cells [26], and less critical in platelets, astrocytes, and sGC-transfected HEK cells [27–29]. This differential in the level of involvement in regulating $[cGMP]_i$ seems to be cell- and tissue-type specific. Therefore, the maintenance and monitoring of $[cGMP]_i$ is imperative for the study of vasomotor reactivity.

Accurate, kinetic measurements of $[cGMP]_i$ have proven difficult in the past. A common methodology used is the cGMP radioimmunoassay, where I^{125}-labelled cGMP is incubated with cell lysates and a specific cGMP antibody is used to extract the label [30, 31]. This method only allows fixed time points to be analyzed, which can make extrapolating fine kinetic details tenuous, and the assay measures total rather than free cGMP, and often requires a PDE inhibitor to increase sensitivity. This method is also indirect, where the quantity of antibody binding is measured rather than cGMP itself.

The greatest advance in cGMP detection has come from the use of fluorescent indicators based on green fluorescent protein (GFP). Two of the most commonly used GFP variants are cyan and yellow (CFP and YFP, respectively). The excitation and emission wavelengths of these two fluorophores make an ideal pair for fluorescence-resonance electron transfer, or FRET [32, 33]. FRET occurs when one fluorophore, the donor, is excited and its emission causes the excitation of the second fluorophore, the acceptor, if the two are in close proximity to one another (20–60 Å) [32, 34, 35]. The first cGMP indicators were termed cyclic GMP *in*dicators using *e*nergy *t*ransfer or “cygnets” and consisted of a fragment of PKGIα sandwiched between ECFP and EYFP or the pH-stable citrine [36, 37]. Cygnets were shown to be capable of monitoring both spatial and temporal changes in cGMP [36, 38]. Another, more recent, FRET-based cGMP indicator was created using the cGMP-binding GAF domains of PDE2 and PDE5 [39]. Although FRET-based indicators were pioneering developments in their ability to monitor intracellular second messengers, they do not have the capacity to monitor small spatial changes as would be required with high-speed confocal microscopy. Serendipitously, circular permuted EGFP (cpEGFP) provided a unique opportunity to create a single wavelength biosensor, although cpEGFPs have a decreased total fluorescence intensity [40]. The Ca^{2+} indicator, G-CaMP, was the first of this single fluorophore-based indicator where the fluorophore was sandwiched between a calmodulin and an M13 domain [41]. A further advance of this principal design was the developments of *fl*uorescent *in*dicators of *cG*MP (FlincGs). FlincGs include fragments of the PKGIα regulatory domain containing the two cGMP-binding domains fused to cpEGFP on the N-terminus [42]. Particularly, the variant δ-FlincG exhibited high cGMP specificity, rapid binding, and dissociation kinetics and the capacity for confocal imaging, providing the first evidence for distinct spatial localization of cGMP signals in VSM cells [26, 42]. δ-FlincG has provided vertical progress by granting the most accurate method for cGMP detection applicable to many different cell types [26, 28, 42–47]. FlincG-type biosensors have helped to overcome the two major roadblocks in determining VSM signaling: (1) smooth muscle cells lose the expression of many smooth muscle-specific markers with culturing [38], and (2) nitric oxide has an extremely short biological half-life [48–50], which creates an experimental dilemma whereby small signaling events may be missed with traditional immunoblotting or kinase assays. Using FlincGs have allowed us to elicit the separate localization of the cytosolic NO- and membrane-restricted natriuretic peptide (NP)-induced cGMP in VSM cells, as well as create cells capable of detecting picomolar concentrations of cGMP produced by sGC [28, 42]. We have also determined the temporal kinetic relationship of NO and NO-induced cGMP, and which PDEs are critical in its maintenance in VSM [26].

A critical aspect to our careful analysis using FlincGs has been the cellular model with which we have chosen to study. VSM cells have a tendency to lose many smooth muscle-specific markers (PKG, PDE5, sGC, myosin heavy chain) over time in culture [38]. To ensure our cultured VSM cells are as close to in vivo cells as possible, we have developed an isolation assay to maintain their integrity. In combination with the adenovirus-transfected FlincG, we have developed an incredibly sensitive readout system for a major intracellular signaling molecule in vessel musculature.

2 Materials

2.1 Tools

1. Two forceps (Dumont #5 or #55 works well), small spring scissors, and large dissection scissors (mouse) or large spring scissors and large chicken bone kitchen shears (rat).
2. 0.22 μM syringe filters, 5 mL luer-lock syringes, 60 mm tissue culture dishes, 9″ glass Pasteur pipettes, automatic pipettor.

3. Bioptechs Delta T4 culture dishes with clear glass bottoms. These dishes work with the Bioptechs Delta T4 temperature control system mounted on the microscope stage (*see* **Note 1**).

2.2 Microscope Setup

1. Whether using epifluorescence or confocal microscopy, the stage should be equipped with a dish warmer, such as the Delta T4 culture system. Maintaining 37 °C is critical for proper enzymatic and biochemical kinetics.
2. Since δ-FlincG can be used as a single-excitation biosensor, epifluorescence imaging requires a mercury-halide lamp (X-CITE 120, EXFO Photonics, Toronto, ON) coupled with a single 480 nm excitation filter and 535 nm emission filter. Confocal imaging can be performed with a 488 nm laser, collecting the emission above 510 nm.

2.3 Enzyme Digestion Solutions

1. Digestion solution #1: Dissolve 175 U/mL Collagenase Type 2 (Worthington Biochemical) and 1.25 U/mL Elastase (Worthington Biochemical) into 5 mL Hank's Balanced Salt Solution (with calcium and magnesium, without phenol red, HBSS, Cellgro, #21-023-CV) for digesting rat aortae. For digesting up to 8 mouse aortae, use 2.5 mL total solution. Sterile filter solution using a 0.2 μm syringe filter. Keep on ice until use.
2. Digestion solution #2: Prepare on second day. Dissolve 175 U/mL collagenase and 2.5 U/mL elastase into 5 mL HBSS. Sterile filter. Keep on ice. Again, for digesting mouse aortae, use 2.5 mL total solution.

2.4 Anesthetics

1. For rat euthanasia, expose to 100 % CO_2 for 5 min, then inject with 1 mL pentobarbital sodium (50 mg/mL) for rats 250–350 g.
2. For mouse euthanasia, inject a 25 g mouse with 100 μL ketamine/xylazine in 0.9 % sterile saline solution (100 mg/kg ketamine, 10 mg/kg xylazine).

2.5 VSM Cell Culture Medium

1. Rat VSM cell culture: Dulbecco's Modified Eagle Medium (DMEM) with high glucose, L-glutamine, and no sodium pyruvate (Invitrogen #11965-092) supplemented with 10 % bovine growth serum (BGS) and 1× penicillin/streptomycin (100 U pen/0.1 mg strep). Rat cells do not require fetal bovine serum (FBS) (*see* **Note 2**).
2. Mouse VSM cell culture: DMEM supplemented with 10 % FBS (low grade is sufficient) and 1× pen/strep. Sterile filtering is optional, but recommended.

2.6 δ-FlincG Adenovirus

1. Adenoviral-δ-FlincG was prepared using the ViraPower™ Adenoviral Expression System (Invitrogen). Viral supernatant (10^7–10^9 per mL titer), and not purified virus, was kept in frozen aliquots at −80 °C. Freeze–thaw cycles should be avoided to maintain viral integrity.

2.7 Preparation of Imaging Buffer

1. HBSS was supplemented with 10 mM 2-[[1,3-dihydroxy-2-(hydroxymethyl)propan-2-yl]amino]ethanesulfonic acid (TES, pH 7.4) and 1 g/liter D-glucose and kept on ice. Before imaging, equilibrate to 37 °C. 10 mM HEPES (pH 7.4) can also be used in place of TES buffer.

2.8 Imaging Compounds

1. NONOate family of NO donors (PROLI/NO, MAHMA/NO, DEA/NO, Spermine/NO, DETA/NO). These should be bought in 10 mg aliquots and stored at −80 °C, and kept on ice until dissolved in ice cold 10 mM NaOH to a stock of 50 mM.
2. CPTIO (2-(4-carboxyphenyl)-4,5-dihydro-4,4,5,5-tetramethyl-1H-imidazol-1-yloxy-3-oxide), an NO scavenger, should be dissolved in DMSO to a stock of 50 mM. It is deep blue in color, and can be stored at −20 °C. CPTIO is generally used at 50 or 60 μM working solution.
3. ODQ (1H-[1,2,4]oxadiazolo[4,3-a]quinoxalin-1-one), an inhibitor of sGC, should be dissolved fresh on the day of use in DMSO, and kept on ice as a 10 mM stock. ODQ can be used in a working solution at 10 μM.
4. 8-Br-cGMP, a cGMP analog, can be dissolved in water and stored at −20 °C as a 100 mM stock, and used at 50 μM for the working solution.

3 Methods

The following protocols instruct the step-by-step isolation of aortic VSM cells from both rats and mice. Isolation of aortae is most successful from young adult animals (approximately 8 weeks).

3.1 Isolating VSM Cells

1. Day 1: Prepare digestion solution #1. Inject peritoneally a sublethal dose of anesthetic (pentobarbital or xylazine/ketamine) in accordance with animal welfare guidelines. Spray fur with 70 % ethanol. Laying the mouse/rat on its back, make an incision mid-torso, just below the diaphragm. Reposition the animal on its side, so the tail is to the right, and legs facing towards you (Fig. 1a). Enlarge the incision towards the spine. Locate the kidney, and sever the renal artery. Prop the animal up to exsanguinate (*see* **Note 3**).

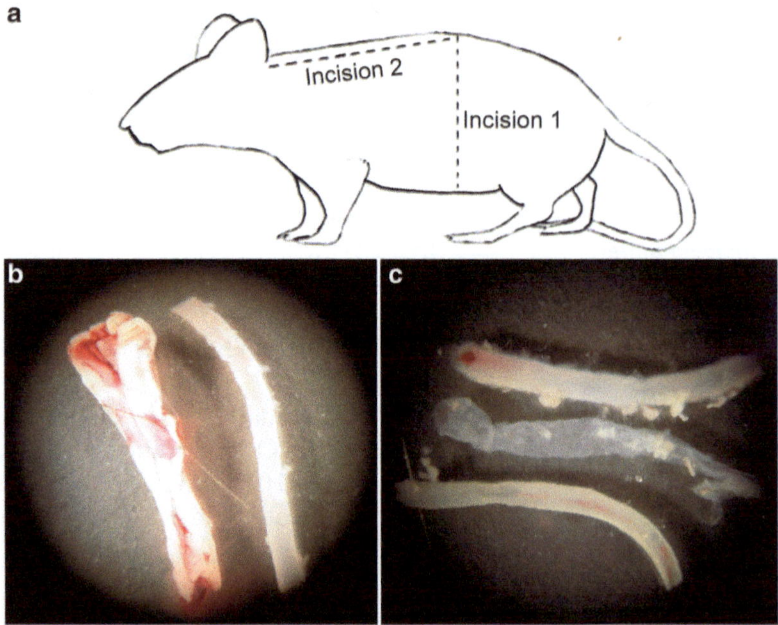

Fig. 1 **(a)** Diagram of the lateral incisions for aortic dissection. The first incision runs from the mid-torso to the spine. The second incision follows the spine and dissects the rib cage from the vertebrae, up to the clavicle. (**b**) Stereoscope view of a freshly dissected rat aorta (*left*) and after removal of extraneous tissue (*right*). (**c**) Following the initial digestion step, the aorta will appear "fuzzy" (*upper* vessel). The adventitia can be teased apart and removed (*middle*), leaving the tunica media which consists mainly of smooth muscle cells (*bottom*)

2. Position the rat back in the side-lying position for a lateral thoracotomy. Using the large shears, cut through the ribcage along the spine up to the clavicle, 90° from the original incision. For a mouse, position on the belly, spine directly up, and make the same incision using the dissection scissors. Cut all the way through the clavicle.
3. Using the spring scissors, cut away the diaphragm and plural membrane. The aorta lies along the spine, sandwiched by a strip of fat above and below the vessel. Keeping the scissors parallel to the spine and using small incisions, cut the fat to release the aorta from the spine. The aorta will relax downward (*see* **Note 4**).
4. Cut across the aorta at the base of the diaphragm and just above the heart, at the aortic arch. Carefully pinching only the very end with the forceps, place the aorta in cold HBSS. Keep on ice (*see* **Note 5**).
5. Place the aorta in a 60 mm dish of cold HBSS. Refreshing the media with cold HBSS during the process helps to reduce the viscosity of the solution, making for an easier dissection.

Using a stereoscope and using small spring scissors, carefully trim away any fat attached to the aorta. The vessel will maintain a tube shape, aiding in the procedure (Fig. 1b). Try not to nick the aorta itself; this makes the digestion removal more difficult.

6. Transfer the aorta into a 60 mm dish with digestion solution #1 (35 mm dish for mouse aortae). Up to 2 rat aortae and 4 mouse aortae can be digested simultaneously. Place in a rocking 37 °C incubator for 25 min. The aorta should appear "fuzzy." If not, keep incubating and monitor in 5 min intervals.
7. Transfer the aorta to HBSS. Under the stereoscope, gently tease one end of the aorta to separate the adventitia and media layers using two forceps. The adventitia appears as a white wispy covering, while the media emerges as a solid beige tube. Once the adventitia is loosened around the entire base of the media, hold the media with one set of forceps, and the adventitia in the other. Gently, in one smooth motion, gently remove the adventitia. The adventitia should peel off like a sock (Fig. 1c). Never pull on the media. It is paramount that during the entire procedure the media remains relaxed! If force is required, transfer the aorta back to the digestion solution and incubate for another 5 min. Repeat until the adventitia is completely removed.
8. Remove the media and place in a new dish with warm VSM culture medium. Let the tissue recover overnight at 37 °C, 5 % CO_2 in a humidified tissue culture incubator.
9. Day 2: Prepare digestion solution #2. Transfer the aortic media to a fresh dish of HBSS. Swirl to rinse. Transfer to digestion solution #2 with sterile forceps. Using small spring scissors, cut into 1–2 mm rings. Uniform sizes promote an even digestion.
10. Incubate at 37 °C, rocking for 2.5 h (45 min for mouse). The rings should appear very fuzzy and wispy.
11. Using a glass Pasteur pipette in an electric pipettor, gently triturate the solution four times. The shear forces should facilitate breaking apart the pieces of media and to dissociate individual cells. If there is little to no change, incubate for 5–10 min longer. Repeat trituration. Add 10 mL culture medium and transfer to a 15 mL conical tube.
12. Centrifuge at $130 \times g$ for 5 min. Aspirate the media, careful to leave the pellet undisturbed.
13. Add 1.5 mL BGS to the tube and resuspend the pellet by gentle trituration. Add 13.5 mL culture medium. Plate the cell suspension in 1 mL aliquots to 15 Delta T4 dishes that have been pre-rinsed with culture medium. For mouse cells,

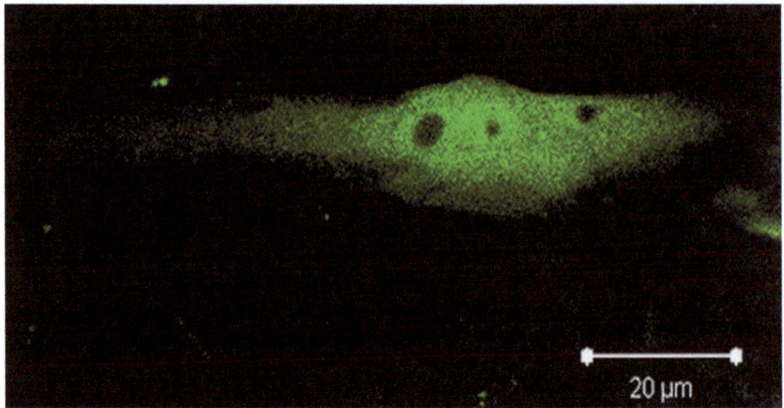

Fig. 2 Epifluorescent image of a δ-FlincG-transfected (100 µL 107 per mL titer adenovirus for 18 h), non-passaged VSM cell taken during live-cell imaging using a 40× objective, D480/20 m excitation filter, 505drxr dichroic mirror, and D535/30 m emission with a mercury-halide lamp (X-CITE 120; EXFO Photonics, Toronto)

add 600 µL FBS for the pellet resuspension and add 5.5 mL culture medium. Plate on 6 Delta T4 dishes.

14. Incubate cells at 37 °C, 5 % CO_2 in a tissue culture incubator. After 24 h, change the media careful not to disturb the loosely attached cells, replacing with fresh 10 % BGS culture medium. Cells will appear mostly rounded, but attached. Some will start to elongate.
15. Two days after plating, cells will appear elongated, with few processes. They have not yet begun to proliferate, which will ensure these VSM cells are as phenotypically close to their in vivo counterparts as possible.

3.2 Adenoviral δ-FlincG Infection

1. Thaw an aliquot of adenoviral supernatant on ice.
2. After 3 days in culture (2 days after plating), add 1 mL fresh VSM culture medium, adding 100 µL (10^7 per mL titer) adenovirus. Incubate for 18 h at 37 °C.
3. Remove the adenovirus, and add fresh culture medium. Incubate for 2–3 h at 37 °C. Prepare imaging buffer (*see* **Note 6**).

3.3 Epifluorescent Imaging

1. Keep the cells in a tissue culture incubator until imaging. For each dish, remove from the incubator and replace the culture medium with warm 1 mL imaging buffer just prior to imaging. At 37 °C, scan the plate for a group of well-transfected, healthy looking VSM cells (Fig. 2). These cells should have a smooth elongated appearance, with no processes and a mid-range of brightness. Overexpression of the indicator is usually toxic to the cells (rounding followed by apoptosis), while low-transfected cells are difficult to monitor and analyze.

2. Using imaging software such as Metamorph, draw several regions, approximately 10 % of the visible cytosolic area in size, on each cell to collect data from. Multiple regions allow the determination of specific local or global events.
3. Collect at least 2 min of baseline. A good baseline should be relatively smooth, with little variation, with an F/F_0 ratio of around 1.0. Most cells will have a slight decrease in their fluorescence within the first 30 s (*see* **Note 7**). Add compounds such as NO donors and cGMP analogs as 1:1,000 dilutions directly into the imaging buffer (*see* Subheading 2.7 for stock concentrations). Using a P1000 pipette, gently mix the buffer, careful not to touch the dish. Any additions should be mixed within 5–10 s to allow for accurate assessment of cGMP fluctuations.
4. To control for the transfection efficiency, and to assure cell viability and health, cGMP analogs, such as 8-Br-cGMP, should be employed. These analogs are cell permeable, and will give a delayed, slow, and steady rise in fluorescence. This increase is diffusion controlled, and maximal responses seen maybe lower than those seen with other, receptor-mediated ligands (*see* Fig. 3).
5. To determine the endogenous sGC activity, NO donors are ideal, because they release only one or two equivalents of nitric oxide and no other small molecules that can potentially harm cells. The NONOate family of donors (PROLI/NONOate, MAHMA/NONOate, DEA/NONOate, Spermine/NONOate, PAPA/NONOate, DETA/NONOate) have a variety of increasing NO release rates (1.8 s to 20 h; [51]). NONOates should be diluted in 10 mM NaOH to desired working concentrations and kept on ice to prevent NO release. Once added to the 37 °C, pH 7.4 imaging buffer, the donor will release NO, therefore, quick mixing is paramount. Exact concentrations of NO can be delivered if the NONOate donor is coupled with the NO scavenger, CPTIO [52, 53]. Mathematical software can be applied to calculate the concentrations required. For example, a 5 nM quick pulse of NO can be achieved with 50 μM CPTIO pre-incubated for 3 min, and 200 nM MAHMA/NO (Fig. 4a). Likewise, a steady state application of 5nM NO is achieved by 60 μM CPTIO 3 min pre-incubation and 8 μM Spermine/NO (Fig. 4b), as we have reported recently [26].
6. Other pharmacological inhibitors of the sGC/cGMP/PKG pathway can be useful to assess cGMP dynamics. The sGC inhibitor, ODQ, is very effective at completely blocking basal and stimulated cGMP production by oxidizing the heme group

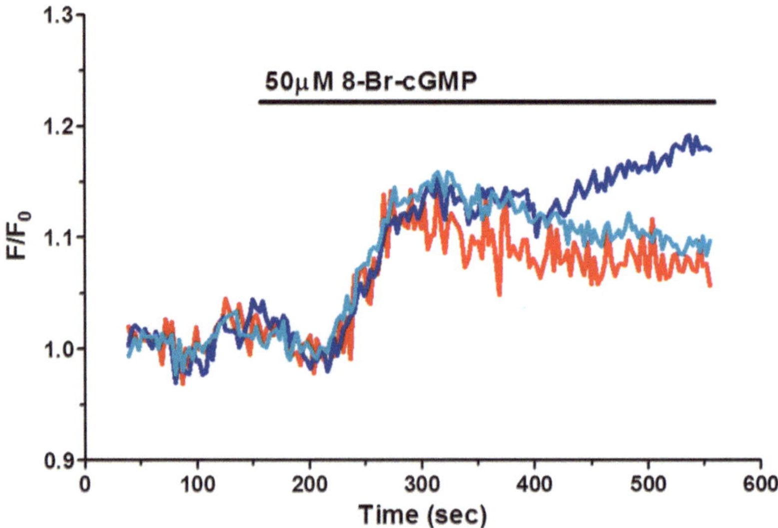

Fig. 3 δ-FlincG-detected cGMP from multiple regions within a single VSM cell (not shown). A single, non-passaged, FlincG-transfected VSM cell was imaged and then exposed to the cGMP analog, 8-Br-cGMP, resulting in a slow but steady rise in intracellular cGMP that remains elevated throughout analog application. The rise in cGMP is measured by FlincG fluorescence increase over baseline (F/F_0)

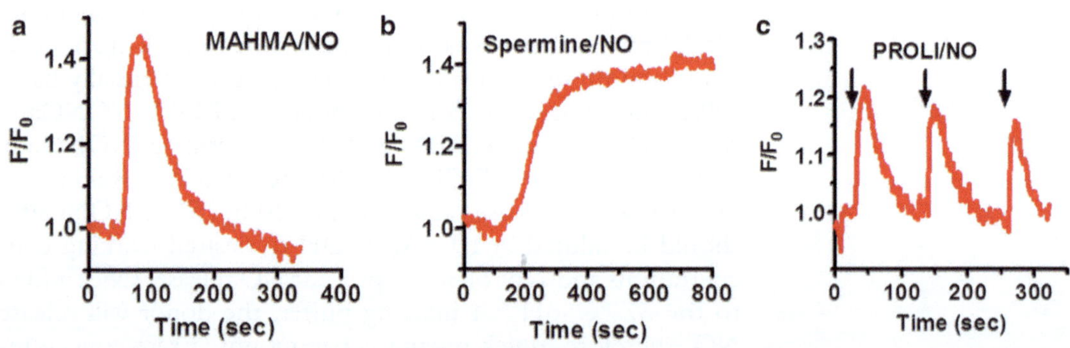

Fig. 4 Average traces of common NO-mediated $[cGMP]_i$ responses. NONOate family donors were coupled with the NO scavenger CPTIO to control NO concentration. (**a**) Transient cGMP increases upon 5 nM pulsed NO utilized by 200 nM MAHMA/NO and 50 μM CPTIO. (**b**) Sustained cGMP upon 5 nM clamped NO utilized by 8 μM Spermine/NO and 60 μM CPTIO. (**c**) Multiple, cGMP transients upon repeated 5 nM NO pulses utilized by 200 nM MAHMA/NO and 50 μM CPTIO. Figures adapted from [26]

of sGC [54–56]. However, this compound must be dissolved in fresh DMSO the day of the experiment, and cannot be stored in solution form.

3.4 Confocal Imaging

1. Confocal imaging with a water-dipping objective provides the most spatially resolute traces and movies. As with epifluorescent imaging, replace the culture medium with pre-warmed imaging buffer, and insert into the 37 °C dish warmer on the

microscope stage. Focus, and scan for an ideally transfected VSM cell. Cell shape is important here as well. A cell with long processes tends to move out of the focal plane easily, skewing the data collected. Refocus just above the bottom membrane attached to the glass for optimal results.

2. Collect at least 45 s of baseline at an acquisition rate of 250 ms, and then add compounds in 1:1,000 mixed with a P100 pipette (*see* Subheading 2.7 for stock concentrations). A careful mixing is important, or the focus will be lost and the experiments are forfeited.

3.5 Data Analysis

1. The large series of images can then be imported into a commercially available analysis software such as Metamorph or custom-written software such as SparkAn (courtesy of A. Bonev and M. Nelson at UVM). Within these software packages, small regions can be drawn on each cell to collect the fluorescence intensity. Region placement is critical for accurate depiction of the cell's responses. For epifluorescence, regions should be well within the cytosol and not too close to the edges or nucleus. Confocal imaging allows the opportunity to distinguish between edge effects and the cytosolic events, so many regions in various locations should be analyzed. Cells tend to move during stimulation; therefore adjustment of the region location may be required to maintain the signal (*see* **Note 7**).
2. Once the fluorescence intensities for each region have been collected, the data can be normalized to the initial background, reported as F/F_0. These calculated values can be graphed with the baseline at 1 or, if preferred, as a percentile.
3. Changes in $[cGMP]_i$ can be calculated simply by subtracting the total response intensity from the initial baseline, or the baseline just prior to the stimulus added. This generally gives a more accurate measurement, as the baseline may not be steady at 1 (or 0) for the duration of the experiment.
4. Transient changes in $[cGMP]_i$ can be analyzed by several parameters. The lag time is described as the time from stimulus addition to the first 5 % of the total response elicited. The percent response is total fluorescence intensity achieved minus the baseline, divided by the baseline, multiplied by 100 %. The response time is the time from 5 to 95 % total response. The tau (τ) factor is the degradation parameter. This is described as the time from 95 % response to 5 % of the new baseline. The peak width ($P_{1/2}$) is measured as the time difference at 50 % of the peak height. The area under the curve of each transient peak can also be an important measurement of total $[cGMP]_i$ change (*see* **Note 8**).

4 Notes

1. Alternatively, punch dishes with coverslip glass bottoms (MatTek or homemade) can be used. The drawback is a lack of temperature control and stability while imaging. The Bioptechs Delta T4 culture system provides a stable stage with heating capabilities to keep the cells in a more physiological state during the experiment. Maintenance of 37 °C is critical for the proper NO release kinetics of NO donors.
2. Culture medium should be used until a slight purple shade is noticed. Dispose and prepare fresh media. Old media yield poor growing conditions for VSM cells.
3. Proper exsanguination allows for an easier dissection of the thoracic aorta. Not only is the cavity empty of blood, but also the aorta contains fewer blood clots, which mitigates subsequent digestion.
4. During the initial excision, it is important not to stretch or pull on the aorta. Any stretch severely reduces the viability of the VSM cells after digestion.
5. Once removed from the animal, an aorta can be kept on ice for 1–2 h.
6. Primary VSM cells tend to only tolerate a single biosensor or dye added at one time. Co-transfections or additions of dyes such as Fura-2 or Fluo-4 for calcium detection or Daf-2 for NO measurements appear to be toxic to these cells (rapid apoptosis).
7. Sometimes imaging traces have a tendency to drift downwards over time. Realigning any response to the baseline just prior to adding each compound should be a general practice. This baseline should be relatively even to ensure a proper F/F_0 ratio.
8. Data analysis is best done on an individual cell basis, and then averaged to give mean traces. Not all cells respond identically. Some cells may need to be excluded from the total analysis based on several factors: health of the animal, cell health, total fluorescence intensity, cell shape, movement, morphology changes, loss of focal plane, etc. However, extreme caution should be used when dismissing individual cells. Only strict guidelines and outlier statistical tests will assure proper and unbiased data analysis.

Acknowledgments

The authors would like to thank Drs. Carolyn Sawyer, Sharon Cawley, and Lydia Nausch for their help in perfecting this technique. Support was provided by NIH grants HL68991 (W.R.D.) and T323 HL07944 (K.F.H.), and the Totman Trust for Biomedical Research.

References

1. Friebe A, Koesling D (2003) Regulation of nitric oxide-sensitive guanylyl cyclase. Circ Res 93(2):96–105
2. Kuhn M (2003) Structure, regulation, and function of mammalian membrane guanylyl cyclase receptors, with a focus on guanylyl cyclase-A. Circ Res 93(8):700–709
3. Conti M, Beavo J (2007) Biochemistry and physiology of cyclic nucleotide phosphodiesterases: essential components in cyclic nucleotide signaling. Annu Rev Biochem 76:481–511
4. Francis SH, Blount MA, Zoraghi R et al (2005) Molecular properties of mammalian proteins that interact with cGMP: protein kinases, cation channels, phosphodiesterases, and multidrug anion transporters. Front Biosci 10:2097–2117
5. Kemp-Harper B, Schmidt HH (2009) cGMP in the vasculature. Handb Exp Pharmacol 191:447–467
6. Alverdi V, Mazon H, Versluis C et al (2008) cGMP-binding prepares PKG for substrate binding by disclosing the C-terminal domain. J Mol Biol 375(5):1380–1393
7. Scholten A, Fuss H, Heck AJ et al (2007) The hinge region operates as a stability switch in cGMP-dependent protein kinase I alpha. FEBS J 274(9):2274–2286
8. Surks HK, Mochizuki N, Kasai Y et al (1999) Regulation of myosin phosphatase by a specific interaction with cGMP-dependent protein kinase Ialpha. Science 286(5444):1583–1587
9. Tang KM, Wang GR, Lu P et al (2003) Regulator of G-protein signaling-2 mediates vascular smooth muscle relaxation and blood pressure. Nat Med 9(12):1506–1512
10. Wooldridge AA, MacDonald JA, Erdodi F et al (2004) Smooth muscle phosphatase is regulated in vivo by exclusion of phosphorylation of threonine 696 of MYPT1 by phosphorylation of serine 695 in response to cyclic nucleotides. J Biol Chem 279(33):34496–34504
11. Sun X, Kaltenbronn KM, Steinberg TH et al (2005) RGS2 is a mediator of nitric oxide action on blood pressure and vasoconstrictor signaling. Mol Pharmacol 67(3):631–639
12. Schlossmann J, Ammendola A, Ashman K et al (2000) Regulation of intracellular calcium by a signalling complex of IRAG, IP3 receptor and cGMP kinase Ibeta. Nature 404(6774): 197–201
13. Geiselhoringer A, Werner M, Sigl K et al (2004) IRAG is essential for relaxation of receptor-triggered smooth muscle contraction by cGMP kinase. EMBO J 23(21):4222–4231
14. Lalli MJ, Shimizu S, Sutliff RL et al (1999) [Ca2+]i homeostasis and cyclic nucleotide relaxation in aorta of phospholamban-deficient mice. Am J Physiol 277(3 Pt 2):H963–H970
15. Sausbier M, Schubert R, Voigt V et al (2000) Mechanisms of NO/cGMP-dependent vasorelaxation. Circ Res 87(9):825–830
16. Ellerbroek SM, Wennerberg K, Burridge K (2003) Serine phosphorylation negatively regulates RhoA in vivo. J Biol Chem 278(21): 19023–19031
17. Weber S, Bernhard D, Lukowski R et al (2007) Rescue of cGMP kinase I knockout mice by smooth muscle specific expression of either isozyme. Circ Res 101(11):1096–1103
18. Rybalkin SD, Bornfeldt KE, Sonnenburg WK et al (1997) Calmodulin-stimulated cyclic nucleotide phosphodiesterase (PDE1C) is induced in human arterial smooth muscle cells of the synthetic, proliferative phenotype. J Clin Invest 100(10):2611–2621
19. Rybalkin SD, Rybalkina IG, Feil R et al (2002) Regulation of cGMP-specific phosphodiesterase (PDE5) phosphorylation in smooth muscle cells. J Biol Chem 277(5):3310–3317
20. Thomas MK, Francis SH, Corbin JD (1990) Substrate- and kinase-directed regulation of phosphorylation of a cGMP-binding phosphodiesterase by cGMP. J Biol Chem 265(25): 14971–14978
21. Corbin JD, Turko IV, Beasley A et al (2000) Phosphorylation of phosphodiesterase-5 by cyclic nucleotide-dependent protein kinase alters its catalytic and allosteric cGMP-binding activities. Eur J Biochem 267(9):2760–2767
22. Francis SH, Poteet-Smith C, Busch JL et al (2002) Mechanisms of autoinhibition in cyclic nucleotide-dependent protein kinases. Front Biosci 7:d580–d592
23. Rybalkin SD, Rybalkina IG, Shimizu-Albergine M et al (2003) PDE5 is converted to an activated state upon cGMP binding to the GAF A domain. EMBO J 22(3):469–478
24. Zoraghi R, Bessay EP, Corbin JD et al (2005) Structural and functional features in human PDE5A1 regulatory domain that provide for allosteric cGMP binding, dimerization, and regulation. J Biol Chem 280(12): 12051–12063
25. Omori K, Kotera J (2007) Overview of PDEs and their regulation. Circ Res 100(3):309–327
26. Held KF, Dostmann WR (2012) Subnanomolar sensitivity of nitric oxide mediated regulation of cGMP and vasomotor reactivity in vascular smooth muscle. Front Pharmacol 3:130

27. Halvey EJ, Vernon J, Roy B et al (2009) Mechanisms of activity-dependent plasticity in cellular no-cGMP signaling. J Biol Chem 284(38):25630–25641
28. Batchelor AM, Bartus K, Reynell C et al (2010) Exquisite sensitivity to subsecond, picomolar nitric oxide transients conferred on cells by guanylyl cyclase-coupled receptors. Proc Natl Acad Sci USA107(51):22060–22065
29. Bellamy TC, Wood J, Goodwin DA et al (2000) Rapid desensitization of the nitric oxide receptor, soluble guanylyl cyclase, underlies diversity of cellular cGMP responses. Proc Natl Acad Sci USA97(6):2928–2933
30. Steiner AL, Parker CW, Kipnis DM (1972) Radioimmunoassay for cyclic nucleotides. I. Preparation of antibodies and iodinated cyclic nucleotides. J Biol Chem 247(4):1106–1113
31. Wehmann RE, Blonde L, Steiner AL (1972) Simultaneous radioimmunoassay for the measurement of adenosine 3′,5′-monophosphate and guanosine 3′,5′-monophosphate. Endocrinology 90(1):330–335
32. Lakowicz JR, Gryczynski I, Gryczynski Z et al (1999) Anisotropy-based sensing with reference fluorophores. Anal Biochem 267(2): 397–405
33. Heim R, Prasher DC, Tsien RY (1994) Wavelength mutations and posttranslational autoxidation of green fluorescent protein. Proc Natl Acad Sci USA91(26):12501–12504
34. Zukin RS, Hartig PR, Koshland DE Jr (1977) Use of a distant reporter group as evidence for a conformational change in a sensory receptor. Proc Natl Acad Sci USA74(5):1932–1936
35. Hahn LH, Hammes GG (1978) Structural mapping of aspartate transcarbamoylase by fluorescence energy-transfer measurements: determination of the distance between catalytic sites of different subunits. Biochemistry 17(12):2423–2429
36. Honda A, Adams SR, Sawyer CL et al (2001) Spatiotemporal dynamics of guanosine 3 ,5 -cyclic monophosphate revealed by a genetically encoded, fluorescent indicator. Proc Natl Acad Sci USA98(5):2437–2442
37. Sawyer CL, Honda A, Dostmann WR (2003) Cygnets: spatial and temporal analysis of intracellular cGMP. Proc West Pharmacol Soc 46:28–31
38. Cawley SM, Sawyer CL, Brunelle KF et al (2007) Nitric oxide-evoked transient kinetics of cyclic GMP in vascular smooth muscle cells. Cell Signal 19(5):1023–1033
39. Nikolaev VO, Gambaryan S, Lohse MJ (2006) Fluorescent sensors for rapid monitoring of intracellular cGMP. Nat Methods 3(1):23–25
40. Baird GS, Zacharias DA, Tsien RY (1999) Circular permutation and receptor insertion within green fluorescent proteins. Proc Natl Acad Sci USA96(20):11241–11246
41. Nakai J, Ohkura M, Imoto K (2001) A high signal-to-noise Ca(2+) probe composed of a single green fluorescent protein. Nat Biotechnol 19(2):137–141
42. Nausch LW, Ledoux J, Bonev AD et al (2008) Differential patterning of cGMP in vascular smooth muscle cells revealed by single GFP-linked biosensors. Proc Natl Acad Sci USA105(1):365–370
43. Isner JC, Maathuis FJ (2011) Measurement of cellular cGMP in plant cells and tissues using the endogenous fluorescent reporter FlincG. Plant J 65(2):329–334
44. Wood KC, Batchelor AM, Bartus K et al (2011) Picomolar nitric oxide signals from central neurons recorded using ultrasensitive detector cells. J Biol Chem 286(50):43172–43181
45. Miller CL, Cai Y, Oikawa M et al (2011) Cyclic nucleotide phosphodiesterase 1A: a key regulator of cardiac fibroblast activation and extracellular matrix remodeling in the heart. Basic Res Cardiol 106(6):1023–1039
46. Chao YC, Cheng CJ, Hsieh HT et al (2010) Guanylate cyclase-G, expressed in the Grueneberg ganglion olfactory subsystem, is activated by bicarbonate. Biochem J 432(2): 267–273
47. Tsai EJ, Kass DA (2009) Cyclic GMP signaling in cardiovascular pathophysiology and therapeutics. Pharmacol Ther 122(3):216–238
48. Cocks TM, Angus JA, Campbell JH et al (1985) Release and properties of endothelium-derived relaxing factor (EDRF) from endothelial cells in culture. J Cell Physiol 123(3):310–320
49. Griffith TM, Edwards DH, Lewis MJ et al (1984) The nature of endothelium-derived vascular relaxant factor. Nature 308(5960): 645–647
50. Hakim TS, Sugimori K, Camporesi EM et al (1996) Half-life of nitric oxide in aqueous solutions with and without haemoglobin. Physiol Meas 17(4):267–277
51. Keefer LK, Nims RW, Davies KM et al (1996) "NONOates" (1-substituted diazen-1-ium-1,2-diolates) as nitric oxide donors: convenient nitric oxide dosage forms. Methods Enzymol 268:281–293
52. Griffiths C, Wykes V, Bellamy TC et al (2003) A new and simple method for delivering clamped nitric oxide concentrations in the physiological range: application to activation of guanylyl cyclase-coupled nitric oxide receptors. Mol Pharmacol 64(6):1349–1356

53. Bellamy TC, Griffiths C, Garthwaite J (2002) Differential sensitivity of guanylyl cyclase and mitochondrial respiration to nitric oxide measured using clamped concentrations. J Biol Chem 277(35):31801–31807

54. Garthwaite J, Southam E, Boulton CL et al (1995) Potent and selective inhibition of nitric oxide-sensitive guanylyl cyclase by 1H-[1,2,4] oxadiazolo[4,3-a]quinoxalin-1-one. Mol Pharmacol 48(2):184–188

55. Schrammel A, Behrends S, Schmidt K et al (1996) Characterization of 1H-[1,2,4] oxadiazolo[4,3-a]quinoxalin-1-one as a heme-site inhibitor of nitric oxide-sensitive guanylyl cyclase. Mol Pharmacol 50(1):1–5

56. Evgenov OV, Pacher P, Schmidt PM et al (2006) NO-independent stimulators and activators of soluble guanylate cyclase: discovery and therapeutic potential. Nat Rev Drug Discov 5(9):755–768

Chapter 9

Methods for Identification of cGKI Substrates

Katharina Salb and Jens Schlossmann

Abstract

The cGMP-dependent protein kinases (cGK), which belong to the family of serine/threonine kinases, exhibit their diverse functions in cells through interaction with a variety of substrate proteins. Several substrates were identified and the interactions studied using different methods inter alia co-immunoprecipitation (Co-IP) and cGMP-agarose affinity purification. In the following chapter, we will describe the preparation of cell or tissue lysates, the procedures of cGMP-agarose affinity purification and co-immunoprecipitation, and finally the separation and analysis of the protein complexes by SDS-PAGE or mass spectrometry.

Key words cGMP, cGMP-dependent protein kinases, cGKI substrate proteins, Affinity purification, Co-immunoprecipitation, cGMP-agarose

1 Introduction

The cGMP-dependent protein kinases (cGK) belong to the family of serine/threonine kinases. They are expressed in a variety of eukaryotes in a multitude of tissues and cells including smooth muscles, platelets, brain, lung, and kidney. Two cGK genes, *prkg1* and *prkg2*, that encode the enzymes cGKI and cGKII were identified in mammals [1]. There exist two isoforms of the cGKI, cGKIα and cGKIβ, which differ in their N-terminal ~100 amino acids containing leucine zipper domains that mediate their interaction with different substrate proteins and the autoinhibitory domains [2]. The N-terminal domain is followed by two tandem cGMP-binding sites that bind cGMP with high and low affinity and by a catalytic domain [2]. The kinases are dimers comprising two identical monomers which are each fully activated by binding of two cGMP molecules [3]. Since the cGK bind cGMP, they can be purified by cGMP-agarose which is cGMP immobilized on an agarose matrix via different spacers. Other cGMP-binding proteins like protein kinase A (PKA) which interacts with cGMP

Thomas Krieg and Robert Lukowski (eds.), *Guanylate Cyclase and Cyclic GMP: Methods and Protocols*, Methods in Molecular Biology, vol. 1020, DOI 10.1007/978-1-62703-459-3_9,

in low affinity or phosphodiesterases (PDEs) are also enriched by this affinity matrix as well as proteins that attach to the cGKs. Therefore, the cGMP-agarose affinity purification has already been used several times to study the interactions between the cGKI, especially the cGKIβ isoform, and its substrate proteins [4, 5]. Additionally, cGK-complexes were frequently analyzed by co-immunoprecipitation (Co-IP) experiments. The interaction between cGKIα and MYPT-1 (myosin-binding subunit of myosin phosphatase) was investigated by immunoprecipitation (IP) with both anti-cGKI and anti-MYPT-1 antibodies [6, 7]. Furthermore, a trimeric complex consisting of the cGKIβ isoform, the inositol trisphosphate receptor I ($InsP_3R$-I) and the $InsP_3R$-associated cGMP kinase substrate (IRAG), was identified by cGMP-agarose and Co-IP [4, 8, 9]. Koller et al. also performed cGMP-agarose and IP experiments and detected phospholamban (PLB) as an additional component of the cGKI macro-signalling complex [5].

As mentioned above, a variety of proteins is bound to cGMP-agarose and therefore Co-IP is presumably more specific for analyzing the interactions between the cGKI isoforms and its substrates. Besides Co-IP, there are several other methods used to study protein–protein interactions: the yeast two-hybrid system, the differential scanning calorimetry, surface plasmon resonance (SPR) studies, or pull-down of affinity-tagged proteins (GST- or His-tagged proteins) [10]. The advantages of Co-IP are that the experiments are relatively inexpensive and very reproducible and that the Co-IP procedure can be combined with affinity tagging [11]. Furthermore, both tissues or cells expressing their endogenous proteins and transfected cells can be analyzed. Epitope-tagged proteins can often be eluted by incubation with competing peptides. This specific elution reduces the amount of contaminating proteins in the eluate [10].

There exist different IP techniques which differ in the type of affinity matrix used to bind the antigen(s) [12]. Usually, protein A or protein G beads—which bind to the F_c region of most antibody classes—are used to capture the antibody/antigen complex. Problems of this method involve contamination of the target protein with the IP antibody after elution or waste of expensive antibodies. Hence, before sample addition, the IP antibody can be covalently linked to the protein A or G resin by using a cross-linker like disuccinimidyl suberate (DSS) [13]. A third possibility is to couple the antibody directly to an activated support via lysine residues. Therefore the agarose matrix is activated by sodium periodate and then connected with an amine-containing ligand, respectively, antibody [13]. This method does not need protein A or G and allows coupling of all antibody species and classes [13].

In this chapter we will describe the cGMP-agarose method and the traditional co-immunoprecipitation as procedures suitable for identifying cGKI substrates.

2 Materials

Prepare all solutions using ultrapure water and analytical grade reagents.

2.1 Preparation of Cell or Tissue Lysates

1. Co-immunoprecipitation experiments can be performed either under denaturing or nondenaturing conditions (*see* **Note 1**). Examples for nondenaturing lysis buffers: 1× Lubrol buffer: 20 mM Tris–HCl/MOPS, pH 8.0, 2 % Lubrol-PX (v/v) (nonaethyleneglycol monododecyl ether; Sigma-Aldrich), 150 mM NaCl, protease inhibitors (*see* **item 4**) [14]; Triton X-100 buffer: 50 mM Tris–HCl, pH 7.4, 1 % (w/v) Triton X-100, 300 mM NaCl, 5 mM EDTA, 0.02 % (w/v) sodium azide, protease inhibitors [15].
2. Detergent-free lysis buffer: 1× PBS buffer containing 5 mM EDTA, 0.02 % (w/v) sodium azide, protease inhibitors [15]; 10× stock solution of PBS buffer: 80 g NaCl, 2 g KCl, 11.5 g $Na_2HPO_4·7H_2O$, 2 g KH_2PO_4 in 1,000 ml H_2O.
3. Denaturing lysis buffer (used for platelet lysis): 50 mM Tris–HCl, pH 8.0, 17.3 mM SDS, 1 mM DTT [16].
4. Protease inhibitors: use protease inhibitor mix (e.g., complete cocktail tablets; Roche) or the single components (e.g., 0.5 μg/ml leupeptin in H_2O, 1 mM benzamidine in H_2O, 0.3 mM PMSF in isopropanol).
5. Add phosphatase inhibitors if the phosphorylation state of the proteins should be conserved during protein binding: use a commercial available phosphatase inhibitor mix (PhosStop; Roche) or the single components in the following concentrations: 120 nM okadaic acid in DMSO, 50 mM NaF in H_2O, 0.2 mM Na-orthovanadate in H_2O [16].

2.2 cGMP-Agarose Affinity Purification

1. 8-AET-cGMP-Agarose (Biolog): supplied in 30 mM Na_2HPO_4 buffer pH 7, containing 0.1 % sodium azide as preservative; The PDE-stable phosphorothioate modifications Rp-/Sp-8-AET-cGMPS which have lower affinities and allow for milder desorption conditions are also available from Biolog. 2′-AHC-cGMP-agarose and Rp-2′-AHC-cGMPS-agarose are more appropriate for binding of PDEs than for binding of cGKs. Ethanolamine-agarose (Biolog) can be used as a control for specific binding of proteins to the cGMP-agarose [4] (*see* **Note 2**).
2. Tubes/columns: cGMP-agarose and immunoprecipitation experiments are typically performed in microcentrifuge tubes. Careful pipetting is required to prevent aspiration of the resin bed volume. To avoid these pipetting problems, commercially available spin columns which have frits that retain the resin beads can be used (e.g., Thermo Scientific) (*see* **Note 3**).

3. Binding/wash buffer: see nondenaturing lysis buffers in Subheading 2.1 (*see* **Note 4**).
4. Elution:
   (a) With cGMP (Biolog, Sigma-Aldrich): e.g., 20 mM cGMP-Na [5] (*see* **Note 5**).
   (b) With 1× SDS sample buffer: *see* Subheading 2.4.1.
   (c) With denaturing elution buffer: 8 M urea in 50 mM ammonium bicarbonate buffer, pH 8.0 [17].

2.3 Co-immunoprecipitation

1. Protein A-sepharose or protein G-sepharose (aqueous ethanol suspension, Sigma-Aldrich; antibody-binding capacity is 6 mg per ml for protein A-sepharose and 2 mg per ml for protein G-sepharose): Protein A from *Staphylococcus aureus* can be used for binding of human and rabbit primary antibodies. Protein G from *Streptococcus* is more appropriate for binding of mouse IgG_1, goat, and sheep IgGs [12] (*see* **Note 6**).
2. Specific polyclonal or monoclonal antibody for immunoprecipitation (*see* **Notes 7** and **8**).
3. Wash buffers for co-immunoprecipitation experiments under nondenaturing conditions: Lubrol buffer: 20 mM Tris–HCl, pH 8.0, 0.4 % (v/v) Lubrol-PX, 150 mM NaCl; Triton X-100 buffer: 50 mM Tris–HCl, pH 7.4.
4. 0.1 % (w/v) Triton X-100, 300 mM NaCl, 5 mM EDTA, 0.02 % (w/v) sodium azide [15] (*see* **Note 9**).
5. Binding buffer for co-immunoprecipitation experiments under nondenaturing conditions: see lysis buffer.
6. Binding under denaturing conditions: the lysate is diluted one to ten with nondenaturing lysis buffer before it is added to the prepared beads (*see* **Note 10**).
7. BSA solution to block unspecific binding sites on the sepharose matrix: 3 % bovine serum albumin in binding buffer.
8. Phosphorylation buffer: 50 mM Mes, pH 6.9, 10 mM NaCl, 1 mM MgAc, 0.4 mM EGTA.
9. Elution: 1× SDS sample buffer.

2.4 SDS-PAGE and Immunoblotting

2.4.1 SDS-PAGE

1. The SDS polyacrylamide gel components according to the method of Laemmli [18, 19]: Tris–HCl buffer 0.6 M: 60 g Tris–HCl, pH 6.8, 40 ml 10 % SDS, add 1,000 ml H_2O; Tris–HCl buffer 1.8 M: 182 g Tris–HCl, pH 8.8, 40 ml 10 % SDS, add 1,000 ml H_2O. For the preparation of the separating and stacking gel solutions, see Table 1. Examples for 7.5, 10 and 11.5 % acrylamide-separating gels are given.
2. Running buffer: 30 g Tris–HCl, pH 8.3, 144 g Glycin, 10 g SDS, add 1,000 ml H_2O.

Table 1
Separating and stacking gel solutions for SDS-PAGE

	Separating gel 7.5 %	Separating gel 10 %	Separating gel 11.5 %	Stacking gel 5 %
H_2O	8.75 ml	7.3 ml	6.5 ml	3.6 ml
30 % acrylamide/0.2 % bisacrylamide	4.25 ml	5.7 ml	6.5 ml	0.83 ml
Tris–HCl buffer 1.8 M pH 8.8	3.6 ml	3.6 ml	3.6 ml	–
Tris–HCl buffer 0.6 M pH 6.8	–	–	–	0.5 ml
10 % SDS	167 μl	167 μl	167 μl	50 μl
TEMED	20 μl	20 μl	20 μl	10 μl
10 % ammonium persulfate	200 μl	200 μl	200 μl	50 μl

The components of a 7.5, 10, and 11.5 % separating gel are listed

3. Tris–Tricine-PAGE is recommended for separation of proteins smaller than 20 kDa [18] (*see* **Note 11**): Separating, spacer, and stacking gel solutions (gel buffer: 3 M Tris–HCl, pH 8.45, 0.3 % SDS), *see* Table 2.
4. Tris–Tricine-PAGE running buffers: anode buffer (0.2 M Tris, pH 8.9) [18] and cathode buffer (0.1 M Tris, pH 8.25, 0.1 M Tricine, 0.1 % SDS).
5. SDS sample buffer (5×): 0.3 M Tris–HCl, pH 6.8, 10 % SDS, 25 % β-mercaptoethanol, 0.1 % bromophenol blue, 45 % glycerol [19].

2.4.2 Staining with Coomassie Blue or Silver

1. Coomassie blue solution: 1.5 g Coomassie blue R-250, 455 ml methanol, 90 ml glacial acetic acid, add 1,000 ml H_2O.
2. Coomassie blue destaining solution: 100 ml glacial acetic acid, 300 ml methanol, add 1,000 ml H_2O.
3. Silver stain kit: ProteoSilver™ Plus Silver Stain Kit (Sigma-Aldrich) with supplied solutions for staining, developing, sensitization, and destaining.

   Additional solutions are:

   (a) Fixing solution: 50 ml ethanol, 10 ml glacial acetic acid, add 100 ml H_2O.

   (b) 30 % ethanol solution.

   Staining utilizes silver nitrate, which binds to proteins under weakly acidic or neutral conditions and is then reduced to metallic silver with formaldehyde at alkaline pH.

Table 2
Separating, spacer, and stacking gel solutions for Tris–Tricine PAGE

	Separating gel (16.5 %)	Spacer gel	Stacking gel
H_2O	–	1.3 ml	3.7 ml
24.25 % acrylamide/0.75 % bisacrylamide	5 ml	2 ml	1 ml
Glycerol	1 ml	–	–
Gel buffer	2.5 ml	1.6 ml	1.55 ml
TEMED	7 µl	6 µl	12 µl
10 % ammonium persulfate	45 µl	20 µl	60 µl

2.4.3 Immunoblotting

1. Membranes: nitrocellulose (0.45 µm pore size) or PVDF membrane (0.45 µm pore size; Millipore).
2. Ponceau S solution: 0.5 g Ponceau S, 1 ml glacial acetic acid, add 100 ml H_2O [20]; alternatively: 0.2 g Ponceau S, 2 ml trichloroacetic acid, add 100 ml H_2O.
3. TTBS (TBS buffer with Tween 20): 100 mM Tris–HCl, pH 7.5, 150 mM NaCl, 0.1 % Tween 20 [20].
4. Luminol reagent: e.g., Pierce ECL Western Blotting Substrate (Thermo Scientific).

3 Methods

It is essential to keep all buffers and tubes cold by using an ice bath and a refrigerated centrifuge.

3.1 Preparation of Cell or Tissue Lysates

1. Attached cells (e.g., COS-7 or HEK293 cells) are rinsed twice with ice-cold PBS and after addition of lysis buffer scraped off the plate. Cells in suspension (e.g., platelets) are collected by centrifugation, washed twice with PBS, and afterwards resuspended in lysis buffer. Mammalian tissues are isolated and immediately snap frozen. References for preparation of cell and tissue lysates: COS-7 cell lysate [8], platelet lysate [16], tissue lysate [14], or smooth muscle membrane proteins [9] (*see* **Note 12**).
2. If proteins from transfected cells are immunoprecipitated, untransfected cells can be used as a control.
3. The protein concentration of the lysate is determined to allow the use of a particular protein amount. If a nondenaturing lysis buffer without SDS, urea, or DTT was used the Bradford protein

assay can be performed [21]. If the lysis buffer contains one of these substances the so-called Lowry test has to be chosen.

4. Solubilization: If the sample was homogenized in detergent-free lysis buffer, the proteins should be solubilized in nondenaturing lysis buffer. The protein sample is diluted with lysis buffer and a protein concentration of, e.g., 2 μg/μl is adjusted. Afterwards, the proteins are kept on ice for 20–30 min, meanwhile the tube is gently inverted every 5 min.
5. Centrifuge for separation of unsoluble proteins: 4 °C, 30,000 × *g*, 20 min; discard the pellet and use supernatant for precipitation (*see* **Note 13**).
6. Before IP, the lysates can be precleared twice for 1 h at 4 °C with protein A/G-sepharose beads (*see* **Note 14**).

3.2 cGMP-Agarose Affinity Purification

1. 15 μl settled beads of cGMP-agarose are used per experiment.
2. For equilibration, wash beads three times with binding buffer (1× Lubrol buffer or Triton X-100 buffer): add approximately 500 μl buffer to the beads, resuspend them by inverting the tube three to four times, and centrifuge at 12,000 × *g* for 30 s. Remove supernatant and add fresh buffer. After the last washing step, remove supernatant completely but carefully.
3. Add protein lysate (preparation described in Subheading 3.1, **step 4**; *see* **Note 15**) plus protease inhibitors (and protein phosphatase inhibitors if phosphorylated proteins are analyzed) and incubate at 4 °C and gentle agitation for at least 2 h or overnight.
4. Beads are washed at least three to four times with 500 μl binding buffer. As much of the washing solution as possible should be removed at each washing step to remove all unbound proteins. Finally, beads can be washed once more with ice-cold PBS to remove detergents that could lead to decreased resolution in SDS-PAGE [15].
5. For SDS-PAGE analysis, attached proteins are eluted with 20 μl elution buffer or SDS sample buffer: Incubate the beads several minutes at 4 °C in elution/SDS buffer and then heat the samples 5 min at 95 °C to denature the proteins (*see* **Note 16**).
6. To load the eluted proteins onto the gel, remove supernatant completely and thereby avoid taking up the agarose beads together with the protein complex containing supernatant.
7. For further analysis of cGK-complexes in cell/tissue lysates, the proteins can be first precleared with cGMP-Agarose, then eluted with cGMP, and then immunoprecipitated with different specific antibodies [5]. This procedure allows a refined identification of cGK-interacting proteins. The preclearing step

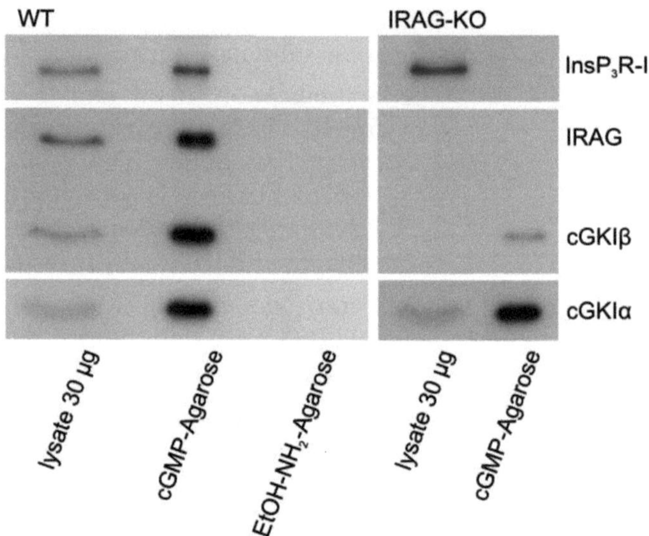

Fig. 1 Immunoblots of cGMP-agarose experiments with murine WT and IRAG-KO platelets: Murine blood was collected by cardiac puncture, then the platelets were isolated by centrifugation, and lysed with 2 % lubrol buffer. The platelet proteins (250 µg per experiment) were added to the agarose beads, incubated overnight, and eluted with SDS sample buffer. $InsP_3R$-I, IRAG, and cGKIα/β were detected in the immunoblot

is carried out as the described cGMP-agarose purification with protein binding for 2 h at 4 °C. After washing the bound proteins are eluted with 2 ml of 20 mM cGMP-Na in 20 mM Tris, pH 8.0, 60 mM NaCl, 0,1 % Lubrol PX, 5 mM EDTA for 1 h at room temperature [5]. cGMP is removed by dialysis against Lubrol buffer (2, 12, and 2 h) and then the proteins are used for Co-IP [5]. To perform a Co-IP after the cGMP-agarose assay is useful because all cGMP-binding proteins are attached to the matrix. It cannot be distinguished between the different complexes of the two cGKI isoforms. Additionally, if the proteins to be analyzed are low expressed in the respective tissue or cell line it is helpful to enrich the proteins before the Co-IP experiment.

8. Figure 1 shows an immunoblot of cGMP-agarose experiments with protein lysate from platelets of WT and IRAG-KO mice which do not express the IRAG protein at all [4]. In the WT sample, $InsP_3R$-I, IRAG, cGKIβ, and cGKIα are bound to cGMP-agarose. Since in the Ethanolamine-agarose experiment none of the proteins was detected, the binding of the proteins to cGMP-agarose was specific. Only cGKIα and cGKIβ were detected in the IRAG-KO sample, because $InsP_3R$-I cannot directly interact with the kinases.

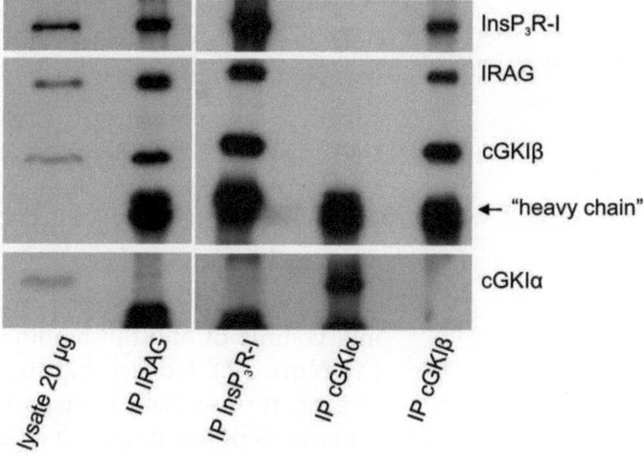

Fig. 2 Immunoblots of co-immunoprecipitation experiments with murine platelets using different antibodies: Protein A-sepharose beads were preincubated for 2 h with specific primary antibodies against IRAG, $InsP_3R$-I, cGKIα, or cGKIβ. Then platelet lysate (250 μg per IP experiment) was added and the samples were incubated overnight. Afterwards, the bound proteins were eluted with SDS sample buffer and separated by SDS-PAGE (11.5 % separating gel). $InsP_3R$-I, IRAG, and cGKIα/β were detected in immunoblot with the same antibodies used for IP. The heavy chain of the antibodies—which are co-eluted with the bound proteins—is detected by the anti-cGKIα/β antibodies

3.3 Co-immunoprecipitation

The interactions between the cGMP-dependent protein kinases (cGK) and their substrate proteins were frequently demonstrated by co-immunoprecipitation (Co-IP) experiments [9, 22, 23]. Furthermore, IRAG-phosphorylation in human platelets was analyzed by autoradiography after ^{33}P in vivo phosphorylation and IRAG immunoprecipitation [16]. Figure 2 shows the trimeric macrocomplex (consisting of $InsP_3R$-I, IRAG, and cGKIβ) in platelets from WT mice isolated by Co-IP. cGKIα does not interact with the other proteins and thus is not a component of this signaling complex.

Traditionally, in a Co-IP experiment the precipitating antibody is first added to the lysate and then this complex is bound to the protein A/G resin. Alternatively, the antibody can be prebound to protein A/G-sepharose before addition of the antigen-containing protein sample. Both methods can lead to good results but prebinding of the antibody allows removal of unbound immunoglobulins and other proteins (e.g., serum proteins) which could be contaminants in the antibody sample. In the following, a basic protocol for an IP experiment with prebinding of the antibody is described:

1. Wash sepharose beads for equilibration three times with wash buffer (~500 μl per sample and washing step). Remove buffer completely after the last wash.
2. Incubate sepharose beads for 30 min at 4 °C on the overhead shaker with 3 % BSA solution to block unspecific binding sites on the matrix.
3. Wash three times with wash buffer to remove the residual unbound BSA.
4. Incubate with the respective monoclonal or polyclonal antibody in a volume of ~500 μl binding buffer for at least 2 h at 4 °C (*see* **Note 17**). Concerning the right amount of antibody, it is important—especially in quantitative studies—that the antibody is in excess of the antigen. This can be ascertained by sequential immunoprecipitation of the sample (*see* **Note 18**) [15]. 1 μg antibody can be used as a starting point for immunoprecipitation experiments with 0.5–1 mg cell or tissue extract [24].
5. Alternatively the antibody can be covalently attached to protein A or G which prevents co-elution of the antibody with the immunoprecipitated protein(s) and thus contamination of the eluted fraction. Contamination will be problematic if the complex proteins have molecular weights similar to the antibody heavy or light chains and if the eluted proteins are analyzed by SDS-PAGE. Additionally, a protein A/G resin with covalently bound antibody can be reused in a second experiment (*see* **Note 19**).
6. Wash again three times with wash buffer (500 μl per washing step).
7. Add protein lysate plus protease inhibitors (and protein phosphatase inhibitors if necessary) in a total volume of approximately 500 μl and incubate for at least 2 h or overnight at 4 °C and gentle agitation. If the phosphorylation state of the proteins should be conserved, the incubation time should be as short as possible to prevent dephosphorylation during protein binding.
8. Beads are washed at least three to four times with wash buffer (500 μl per washing step) to remove all unbound proteins (see Washing of cGMP-Agarose).
9. For SDS-PAGE analysis, elute bound proteins with 20 μl 1× SDS sample buffer.
10. If a protein kinase is co-immunoprecipitated with its substrate proteins, the phosphorylation of the substrates by the kinase can be stimulated. Upon immunoprecipitation of smooth muscle cell proteins with specific antibodies against $InsP_3R$-I, IRAG and cGKI, the beads are phosphorylated in phosphorylation buffer in the presence of 3 μM 8-pCPT-cGMP and 0.1 mM [γ-^{32}P]-ATP (2,000 cpm/pmol) for 2 min at 30 °C. Proteins are eluted by SDS sample buffer and separated by SDS/PAGE and immunoblotting. Protein phosphorylation is then analyzed by autoradiography (*see* Subheading 3.4.3) [5].

3.4 SDS-PAGE Followed by Coomassie Blue or Silver Staining or by Immunoblotting

SDS-polyacrylamide gel electrophoresis (PAGE) can be performed by the methods of Laemmli (10–12.5 % polyacrylamide gels, range: 240–35 kDa) [25] and Schägger (Tricine-PAGE, 16.5 %, range: 35–5 kDa) [26].

3.4.1 SDS-PAGE

1. Firstly, the gels for the SDS-PAGE are prepared. Cast the separating gel, overlay it with isopropanol for a flat surface, and wait for 1 h before casting the stacking gel. Do not forget the comb to form the gel pockets.
2. Fill the gel chamber with running buffer. If a Tricine-PAGE is performed a cathode and an anode running buffer must be used.
3. For the determination of the protein sizes load one lane with protein standard and the others with the prepared protein samples.
4. The proteins are separated for 1.5–2 h at 150 V or—if a Tricine-PAGE is run—for 1 h at 30 V and then for 1.5 h at 150 V.

3.4.2 Staining with Coomassie Blue or Silver

If the gel is not immunoblotted, the proteins can be fixed and stained with Coomassie blue or silver to visualize all proteins eluted from the sepharose matrix [27]. The silver stain method is suggested to be 100-fold more sensitive than Coomassie blue [27].

1. For Coomassie staining the gel is incubated with approximately 50 ml Coomassie blue solution for 20 min on the overhead shaker at room temperature. Thereby, the gel is equally stained blue.
2. To make the proteins visible, the gel has subsequently to be destained. It is incubated with destaining solution until the background is relatively colorless. During this process the destaining solution has to be changed several times.
3. For silver staining we used the ProteoSilver™ Plus Silver Stain Kit (Sigma-Aldrich) and performed the staining according to the manufacturers protocol. In brief, the gel is incubated with 100 ml fixing solution for 20 min, washed with 30 % ethanol solution and 200 ml water for each 10 min, and equilibrated with sensitizer solution for 10 min followed by two washes with water. The gel is then incubated with 100 ml of silver solution for 10 min followed by a 1–1.5 min water wash. Subsequently, developer solution is added for 3–12 min according to the required sensitivity of the staining. Finally, the stop solution is added for 5 min and afterwards gels are washed for 15 min with 200 ml of water.
4. Protein complexes purified by co-immunoprecipitation or cGMP-agarose may be analyzed by mass spectrometry. In brief, the Coomassie blue- or silver-stained proteins are cut out from the gel and analyzed by MALDI-TOF-MS (matrix-assisted laser desorption/ionization–time-of-flight–mass spectrometry) or LC-MS (liquid chromatography-mass spectrometry) [5, 17]. The mass spectrometry allows also the identification of protein phosphorylation sites [28, 29].

3.4.3 Immunoblotting

1. The proteins within the gel are transferred onto a nitrocellulose (0.45 μm pore size) or PVDF-membrane (0.45 μm pore size; Millipore) using a semidry electroblotter for 60 min at 0.6 mA per gel (*see* **Note 20**).
2. To verify transfer efficiency, membranes can be reversibly stained with Ponceau S solution (0.2 % Ponceau S in 2 % trichloroacetic acid).
3. To avoid unspecific binding the membrane is incubated with blocking buffer (5 % nonfat dry milk in TTBS) for 1–2 h at room temperature or overnight at 4 °C (*see* **Note 21**).
4. To detect the protein to be analyzed the membrane is first incubated with the respective primary antibody (2–3 h at room temperature or overnight at 4 °C). After a washing step (with TTBS) it is exposed to the enzyme-conjugated secondary antibody (e.g., goat anti-rabbit IgG for a rabbit primary antibody).
5. Chemiluminescence reaction: The antibody-conjugated horseradish peroxidase (HRP) catalyzes the oxidation of luminol in the presence of H_2O_2 and p-Iodophenol as enhancer [28]. The oxidized luminol substrate emits light which can be detected with photographic films. Therefore, the film is placed onto the membrane for a few seconds to 30 min [28].
6. Radiolabeled proteins are visualized by autoradiography after blotting: an erased imaging plate (IP; Fujifilm) is placed onto the membrane and the radioactivity is detected by an image reader. The cGMP-stimulated phosphorylation of cGK substrates is usually determined by autoradiography [5, 16, 23].

4 Notes

1. A precondition for immunoprecipitation is the availability of the antigen for binding to specific antibodies. Extraction with nondenaturing detergents or in the absence of detergent enables binding of antibodies to epitopes that are exposed on native proteins. If the protein complex is unsoluble in nondenaturing detergents or if the epitope is hidden within the protein structure, the proteins must be extracted under denaturing conditions.
2. The gel matrix and the spacer of ethanolamine-agarose can be purchased identical to the other cyclic nucleotide gels offered by Biolog. Therefore, it can be used as a control for unspecific adsorption effects.
3. Thermo Scientific offers several IP kits, inter alia the Classic Immunoprecipitation Kit. This kit contains Protein A/G Plus Agarose, lysis/wash buffer, binding and elution buffers, spin columns, and collection tubes. The spin columns prevent resin loss

and allow therefore more reproducible IP results. The immune complex can be eluted with non-reducing sample buffer or low-pH elution buffer (pH 2.8) which contains primary amine.

4. If there is no interaction detectable, it is helpful to use less stringent conditions (reduced salt and/or non-ionic detergent) or use smaller wash volumes to reduce undesired protein removal [24].
5. The cGMP-agarose affinity purification method and co-immunoprecipitation experiments can be combined. Koller et al. [5] first purified microsomal membrane proteins by cGMP-Agarose, eluted the bound proteins with 20 mM cGMP-Na, and then performed immunoprecipitations with different antibodies (e.g., against $InsP_3R$-I, IRAG, or cGKI).
6. Protein A is less expensive than protein G, but protein G binds a greater variety of antibody classes [12, 15]. When the IP antibody is directly bound to an activated matrix, even chicken IgY can be used [13].
7. Polyclonal antibodies, which are mixtures of different antibodies, can be produced against whole proteins, protein fragments, or peptides. Antibodies against whole proteins frequently recognize several epitopes which increases the chance for interaction. But polyvalent antibodies can cross-react with other proteins resulting in false protein binding and higher background. These problems are less by use of anti-peptide antibodies. Unpurified antisera can contain antibodies against other antigens, but affinity purification can cause loss of affinity. Monoclonal antibodies, which are produced only by clones of one B-cell, exhibit the highest specificity and titer, but low-affinity ones are possibly not capable of preserving the interaction with the antigen during the several washing steps [15].
8. Some polyclonal or monoclonal antibodies bind very weakly to protein A or protein G. These antibodies can be coupled to the sepharose matrix by using an intermediate rabbit antibody against the respective immunoglobulins [15].
9. If cells/tissues were lysed with 2 % Lubrol buffer use 0.4 % Lubrol buffer for washing. The Triton X-100 wash buffer is appropriate after lysis with 1 % Triton X-100.
10. Through the dilution with nondenaturing lysis buffer, the SDS concentration of the cell/tissue lysate is reduced to 0.1 % which is the recommended concentration for a Co-IP experiment.
11. The separation of proteins and peptides in the range of 5–20 kDa can be achieved by the Tris–tricine method or by a system using increased buffer concentrations. The method presented here is the modified Tris–tricine method with a 10 % spacer gel between the stacking and resolving gel which allows the separation of peptides below 5 kDa [18].

12. The homogenization methods used for proteomics can be divided into the following major categories: mechanical, ultrasonic, pressure, freeze-thaw, and osmotic or detergent lysis [29, 30]. Mechanical homogenization is widely used for various tissues and cells. A problem could be the loss of activity, particularly when the investigated material is heat insensitive and the cooling during processing is ineffective [29]. Ultrasonic homogenization is mainly used to break up small pieces of soft tissues (brain, blood, liver). Pressure homogenization is appropriate for eukaryotic cells as well as for microorganisms in suspension. Freeze-thaw homogenization is effective towards a multitude of cells and could be additionally used after mechanical or ultrasonic procedures [29].
13. Only soluble proteins can be used for cGMP-agarose and co-immunoprecipitation experiments to avoid that protein aggregates are bound to the gel matrix.
14. A preclearing step is not essential but reduces unspecific binding of proteins to the respective matrix. Only the proteins that are not bound to the antibody-free sepharose are then used for the following immunoprecipitation experiment.
15. The amount of protein applied to the matrix can be critical: If too little protein is used, only a small amount of protein is bound and the interaction can hardly be detected. If too much protein is used, the distinct ligands compete for the binding sites and therefore weak interactions cannot be detected [31].
16. Non-immunoprecipitated samples should not be stored in SDS sample buffer at room temperature. They should be first heated at 95 °C to inactivate proteases, because endogenous proteases are very active in SDS sample buffer [18].
17. To determine unspecific binding of sample proteins to the sepharose matrix, incubate one tube (control) directly with tissue/cell lysate (without addition/binding of antibody).
18. In sequential immunoprecipitation, the supernatant of the first IP is used for a second experiment. If the second IP yields only a small amount (<10 %) of the antigen isolated in the first IP, the antibody titer is appropriate [15].
19. Dimethyl pimelimidate (20 mM) can be used for cross-linking an antibody with protein A [32, 33]. The beads are incubated with the antibody for 2 h at 4 °C. Afterwards, the protein A sepharose is washed with 0.1 M borate buffer and once with 0.2 M triethylamine (pH 8.2) and subsequently incubated with 20 mM dimethyl pimelimidate dihydrochloride (Pierce) in 0.2 M triethylamine (pH 8.2) for 45 min. The reaction is stopped by addition of ethanolamine and the beads are washed in 0.1 M borate buffer and finally resuspended in binding/lysis buffer [33]. DSS is also used as a cross-linker for covalent attachment of antibodies to protein A or G [12].

20. Nitrocellulose and PVDF membranes are both commonly used to bind transferred proteins. Nitrocellulose is less favorable since the proteins are not covalently bound. PVDF is more advantageous because of its high binding capacity, physical strength, and chemical stability [34]. The advantages of semi-dry blotting versus wet transfer are that several gels can be blotted simultaneously and that simple carbon blocks can be used as electrodes [34]. Tank blotting should be used for proteins that require long blotting times for efficient transfer [20].
21. If primary antibodies which show a high unspecific binding are used, blocking overnight is recommended.

Acknowledgments

This work was supported by the German Research Council (DFG) and the collaborative research center SFB699.

References

1. Hofmann F, Feil R, Kleppisch T et al (2006) Function of cGMP-dependent protein kinases as revealed by gene deletion. Physiol Rev 86:1–23
2. Casteel DE, Smith-Nguyen EV, Sankaran B et al (2010) A crystal structure of the cyclic GMP-dependent protein kinase I{beta} dimerization/docking domain reveals molecular details of isoform-specific anchoring. J Biol Chem 285:32684–32688
3. Wall ME, Francis SH, Corbin JD et al (2003) Mechanisms associated with cGMP binding and activation of cGMP-dependent protein kinase. Proc Natl Acad Sci USA 100:2380–2385
4. Desch M, Sigl K, Hieke B et al (2010) IRAG determines nitric oxide- and atrial natriuretic peptide-mediated smooth muscle relaxation. Cardiovasc Res 86:496–505
5. Koller A, Schlossmann J, Ashman K et al (2003) Association of phospholamban with a cGMP kinase signaling complex. Biochem Biophys Res Commun 300:155–160
6. Given AM, Ogut O, Brozovich FV (2007) MYPT1 mutants demonstrate the importance of aa 888-928 for the interaction with PKGIalpha. Am J Physiol Cell Physiol 292:C432–C439
7. Surks HK, Mochizuki N, Kasai Y et al (1999) Regulation of myosin phosphatase by a specific interaction with cGMP-dependent protein kinase Ialpha. Science 286:1583–1587
8. Ammendola A, Geiselhoringer A, Hofmann F et al (2001) Molecular determinants of the interaction between the inositol 1,4,5-trisphosphate receptor-associated cGMP kinase substrate (IRAG) and cGMP kinase Ibeta. J Biol Chem 276:24153–24159
9. Schlossmann J, Ammendola A, Ashman K et al (2000) Regulation of intracellular calcium by a signalling complex of IRAG, IP3 receptor and cGMP kinase Ibeta. Nature 404:197–201
10. Berggard T, Linse S, James P (2007) Methods for the detection and analysis of protein–protein interactions. Proteomics 7:2833–2842
11. Dwane S, Kiely PA (2011) Tools used to study how protein complexes are assembled in signaling cascades. Bioeng Bugs 2:247–259
12. Kaboord B, Perr M (2008) Isolation of proteins and protein complexes by immunoprecipitation. Methods Mol Biol 424:349–364
13. Qoronfleh MW, Ren L, Emery D et al (2003) Use of immunomatrix methods to improve protein–protein interaction detection. J Biomed Biotechnol 2003:291–298
14. Geiselhoringer A, Werner M, Sigl K et al (2004) IRAG is essential for relaxation of receptor-triggered smooth muscle contraction by cGMP kinase. EMBO J 23:4222–4231
15. Bonifacino JS, Dell'Angelica EC, Springer TA (2001) Immunoprecipitation. Curr Protoc Mol Biol Chapter 10, Unit 10 16
16. Antl M, von Bruhl ML, Eiglsperger C et al (2007) IRAG mediates NO/cGMP-dependent inhibition of platelet aggregation and thrombus formation. Blood 109:552–559

17. Margarucci L, Roest M, Preisinger C et al (2011) Collagen stimulation of platelets induces a rapid spatial response of cAMP and cGMP signaling scaffolds. Mol Biosyst 7: 2311–2319
18. Gallagher SR (2006) One-dimensional SDS gel electrophoresis of proteins. Curr Protoc Mol Biol Chapter 10:Unit 10.2A
19. Kurien BT, Scofield RH (2009) Nonelectrophoretic bidirectional transfer of a single SDS-PAGE gel with multiple antigens to obtain 12 immunoblots. Methods Mol Biol 536:55–65
20. Gallagher S, Winston SE, Fuller SA et al (2008) Immunoblotting and immunodetection. In: Frederick M. Ausubel et al (eds) Current protocols in molecular biology. Chapter 10, Unit 10 18
21. Bradford MM (1976) A rapid and sensitive method for the quantitation of microgram quantities of protein utilizing the principle of protein-dye binding. Anal Biochem 72: 248–254
22. Tang KM, Wang GR, Lu P et al (2003) Regulator of G-protein signaling-2 mediates vascular smooth muscle relaxation and blood pressure. Nat Med 9:1506–1512
23. Wilson LS, Elbatarny HS, Crawley SW et al (2008) Compartmentation and compartment-specific regulation of PDE5 by protein kinase G allows selective cGMP-mediated regulation of platelet functions. Proc Natl Acad Sci USA 105:13650–13655
24. Elion EA (2006) Detection of protein–protein interactions by coprecipitation. Curr Protoc Mol Biol Chapter 20, Unit20 25
25. Laemmli UK (1970) Cleavage of structural proteins during the assembly of the head of bacteriophage T4. Nature 227:680–685
26. Schagger H, von Jagow G (1987) Tricine-sodium dodecyl sulfate-polyacrylamide gel electrophoresis for the separation of proteins in the range from 1 to 100 kDa. Anal Biochem 166(2):368–379
27. Gauci VJ, Wright EP, Coorssen JR (2011) Quantitative proteomics: assessing the spectrum of in-gel protein detection methods. J Chem Biol 4(1):3–29
28. Kricka LJ (1991) Chemiluminescent and bioluminescent techniques. Clin Chem 37(9): 1472–1481
29. Bodzon-Kulakowska A, Bierczynska-Krzysik A, Dylag T et al (2007) Methods for samples preparation in proteomic research. J Chromatogr B Analyt Technol Biomed Life Sci 849:1–31
30. Chaiyarit S, Thongboonkerd V (2009) Comparative analyses of cell disruption methods for mitochondrial isolation in high-throughput proteomics study. Anal Biochem 394:249–258
31. Phizicky EM, Fields S (1995) Protein–protein interactions: methods for detection and analysis. Microbiol Rev 59:94–123
32. Schneider C, Newman RA, Sutherland DR et al (1982) A one-step purification of membrane proteins using a high efficiency immunomatrix. J Biol Chem 257:10766–10769
33. Dickson C (2008) Protein techniques: immunoprecipitation, in vitro kinase assays, and Western blotting. Methods Mol Biol 461:735–744
34. Kurien BT, Scofield RH (2006) Western blotting. Methods 38(4):283–293

Chapter 10

Approaches for Monitoring PKG1α Oxidative Activation

Joseph Robert Burgoyne and Philip Eaton

Abstract

cGMP-dependent protein kinase, also known as protein kinase G (PKG), is activated independently of cGMP by a novel thiol-reactive mechanism involving the formation of an intermolecular disulfide. This oxidative modification within PKG is generally not detected by conventional Western immunoblot analysis due to the experimental conditions used. Here, we describe the proteomic approach that lead to PKG being identified as a kinase susceptible to oxidant-dependent disulfide dimer formation, these methods being applicable for the identification of other disulfide bound protein complexes. In addition a nonreducing Western immunoblot method for routinely measuring PKG oxidation in complex protein mixtures generated from cell lysates or tissue homogenates is also described.

Key words Protein kinase G, Oxidant, Hydrogen peroxide, EDHF, SDS-PAGE, Disulfide

1 Introduction

The regulation of blood pressure is critical for maintaining health with hypertension increasing the risk of heart attacks, aortic aneurysms, peripheral artery disease, stroke, and kidney failure [1]. At a cellular level the contractile machinery of the smooth muscle cells that controls blood pressure by dilating vessels is regulated by three major pathways. Firstly, prostaglandins (such as PGE_2) formed by cyclooxygenase in endothelial cells diffuse into smooth muscle cells where they stimulate cAMP formation [2]. Elevations in cAMP activate the cAMP-dependent protein kinase, also known as protein kinase A (PKA), leading to downstream phospho-dependent smooth muscle relaxation. Another process that is considered a primary regulator of blood pressure is the formation of nitric oxide in endothelial cells by nitric oxide synthase stimulated by the ensuing increase in Ca^{2+} generated by sheer stress or following the binding of vasodilatory ligands (such as bradykinin or acetylcholine) to specific cell surface receptors [3]. Endothelial nitric oxide diffuses into smooth muscle cells where it binds to and stimulates the

Thomas Krieg and Robert Lukowski (eds.), *Guanylate Cyclase and Cyclic GMP: Methods and Protocols*,
Methods in Molecular Biology, vol. 1020, DOI 10.1007/978-1-62703-459-3_10, © Springer Science+Business Media, LLC 2013

activity of the heme-bound guanylate cyclase, which in turns generates cyclic guanosine monophosphate (cGMP). The cyclic nucleotide cGMP is a natural ligand for PKG, which upon binding stimulates kinase activity. Like PKA, PKG phosphorylates numerous targets to induce vasodilation by decreasing intracellular Ca^{2+} and myofilament Ca^{2+} sensitivity to alleviate Ca^{2+}-dependent vasoconstriction in smooth muscle cells [4]. The third known pathway for mediating vessel dilation involves the formation of a substance termed endothelial derived hyperpolarizing factor (EDHF), which as the name suggests is generated by endothelial cells and diffuses into smooth muscle cells where it induces membrane hyperpolarization, inhibiting the cellular import of Ca^{2+} [5]. The identity of EDHF remains controversial, with evidence in support of several different candidates that include K^{+}, epoxyeicosatrienoic acids (EETs), and H_2O_2. Although the true identity of EDHF remains elusive it is at least in part mediated by H_2O_2 as the enzyme catalase [6], which decomposes this oxidant, is able to markedly attenuate the ability of EDHF to mediate vessel dilation. In addition it is reported that PKG1α transduces the EDHF response by undergoing direct oxidation [7]. The formation of disulfide bond within the N-terminus of PKG1α directly activates this kinase independently of cGMP by considerably increasing its affinity for substrate [8]. This phenomenon explains how the EDHF response is likely mediated and in addition is potentially an important functional process that occurs under conditions where the cellular reducing/oxidant (redox) capacity is altered.

The cellular proteome contains many redox-sensitive proteins that change their function when they become oxidized, allowing them to transduce changes in cellular oxidants to a functional response. Protein thiols can become oxidatively modified in a number of ways, one of which involves the formation of intermolecular disulfides. This structural modification can directly alter the activity or interactions of the oxidatively modified protein, providing a mode of redox regulation of function. In addition to PKG1α many other targets that form intermolecular disulfides during oxidative stress have been identified including PKA regulatory subunit 1 (PKARI) [9], NEMO [10], the C-terminal catalytic domain of receptor protein–tyrosine phosphatase alpha [11], KEAP1, and ATP synthase [12, 13].

Here we describe a diagonal SDS-polyacrylamide gel electrophoresis (SDS-PAGE) technique for detecting proteins that form intermolecular disulfides. This technique relies on the difference in migration of proteins on nonreducing SDS-PAGE compared with under reducing conditions where the disulfide holding the complex together is lost. The principal being that proteins bound together by a disulfide will run at a higher, combined molecular weight when separated under nonreducing conditions but when then run in a second dimension on a reducing gel will resolve at their natural

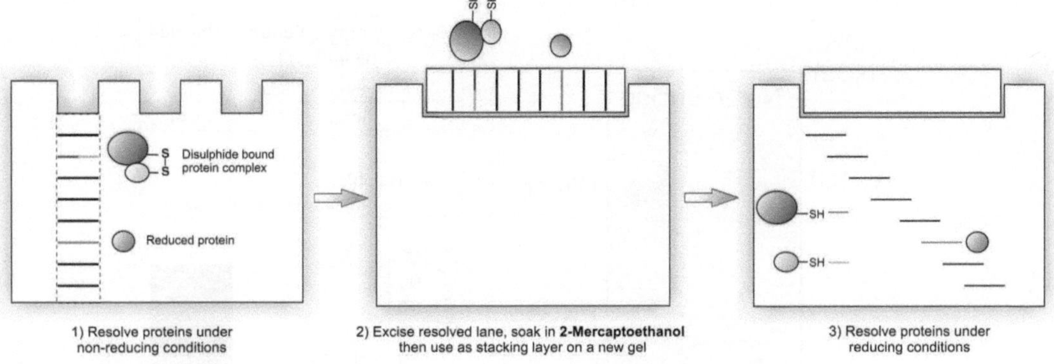

Fig. 1 Principle behind diagonal gel SDS-PAGE for detecting disulfide bound complexes

individual molecular weights [14]. Therefore proteins that form a disulfide complex will migrate off the diagonal plane of the gel after being resolved in the second dimension under reducing conditions (*see* Fig. 1), making them easy to identify following gel staining using mass spectrometry. In addition to describing this technique we also provide a detailed protocol for monitoring PKG1α intermolecular disulfide formation using nonreducing SDS-PAGE. The principal behind this technique is that by lysing cells or tissue into an alkylating buffer the oxidation status of PKG1α can be preserved by chemically blocking reduced cysteine thiols preventing artificial air oxidation. The oxidation status of PKG1α can then be determined using nonreducing SDS-polyacrylamide gel electrophoresis to prevent loss of protein oxidation that would occur under reducing conditions. The reduced form of PKG1α resolves at 75 kDa while the oxidized disulfide bound dimer resolves at twice the weight at 150 kDa (*see* Fig. 2).

2 Materials

2.1 SDS-Polyacrylamide Gel Electrophoresis

1. 12.5 % resolving gel buffer: 4.2 ml of 30 % acrylamide (37.5:1 acrylamide to bisacrylamide), 3.8 ml 1 M Tris–HCl buffer pH 8.8 (*see* **Note 1**), 1.85 ml distilled water, 100 μl of 10 % SDS, and 6 μl *N,N,N′,N′*-tetramethylethylenediamine (TEMED). Add 10 % ammonium persulfate (APS) just before use (*see* **Note 2**).
2. 8 % resolving gel buffer: 2.67 of 30 % acrylamide (37.5:1 acrylamide to bisacrylamide), 3.8 ml 1 M Tris–HCl buffer pH 8.8, 3.38 ml distilled water, 100 μl SDS, and 6 μl TEMED. Add 10 % ammonium persulfate (APS) just before use (*see* **Note 2**).
3. Stacking gel buffer: 0.9 ml of 30 % acrylamide (37.5:1 acrylamide to bisacrylamide), 0.72 ml 1 M Tris–HCl buffer pH 6.8 (*see* **Note 1**), 4.2 ml distilled water, 60 μl of 10 % SDS, and 10 μl TEMED. Add 10 % APS just before use (*see* **Note 2**).

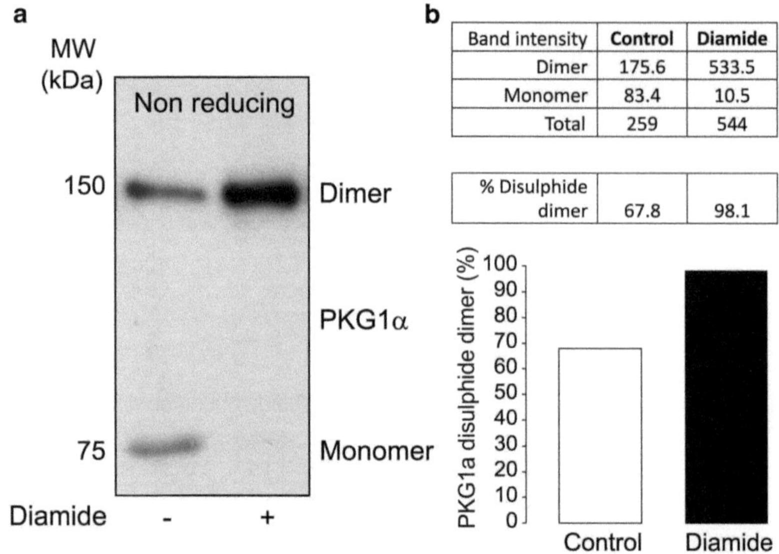

Band intensity	Control	Diamide
Dimer	175.6	533.5
Monomer	83.4	10.5
Total	259	544

Fig. 2 (**a**) Immunoblot for PKG1α generated from isolated cardiac myocytes treated with or without the disulfide inducing compound diamide. These samples where run on a nonreducing SDS-PAGE gel allowing detection of the disulfide dimer. The monomeric form of PKG1α is seen as a 75 kDa band whereas the oxidized disulfide bound dimer migrates at 150 kDa. (**b**) Quantification for PKG1α disulfide dimerization in control- and diamide-treated samples

4. 2× Nonreducing alkylating sample buffer (*see* **Note 3**): 100 mM Tris–HCl buffer pH 6.8, 4 % SDS (volume/volume), 20 % glycerol (volume/volume), 0.01 % bromophenol blue (weight/volume), and 100 mM maleimide (Sigma).
5. 1× reducing sample buffer: 50 mM Tris–HCl buffer pH 6.8, 2 % SDS (volume/volume), 10 % glycerol (volume/volume), 0.005 % bromophenol blue (weight/volume), and 5 % of 14.3 M 2-mercaptoethanol.
6. 10× Tris-Base running buffer: 144 g of glycine and 30 g of Tris-Base in 1 l of deionized water (*see* **Note 4**).
7. 1× Tris-Base + 1 % SDS running buffer: 100 ml of 10× Tris-Base running buffer and 10 ml of 10 % SDS in 1 l of deionized water.
8. Krebs buffer: 6.93 g of NaCl, 0.35 g of KCl, 0.16 g of KH_2PO_4, 2.1 g $NaHCO_3$, 0.30 g $MgSO_4$ and 2 g of glucose. Make up to 800 ml with deionized H_2O_2 then bubble with 5 % CO_2/95 % O_2 for 10–15 min before adding 0.21 g of $CaCl_2$ (*see* **Note 5**). Finally make volume up to 1 l with deionized H_2O_2 then filter through a 5 μM filter membrane into a conical flask.
9. Tissue homogenization buffer: 100 mM Tris–HCl pH 7.4, protease inhibitors (Roche, protease Inhibitors Complete EDTA free), and 100 mM maleimide.

10. Colloidal Coomassie Blue: 100 g of ammonium sulfate, 200 ml ethanol, 30 ml *ortho*-phosphoric acid in 980 ml of H_2O_2, and then add 20 ml of 5 % Coomassie Blue G-250 (weight/volume). Directly before use shake Colloidal Coomassie to get a homogeneous solution then take 40 ml and mix with 10 ml of methanol.

2.2 Immunoblotting

1. Polyvinylidene difluoride (PVDF) membrane: Cut PVDF membrane into 9 × 7 cm pieces.
2. Transfer buffer: 10 ml of 10 × Tris-Base running buffer stock (see above), 100 μl of 10 % SDS, and 20 ml of 100 % methanol, made up to a final volume of 100 ml with deionized water.
3. Phosphate-buffered saline and 1 % Tween (PBS-T): 10 mM phosphate buffer pH 7.4, 2.7 mM potassium chloride, 137 mM sodium chloride, and 1 % Tween-20.
4. Blocking solution: 5 % milk powder in PBS-T.
5. PKG1α antibody: diluted 1:1,000 (Santa Cruz, E-15) into blocking solution (*see* **Note 6**).
6. Secondary anti-goat antibody: diluted 1:1,000 (Dako) into blocking solution.
7. Enhanced chemiluminescence (ECL): Mix equal volume of each solution together just before use (Pierce).
8. ECL Film: 18 × 24 cm Amersham Hyperfilm ECL.

3 Methods

3.1 Tissue and Cell Lysate Preparation

3.1.1 Preparation of Denatured Lysate from Cultured Cells

1. For experiments using cultured cells rinse in PBS (2 ml/well for a 6-well plate) before lysing directly into an appropriate volume of 1× nonreducing alkylating sample buffer (*see* **Note 7**). The number of cells and the size of the dish should be varied depending on the cell type and experimental conditions so that there is enough protein to be visualized using diagonal gel electrophoresis or for the detection of PKG1α.

3.1.2 Preparation of Denatured Lysate from Tissue/Organs

1. Excised tissue/organs should be rinsed briefly in Krebs solution to remove excess blood before being rapidly snap frozen in liquid nitrogen to be stored until needed for homogenization. If tissues are ex vivo perfused then they should be immediately snap frozen at the end of the protocol.
2. Tissue/organs should be rapidly homogenized using a Polytron handheld homogenizer into tissue homogenization buffer to yield a 10 % (weight/volume) homogenate. Once homogenized place the required amount of protein mixture into an

equal volume of 2× nonreducing alkylating sample buffer with the remaining homogenate being snap frozen and then stored either in liquid nitrogen or a −80 °C freezer.

3.2 Diagonal Gel Electrophoresis for the Detection of Disulfide Bound Protein Complexes

For simplicity this protocol is described for comparing two samples, one a control tissue and the other from tissue under oxidative stress in which interprotein disulfides may have formed. However the same method can also be used to compare multiple samples but additional gels then have to be cast.

1. Prepare solution for two 12.5 % resolving gels (recipe provided in materials section is enough for one gel) then add 180 μl of APS and cast gels within 15 × 10.6 × 5.3 cm gel cassettes (Bio-Rad, Criterion gel cassettes). Allow sufficient space for the stack and carefully overlay the resolving gel with a layer of isobutanol/H_2O with enough to cover the gel surface (*see* **Note 8**).
2. Prepare two volumes of stacking gel buffer and once the resolving gel has set remove the isobutanol by carefully rinsing the gel surface with water. Take the stacking gel buffer and add 100 μl of APS then pour onto the surface of the resolving gel to fill the remaining space. Quickly place an 18-well comb into the top of one cassette and an IPG + 1-well comb (this is a comb that generates a large lane designed for an immobilized pH gradient (IPG) gel strip and an extra small lane for loading a protein standard) or equivalent into the other.
3. Heat samples at 95 °C for 5 min then centrifuge at 25,000 × *g* for 10 min. During this time prepare 1× running buffer + 1%SDS, and then once the stack has set assemble gel running apparatus using only the 18-well gel. Carefully remove the comb from the 18-well gel, place into gel apparatus, and submerge in running buffer, then load samples into two individual wells separated by a lane containing 5 μl of protein standard (Bio-Rad).
4. Resolve proteins by running gel apparatus at 200 V until the dye front has just run off the gel (approximately 40 min). Remove gel cassette and prise open to reveal the gel, then place into an open container and submerge in 1× reducing sample buffer.
5. After being submerged for 5 min place the gel onto the surface of a clean glass plate (*see* **Note 9**). Remove the stacking layer and using a scalpel carefully cut out each lane in which the samples have been resolved using the protein standard as a guide (*see* **Note 10**). Once the gel strips have been cut-out, to prevent them from drying out they can then be submerged once again in 1× reducing sample buffer while being careful to remember their orientation.
6. Take the second gel that was cast and remove the IPG + 1-well comb, place into gel apparatus, and submerge in running buffer.

Place 5 μl of protein standard into the single small left hand well. Then take one of the gel strips and carefully place into the large IPG-well, gently pressing down using fingers to edge it into the well and so that it is aligned against one side of the gel. Once it is firmly in place and flat against the resolving layer take the second strip and do the same so the well contains both strips of gel side by side (*see* **Note 11**).

7. Place a layer of 1× reducing sample buffer over the top surface of each strip then run the gel apparatus at 200 V until the dye front has just run off the gel.
8. Once the samples have resolved remove the cassette, prise open, remove the gel, and soak in an open container for 5 min in deionized water.
9. Remove the water then submerge the gel in 25 ml of Colloidal Coomassie Blue and seal the container with a sheet of Clingfilm. Place the container onto a platform shaker overnight at room temperature at a speed that allows enough agitation of the solution so that gel is able to move but remains submerged.
10. The following day pour away the Colloidal Coomassie Blue and rinse the gel in deionized water. Then wash the gel on a platform shaker at room temperature, replacing the water at regular intervals until the gel has destained to a suitable extent (*see* **Note 12**).
11. The gel can then be visualized using a light box to identify any novel differences in the presence or intensity of bands between the two samples that have resolved off the diagonal of the gel. These bands can be removed using a sterile scalpel and sent for mass spectrometry analysis to identify the proteins found in each.

3.3 Nonreducing Gel Electrophoresis for Detecting PKG1α Oxidation

Both cultured cells and animal tissue that express PKG1α can be assessed for its oxidation. The PKG1α isoform is known to be replete in smooth muscle and HEK293 cells and also in cardiac, neuronal, liver, and lung tissue. The oxidation of PKG1α can be analyzed in tissue or cells that have been directly treated in vitro or from tissue of animals that have undergone an intervention. For example, cultured cells can be treated with H_2O_2 (50–1,000 μM for 10 min) or diamide (50–1,000 μM for 10 min) to induce PKG1α disulfide dimerization. Cells and tissue can then be processed as described in Subheading 3.1 and analyzed as described below.

1. Prepare solution for an 8 % resolving gel (recipe in materials section is enough for two gels) then add 90 μl of APS and cast gel by assembling a 0.75 mm base plate (8.6×6.7 cm) with a cover plate using a casting frame and casting stand. Allow sufficient space for the stack and carefully overlay the resolving gel with a layer of isobutanol/H_2O with enough to cover the gel surface (*see* **Note 8**).

2. Prepare stacking gel buffer and once the resolving gel has set remove the isobutanol by carefully rinsing the gel surface with water. Take the stacking gel buffer and add 50 μl of APS then pour onto the surface of the resolving gel to fill the remaining space. Quickly place a 15-well comb into the stacking buffer.
3. Heat samples at 95 °C for 5 min then centrifuge at 25,000 × *g* for 10 min. During this time prepare 1× running buffer + 1 % SDS and once the stack has set assemble gel running apparatus. Carefully remove the comb from the 15-well gel, place into gel apparatus, and submerge in running buffer, then load 5 μl of protein standard (Bio-Rad) into the 1st well followed by each of the samples.
4. Resolve proteins by running gel apparatus at 180 V until the dye front has just run off the gel (approximately 50 min). During this time prepare membrane and blotting paper for Western blotting.
5. Soak PVDF membrane in 100 % methanol for 5 min then submerge in transfer buffer and place onto a flatbed shaker until needed for transfer (*see* **Note 13**).
6. Just before the gel has finished resolving prepare semi-dry blotter (Bio-Rad) by removing the surface plate, then place six sheets of blotting paper (cut to dimensions that are slightly larger than the PVDF membrane) onto each of the two surface plates. Thoroughly wet each pile of blotting paper and remove air bubbles between sheets by carefully flattening each pile using a roller.
7. Once the gel has resolved remove the gel cassette and prise open to reveal the gel. Carefully place a piece of wetted PVDF onto the surface of the gel and prise the gel from the glass plate using a gel releaser (Bio-Rad). Take the gel and place onto the surface of the wetted blotting paper on the inner plate of the semi-dry blotter, so that gel is at the bottom and the membrane is at the top.
8. Place the blotting paper from the outer plate onto the surface of the membrane on the inner plate and carefully flatten using a roller to remove any residual bubbles. Dampen the surface of the outer plate with transfer buffer then fix into place on top of the inner plate to seal the apparatus. Start the transfer by running the apparatus at a constant 10 V, but delimiting the current to a maximum of 0.25 mA per membrane, for 35 min.
9. Once the transfer is complete submerge the membrane in 20 ml of blocking solution and leave agitating overnight at 4 °C on a flatbed shaker.
10. The following day remove the blocking solution and add 10 ml of fresh blocking solution containing PKG1α antibody. Incubate for 1 h at room temperature on a flatbed shaker at room temperature (*see* **Note 14**).

11. After 1 h remove the antibody solution and rinse the membrane three times at regular intervals with 20 ml of PBS-T over an hour long period. Then place the membrane in 10 ml secondary anti-goat antibody for an hour then wash again for another hour as before.
12. Prepare the membrane for development by incubating in 5 ml of ECL solution for 5 min and then attach to a film cassette within a sealed pouch (polished binder pockets). Take the film cassette into a dark room and develop using film and a developer.
13. The extent of PKG1α oxidation can be determined in each sample by measuring the relative intensity of monomeric (75 kDa) and disulfide bound PKG1α (150 kDa) using a gel scanner and appropriate analysis software. The % disulfide dimer = (intensity of 150 kDa band/the sum of the intensity of both the 75 and 150 kDa band) × 100.

4 Notes

1. The 1 M Tris–HCl buffer pH 8.8 and pH 6.8 used for making the resolving and stacking gel, respectively, is generated by mixing 121.14 g of Tris-Base with 1 l of deionized water. However it is advisable to dissolve the Tris-Base in less than 1 l of deionized water first then pH the solution using concentrated HCl. Once the correct pH is reached then make up the remaining liter with deionized water. In addition it is also advisable to store the 1 M Tris–HCl buffers at 4 °C to reduce the risk of contaminant growth.
2. We find that 10 % solutions of APS can be stored in the fridge for several weeks without any loss in its ability to polymerize gels.
3. Maleimide is an essential component of the sample and tissue homogenization buffer as it binds to free (reduced) cysteine thiols preventing them from becoming artificially air oxidized during sample preparation.
4. 10× Tris-Base can be stored for long periods (months) at room temperature in a sealed bottle.
5. It is important to bubble the Krebs with 5 % CO_2/95 % O_2 as it reduces the pH to 7.4 and prevents precipitation of calcium phosphate when subsequently added.
6. There are several commercial antibodies that recognize PKG1α; however care needs to be taken as some do not recognize the disulfide dimer as the epitope of the antibody is within the region in which the disulfide forms. Therefore it is highly recommended that the PKG1α antibody from Santa Cruz (E-15) is used when immunoblotting for the oxidation status of this kinase.

7. It is important to rinse cultured cells with PBS before lysing into sample buffer to remove any residual fetal bovine serum from the culture medium as this contains antibodies that will disrupt the immunoblotting process.
8. A solution of isobutanol mixed with water for overlying gels can be stored at room temperature for several months. It is advisable to mix this solution before using it to overlay blots as the isobutanol will separate out from the water giving two separate layers of solution.
9. When the gel is placed onto a clean glass plate keep it wet by covering in 1× reducing sample buffer to stop it from drying out.
10. When cutting out each lane it may help to use a second clean glass plate, which can be placed on top of the gel aligned against each lane to aid cutting out the lanes by acting as a guide.
11. It is recommended that each gel strip is trimmed slightly at the top (just above 250 kDa mark) to help aid in placing the two gel strips into the IPG lane of a new gel. In addition it is advisable to place both gel strips in the same orientation, either from low to high molecular weight or vice versa to help with post analysis after staining with Colloidal Coomassie Blue.
12. Gels should be destained until there is a good contrast between bands and background staining. Over-destaining can lead to loss of bands; however if this occurs gels can be restrained again overnight.
13. Once the PVDF has been soaked in transfer buffer it is imperative that it remains damp and does not dry out at any later stage during the protocol. Therefore when placed onto a flatbed shaker it is important to verify that there is sufficient agitation and buffer to keep the entire membrane submerged.
14. The PKG1α antibody from Santa Cruz (E-15) can be reused up to three times and should be stored at −20 °C in 5 % milk between each use.

Acknowledgments

We would like to acknowledge support from the Medical Research Council, the British Heart Foundation, the Leducq Foundation, and the Department of Health via the NIHR cBRC award to Guy's and St Thomas' NHS Foundation Trust. Also JR Burgoyne is supported by a Sir Henry Wellcome postdoctoral fellowship from The Wellcome Trust (sponsor reference 085483/Z/08/Z).

References

1. Germino FW (2009) The management and treatment of hypertension. Clin Cornerstone 9(Suppl 3):S27–S33
2. Tanaka Y, Yamaki F, Koike K, Toro L (2004) New insights into the intracellular mechanisms by which PGI2 analogues elicit vascular relaxation: cyclic AMP-independent, Gs-protein mediated-activation of MaxiK channel. Curr Med Chem Cardiovasc Hematol Agents 2(3):257–265
3. Yetik-Anacak G, Catravas JD (2006) Nitric oxide and the endothelium: history and impact on cardiovascular disease. Vascul Pharmacol 45(5):268–276. doi:10.1016/j.vph.2006.08.002
4. Morgado M, Cairrao E, Santos-Silva AJ, Verde I (2012) Cyclic nucleotide-dependent relaxation pathways in vascular smooth muscle. Cell Mol Life Sci 69(2):247–266. doi:10.1007/s00018-011-0815-2
5. Garland CJ, Hiley CR, Dora KA (2011) EDHF: spreading the influence of the endothelium. Br J Pharmacol 164(3):839–852. doi:10.1111/j.1476-5381.2010.01148.x
6. Shimokawa H (2010) Hydrogen peroxide as an endothelium-derived hyperpolarizing factor. Pflugers Arch 459(6):915–922. doi:10.1007/s00424-010-0790-8
7. Prysyazhna O, Rudyk O, Eaton P (2012) Single atom substitution in mouse protein kinase G eliminates oxidant sensing to cause hypertension. Nat Med 18(2):286–290. doi:10.1038/nm.2603
8. Burgoyne JR, Madhani M, Cuello F, Charles RL, Brennan JP, Schroder E, Browning DD, Eaton P (2007) Cysteine redox sensor in PKGIα enables oxidant-induced activation. Science 317(5843):1393–1397. doi:10.1126/science.1144318
9. Brennan JP, Bardswell SC, Burgoyne JR, Fuller W, Schroder E, Wait R, Begum S, Kentish JC, Eaton P (2006) Oxidant-induced activation of type I protein kinase A is mediated by RI subunit interprotein disulfide bond formation. J Biol Chem 281(31):21827–21836. doi:10.1074/jbc.M603952200
10. Herscovitch M, Comb W, Ennis T, Coleman K, Yong S, Armstead B, Kalaitzidis D, Chandani S, Gilmore TD (2008) Intermolecular disulfide bond formation in the NEMO dimer requires Cys54 and Cys347. Biochem Biophys Res Commun 367(1):103–108. doi:10.1016/j.bbrc.2007.12.123
11. van der Wijk T, Overvoorde J, den Hertog J (2004) H_2O_2-induced intermolecular disulfide bond formation between receptor protein-tyrosine phosphatases. J Biol Chem 279(43):44355–44361. doi:10.1074/jbc.M407483200
12. Fourquet S, Guerois R, Biard D, Toledano MB (2010) Activation of NRF2 by nitrosative agents and H_2O_2 involves KEAP1 disulfide formation. J Biol Chem 285(11):8463–8471. doi:10.1074/jbc.M109.051714
13. Wang SB, Foster DB, Rucker J, O'Rourke B, Kass DA, Van Eyk JE (2011) Redox regulation of mitochondrial ATP synthase: implications for cardiac resynchronization therapy. Circ Res 109(7):750–757. doi:10.1161/CIRCRESAHA.111.246124
14. Brennan JP, Wait R, Begum S, Bell JR, Dunn MJ, Eaton P (2004) Detection and mapping of widespread intermolecular protein disulfide formation during cardiac oxidative stress using proteomics with diagonal electrophoresis. J Biol Chem 279(40):41352–41360. doi:10.1074/jbc.M403827200

Chapter 11

Analysis of cGMP Signaling in Adipocytes

Katja Jennissen, Bodo Haas, Michaela M. Mitschke, Franziska Siegel, and Alexander Pfeifer

Abstract

Obesity has reached pandemic dimensions with more than half a billion adults affected worldwide. Detailed knowledge of adipose biology is required for the development of urgently needed novel therapies directed against obesity. Two types of adipose tissue can be distinguished in humans and mice: white adipose tissue (WAT), which primarily stores energy in the form of lipids and has endocrine functions. In contrast, brown adipose tissue (BAT) dissipates energy in the form of heat (thermogenesis). Recent studies in humans demonstrated that BAT not only plays a role for non-shivering thermogenesis in newborns but is also metabolically active in adults. Here, we describe protocols for the generation of cellular models for the analysis of adipogenesis as well as function of brown and white fat. These models are based on the in vitro differentiation of mesenchymal stem cells (MSCs) isolated from adipose tissues. Using specific differentiation protocols, the role of cGMP signaling in both brown as well as white adipocytes can be studied.

Key words Brown adipocytes, White adipocytes, Mesenchymal stem cells, Adipogenesis, In vitro differentiation, Primary cells, Immortalization, Oil Red O staining, Triglyceride quantification, Adipogenic and thermogenic marker expression, cGMP signaling

1 Introduction

Obesity is characterized by excessive accumulation of adipose tissue due to an imbalance between energy intake and consumption. Obesity has reached pandemic dimensions with more than 1.5 billion adults being overweight and approximately 10 % of adults worldwide suffering from obesity [1, 2]. The growing problem of obesity is associated with multiple disorders including cardiovascular diseases, type 2 diabetes, Alzheimer's disease, and cancer [3, 4].

In mammals adipose tissue is divided into two functionally distinct forms. White adipose tissue (WAT) is the major site of energy storage [5], whereas brown adipose tissue (BAT) is a thermogenic organ, densely packed with mitochondria that are able to dissipate energy in the form of heat via uncoupled respiration [6].

Cyclic nucleotide signaling is involved in a variety of physiological processes including neurotransmission, cardiovascular

Thomas Krieg and Robert Lukowski (eds.), *Guanylate Cyclase and Cyclic GMP: Methods and Protocols*, Methods in Molecular Biology, vol. 1020, DOI 10.1007/978-1-62703-459-3_11, © Springer Science+Business Media, LLC 2013

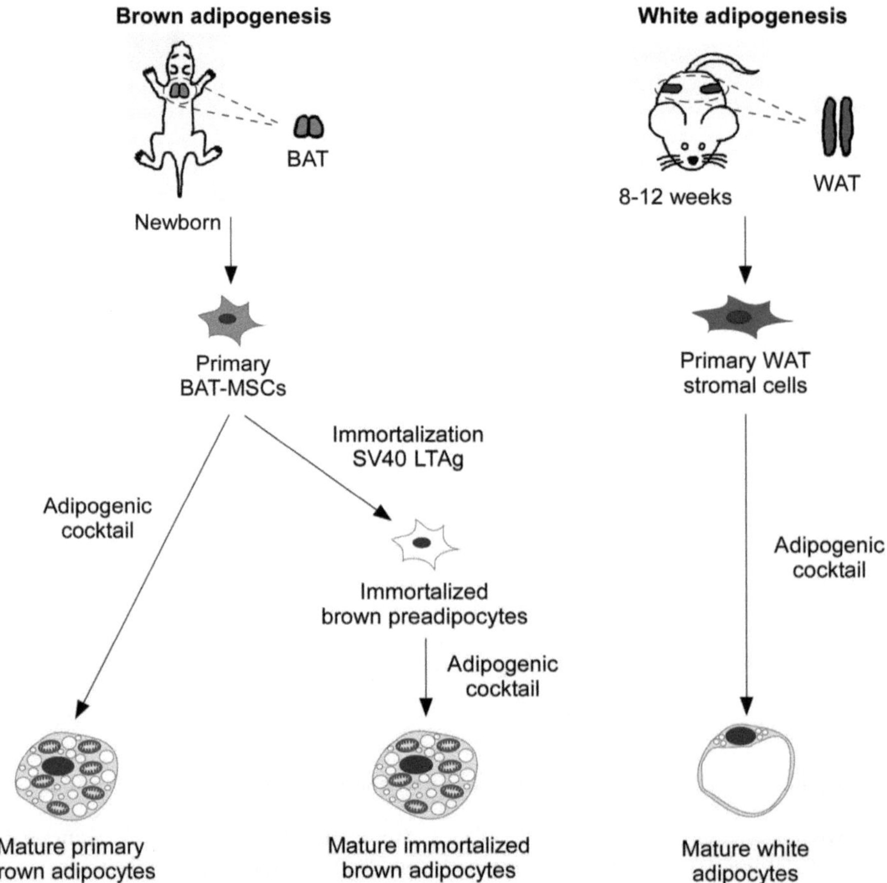

Fig. 1 In vitro models of brown and white adipogenesis derived from explanted adipose tissue. Primary BAT–MSCs from newborn mice can be differentiated to mature brown adipocytes in vitro from either primary MSCs or immortalized cells using different adipogenic cocktails. Primary WAT stromal cells isolated from adult mice can be differentiated to mature white adipocytes by applying a specific adipogenic differentiation protocol

homeostasis, platelet aggregation, and cellular metabolism. In the past, the role of cAMP in adipocyte function has extensively been studied. Recent studies indicate that cGMP signaling is also involved in the regulation of white and brown adipogenesis [7–10].

To study white as well as brown adipogenesis in vitro, several cellular model systems have been developed [11]. Here, we describe in detail protocols for the isolation of mesenchymal stem cells (MSCs) from WAT and BAT as well as protocols that can be used for differentiating MSCs/precursor cells to mature white or brown adipocytes (Fig. 1). These cellular models can be utilized to study the role of cyclic nucleotide signaling cascades in adipogenic differentiation as well as adipocyte function.

2 Materials

Prepare all solutions using ultrapure water. Use 0.22 μm filters for sterile filtration or autoclave solutions as indicated. Sterilized dissection instruments (scissors, forceps), and nylon meshes (30 and 100 μm, Millipore) are required for isolation of BAT–MSCs and WAT stromal cells.

2.1 Standard Buffers/Solutions

Phosphate-buffered saline (PBS), pH 7.4; 70 % ethanol.

2.2 Standard Plastic Labware and Instruments

6-well plates and 10 cm dishes, Falcon tubes, 1.5 ml reaction tubes, 5 ml syringes and cannulas (0.9 × 70 mm; 20 G × 2 ¾).

2.3 Model of Brown Adipogenesis

Use DMEM GlutaMAX™ I + 4,500 mg/l glucose (DMEM) without pyruvate (Invitrogen) for the model of brown adipogenesis.

2.3.1 Isolation of BAT–MSCs

1. BAT collagenase digestion buffer: Dissolve 123 mM NaCl, 5 mM KCl, 1.3 mM $CaCl_2$, 5 mM glucose, and 100 mM *N*-(2-hydroxyethyl)-piperazine-*N*′-2-ethansulfonic acid (HEPES) in water. Adjust pH with NaOH to 7.4, sterile filtrate, and store at 4 °C. Add 1.5 % bovine serum albumin (BSA) and 0.2 % collagenase II (Worthington) directly before use. Sterile filtrate and store at 4 °C.
2. BAT–MSC culture medium (BAT–MSC CM; Table 1): Prepare a 0.1 M HEPES stock solution by dissolving HEPES in DMEM, sterile filtrate, and store at 4 °C. Prepare a 60 μM sodium triiodothyronine stock solution by dissolving 2 mg/ml sodium triiodothyronine in 1 M NaOH and diluting this solution with DMEM up to a concentration of 60 μM sodium triiodothyronine. Store at 4 °C for up to 4 weeks. A 10 mg/ml sodium ascorbate stock solution is prepared by dissolving sodium ascorbate in PBS; sterile filtrate and store at −20 °C. To prepare BAT–MSC culture medium, supplement DMEM with 10 % fetal bovine serum (FBS), 1 % penicillin/streptomycin (P/S), 4 nM insulin, 4 nM sodium triiodothyronine, 10 mM HEPES, and 25 μg/ml sodium ascorbate and store at 4 °C.

2.3.2 Brown Adipogenic Differentiation of Primary BAT–MSCs

BAT–MSC culture medium: Prepare BAT–MSC culture medium as described above (Subheading 2.3.1, **step 2**).

2.3.3 Immortalization of BAT–MSCs

BATi growth medium (BATi GM; Table 1): Add 10 % FBS and 1 % P/S to DMEM and store at 4 °C.

Table 1
Media used for adipogenic differentiation and the final concentrations of adipogenic substances

	Medium	Adipogenic substances Insulin (nM)	T3 (nM)	Asc (μg/μl)	Dexa (μM)	IBMX (mM)	Rosi (μM)	ABP: L-asc (μg/ml) Biotin (μM) Panto (μM)
BAT	BAT–MSC CM	4	4	25	–	–	–	–
	BATi GM	–	–	–	–	–	–	–
	BATi DM	20	1	–	–	–	–	–
	BATi IM	20	1	–	1	0.5	–	–
WAT	WAT GM	–	–	–	–	–	–	–
	WAT IM	172	1	–	0.25	0.5	2.5	L-asc 50 Biotin 1 Panto 17
	WAT MM	172	1	–	–	–	2.5	L-asc 50 Biotin 1 Panto 17

ABP ascorbate, biotin, pantothenate mixture, *Asc* sodium ascorbate, *Biotin* d-biotin, *CM* culture medium, *Dexa* dexamethasone, *DM* differentiation medium, *GM* growth medium, *IBMX* 3-isobutyl-1-methylxanthine, *IM* induction medium, *L-asc* L-ascorbate, *MM* maintenance medium, *Panto* pantothenate, *Rosi* rosiglitazone, *T3* triiodothyronine

2.3.4 Cultivation and Storage of Immortalized Brown Preadipocytes (BATi)

1. Trypsin–ethylene diamine tetraacetic acid (EDTA), 0.05 % (Invitrogen).
2. Freezing medium: Supplement BATi growth medium with 20 % dimethyl sulfoxide (DMSO) directly before use.

2.3.5 Brown Adipogenic Differentiation of BATi

1. BATi differentiation medium (BATi DM; Table 1): Supplement DMEM with 10 % FBS, 1 % P/S, 20 nM insulin, and 1 nM sodium triiodothyronine and store at 4 °C. Use the sodium triiodothyronine stock solution described above (Subheading 2.3.1, **step 2**).
2. BATi induction medium (BATi IM; Table 1): Prepare always freshly before use. Prepare a 50 μM dexamethasone stock solution by dissolving 1 mg/ml dexamethasone in 99 % ethanol (final concentration: 2.5 mM) and diluting this solution with BATi differentiation medium up to a concentration of 50 μM dexamethasone. Prepare a 0.5 M 3-isobutyl-1-methylxanthine (IBMX) stock solution by diluting IBMX in DMSO at 37 °C. To prepare BATi induction medium, warm up BATi differentiation medium to 37 °C and supplement it with 1 μM dexamethasone and 0.5 mM IBMX. BATi differentiation medium

needs to be warmed up to 37 °C before the addition of the IBMX stock solution to prevent precipitation of IBMX.

2.4 Model of White Adipogenesis

Use DMEM GlutaMAX™ I+4,500 mg/l glucose (DMEM) with pyruvate (Invitrogen) for the model of white adipogenesis.

2.4.1 Isolation of WAT Stromal Cells

1. WAT collagenase digestion buffer: Prepare always freshly before use. Prepare a collagenase II stock solution by dissolving 20 mg/ml collagenase II (Worthington) in PBS. Prepare a 10 % BSA stock solution in PBS. To prepare WAT collagenase digestion buffer, dissolve 1.5 mg/ml collagenase II and 0.5 % BSA in DMEM and sterile filtrate.
2. WAT growth medium (WAT GM; Table 1): Add 10 % FBS and 1 % P/S to DMEM and store at 4 °C.

2.4.2 White Adipogenic Differentiation of WAT Stromal Cells

1. WAT induction medium (WAT IM; Table 1): Prepare always freshly before use. Prepare a 10 mM dexamethasone stock solution by dissolving dexamethasone in 99 % ethanol and store aliquots at −20 °C. For short-term storage dilute this stock solution with PBS up to a concentration of 1 mM dexamethasone and store at 4 °C for up to 2 weeks. Prepare and store a 60 μM sodium triiodothyronine stock solution as described above (Subheading 2.3.1, **step 2**) using DMEM with pyruvate. Prepare a 40 mM d-biotin stock solution by dissolving d-biotin in 1 M NaOH and use this solution to prepare an ABP stock solution containing 50 mg/ml L-ascorbate, 1 mM d-biotin, and 17 mM pantothenate dissolved in water and store at −20 °C. Prepare a 25 mM rosiglitazone stock solution by dissolving rosiglitazone in DMSO and store at −20 °C. Prepare a 0.5 M IBMX stock solution by dissolving IBMX in 0.5 M KOH freshly before use. To prepare WAT induction medium, supplement DMEM with 5 % FBS, 1 % P/S, 1 μg/ml insulin (0.172 μM), 0.25 μM dexamethasone, 1 nM triiodothyronine, 1:1,000 ABP, 0.5 mM IBMX, and 2.5 μM rosiglitazone.
2. WAT maintenance medium (WAT MM; Table 1): Prepare always freshly before use. Prepare stock solutions as described above (Subheading 2.4.2, **step 1**). Supplement DMEM with 5 % FBS, 1 % P/S, 1 μg/ml insulin (0.172 μM), 1 nM triiodothyronine, 1:1,000 ABP, and 2.5 μM rosiglitazone.

2.5 Analysis of Adipogenic Differentiation

2.5.1 Oil Red O Staining

1. Oil Red O stock solution: Dissolve 0.5 % Oil Red O in isopropyl alcohol (99 %) to a final concentration of 5 mg/ml with a magnetic stir bar overnight. Store at room temperature.
2. Oil Red O working solution: Dilute Oil Red O stock solution with water up to a final concentration of 3 mg/ml Oil Red O and allow to stand overnight. The next day, filtrate the solution twice through a paper filter.

3. PFA fixative: Dissolve 4 % paraformaldehyde (PFA) in PBS (pH 7.4). Heat for 1 min, cool on ice, and store at −20 °C.

2.5.2 Measurement of Triglyceride (TG) Accumulation

1. TG-assay buffer: Dissolve 150 mM NaCl and 10 mM Tris in water. Adjust pH to 8.0 with HCl and add 0.1 % Triton X-100. Sterile filtrate and store at 4 °C. Before use, add 40 μl/ml Complete® EDTA-free protease inhibitor cocktail (Roche).
2. TG-assay reagent: Mix Free Glycerol Reagent and Triglyceride Reagent (both Sigma-Aldrich) at a ratio of 80:20 %.
3. Glycerol standard solution: 2.5 mg/ml equivalent triolein concentration (Sigma-Aldrich).
4. Coomassie (Bradford) protein assay kit (Pierce) for determination of protein content.

2.5.3 Analysis of Adipogenic and Thermogenic Marker Expression

1. Standard SDS-polyacrylamide-gel electrophoresis (SDS-PAGE)/Western Blot analysis reagents and Coomassie (Bradford) protein assay reagents (Coomassie solution, stacking gel solution, separating gel solution, SDS-PAGE running buffer, transfer buffer, Tris-buffered saline supplemented with Tween-20 (TBST)).
2. Lysis buffer: Dissolve 10 mM Tris–HCl (pH 7.4), 150 mM NaCl, 1 % NP-40, 1 % sodium deoxycholic acid, and 0.1 % sodium dodecyl sulfate (SDS) in water. Sterile filtrate and store at 4 °C. Add the following substances freshly before use: 40 μl/ml Complete® EDTA-free protease inhibitor cocktail (Roche), 10 mM NaF, and 1 mM Na_3VO_4.
3. 3× Laemmli buffer: Dissolve 125 mM Tris–HCl (pH 6.8), 20 % glycerol, 17 % SDS, and 0.015 % bromophenol blue in water and store at −20 °C. Before use, add 5 % β-mercaptoethanol.
4. Blocking buffer: Dissolve 5 % skimmed milk powder in TBST and store at 4 °C.
5. Primary antibody buffer: Dissolve 1 % skimmed milk powder in TBST and supplement with 0.05 % NaN_3. Store at 4 °C.
6. Secondary antibody buffer: Dissolve 1 % skimmed milk powder in TBST and store at 4 °C.
7. Antibodies: Use the antibodies listed in Table 2.

2.6 Analysis of cGMP Signaling in Preadipocytes and Mature Fat Cells

2.6.1 Analysis of PKG Activity

1. SDS-PAGE/Western blot analysis reagents and buffers: Use reagents and buffers as described above (Subheading 2.5.3).
2. Antibodies: Use the antibodies listed in Table 3.

Table 2
Antibodies for the analysis of adipogenic and thermogenic marker expression

Antigen	Manufacturer	Cat. no.	Dilution	Purpose
aP2	Santa Cruz	sc-18661	1:1,000	Adipogenic marker
C/EBPα	Santa Cruz	sc-61	1:1,000	Adipogenic marker
Cytc	BD Biosciences	556433	1:1,000	Thermogenic marker
PPARγ	Santa Cruz	sc-7273	1:1,000	Adipogenic marker
Tubulin	Dianova	DLN-09992	1:1,000	Loading control
UCP-1	Santa Cruz	sc-6529	1:500	Thermogenic marker

Table 3
Antibodies for the analysis of PKG activity

Antigen	Manufacturer	Cat. no.	Dilution	Purpose
RhoA	Santa Cruz	sc-418	1:1,000	Loading control
RhoA pSer188	Santa Cruz	sc-32954	1:1,000	PKG activity
Tubulin	Dianova	DLN-09992	1:1,000	Loading control
VASP	Cell Signaling	3132	1:1,000	Loading control
VASP pSer239	nanoTools	0047-100/ VASP-16C2	1:1,000	PKG activity

2.6.2 Measurement of Cellular cGMP Levels

cGMP enzyme-linked immunosorbent assay (ELISA): Cyclic GMP EIA Kit ELISA (Cayman).

3 Methods

3.1 Model of Brown Adipogenesis

Isolate BAT–MSCs from interscapular brown fat of newborn mice (Figs. 1 and 2) following a protocol modified from Nechad et al. [12] (*see* **Note 1**).

3.1.1 Isolation of BAT–MSCs

1. Autoclave dissecting instruments (scissors, forceps) and nylon meshes (100 and 30 μm; Millipore).
2. Work under sterile bench conditions. Pre-warm a water bath to 37 °C.
3. Pre-warm BAT–MSC culture medium (about 3 ml per mouse) to 37 °C.
4. Provide a beaker with 70 % ethanol to sterilize the dissecting instruments during the isolation of BAT–MSCs from different animals.

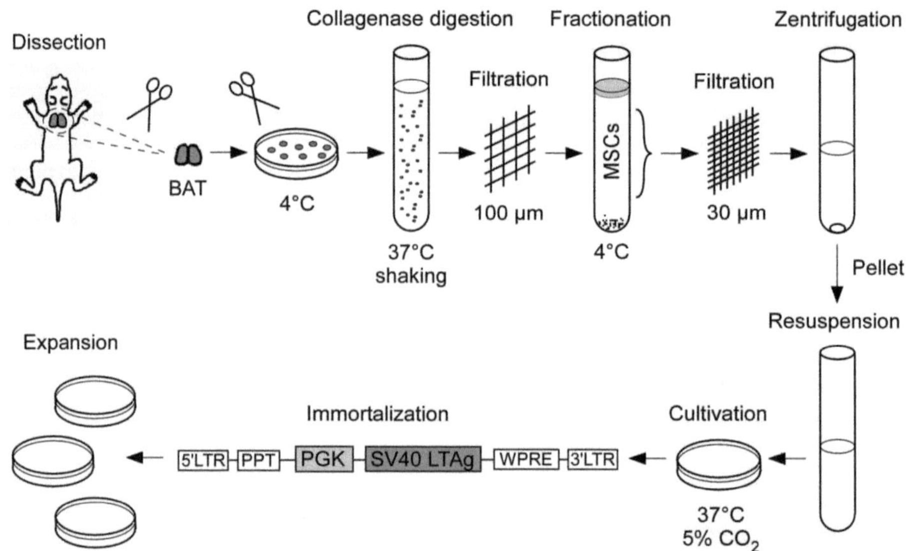

Fig. 2 Isolation of BAT–MSCs from interscapular BAT of newborn mice. BAT pads of newborn mice are dissected and digested with collagenase. BAT–MSCs are separated from mature *brown* adipocytes, seeded on 6-well plates, and can be immortalized followed by expansion, if desired

5. Provide a 6-well plate with 3 ml BAT collagenase digestion buffer per well on ice, sterile 15 ml falcon tubes, 1.5 ml reaction tubes, 5 ml syringes, cannulas, and a dish with 70 % ethanol.
6. Decapitate a newborn mouse and quickly transfer it to the dish with 70 % ethanol to achieve sterile conditions.
7. Remove the two brown fat pads and transfer them to one well of the 6-well plate with BAT collagenase digestion buffer. Chop the fat pads with scissors, transfer the solution to a 15 ml falcon tube, and store it on ice.
8. Cut off the tail tip, transfer it to a 1.5 ml reaction tube, and store it on ice for genotyping.
9. Continue with **steps 6–8** for each mouse pup.
10. Transfer the falcon tubes into the pre-warmed water bath for collagenase digestion. Shake thoroughly by hand every 5 min. Incubate the suspensions for about 30 min until all bigger tissue remnants have disappeared.
11. Suck the suspensions into 5 ml syringes using cannulas (0.9 × 70 mm; 20 G × 2 ¾).
12. Remove the cannulas, fix one 100 μm nylon mesh at each syringe using a cutoff pipette tip, and filtrate the suspension through the nylon mesh into a new 15 ml falcon tube.
13. Incubate the falcon tubes on ice for 30 min.
14. Suck the middle phase of the suspensions into 5 ml syringes using cannulas. Do not use the upper phase (about 0.5 ml),

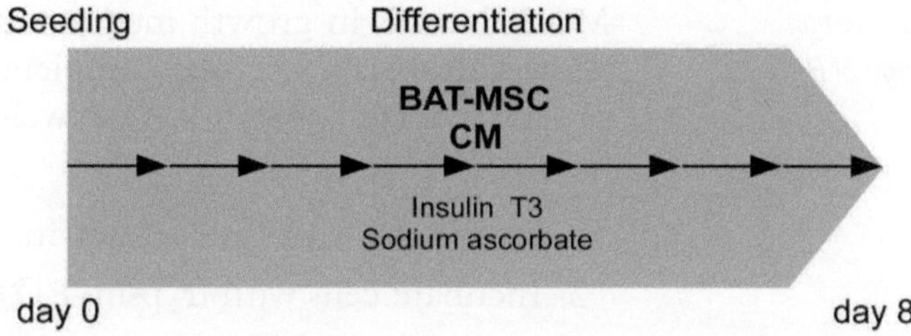

Fig. 3 Protocol of *brown* adipogenic differentiation of primary BAT–MSCs. Medium is changed every day for 8 days (*BAT–MSC CM* BAT–MSC culture medium, *T3* triiodothyronine)

which contains mature adipocytes, and the bottom phase (about 0.5 ml), which contains tissue remnants.

15. Filtrate the suspensions through 30 μm nylon meshes into new 15 ml falcon tubes as described above.
16. Centrifuge at 700 × *g* for 10 min.
17. Resuspend the pellets in 3 ml pre-warmed BAT–MSC culture medium, count the cells, and seed 5.7×10^5 cells per well of a 6-well plate (*see* **Note 2**).
18. Incubate primary BAT–MSCs at 37 °C and 5 % CO_2 for 24 h.
19. Wash cells once with BAT-MSC culture medium to remove dead cells, tissue remnants, and mature fat cells. Maintain cells in BAT-MSC culture medium at 37 °C and 5 % CO_2.

3.1.2 Brown Adipogenic Differentiation of Primary BAT–MSCs

Differentiate primary BAT–MSCs by exchanging BAT–MSC culture medium every day for 8 days (Fig. 3).

3.1.3 Immortalization of BAT–MSCs

Immortalize primary BAT–MSCs (passage 0) 24 h after isolation by transduction with a lentivirus expressing the simian virus 40 (SV40) large T-antigen (L-TAg) under control of the phosphoglycerate kinase (PGK) promoter [7]. Expand immortalized brown preadipocytes (BATi) in BATi growth medium at 37 °C and 5 % CO_2. Do not use the cells beyond passage 4.

1. Thaw and centrifuge virus solution at 16,000 × *g* for 1 min.
2. Remove the amount of supernatant that corresponds to 200 ng viral reverse transcriptase (for 1 well of a 6-well plate) and mix it with 800 μl BATi growth medium.
3. Remove medium from the cells and incubate them with 800 μl viral medium overnight at 37 °C and 5 % CO_2.
4. Remove viral medium from the cells, wash cells three times with PBS, and incubate with BATi growth medium at 37 °C and 5 % CO_2.

3.1.4 Cultivation and Storage of BATi

Maintain cells in growth medium at 37 °C and 5 % CO_2. Do not allow cultures to become completely confluent for expansion. In order to detach cells from the well plate, follow the instructions below:

1. Wash cells with PBS pre-warmed to 37 °C.
2. Incubate cells with trypsin–EDTA for 5 min at 37 °C.
3. Detach cells by resuspending them in BATi growth medium.

To store cells for longer periods of time, trypsinize and resuspend cells in pre-warmed (37 °C) BATi growth medium as described above and freeze them as follows:

1. Centrifuge suspensions for 5 min at $200 \times g$.
2. Resuspend pellets in BATi growth medium mixed with freezing medium at a ratio of 1:1 to achieve a final DMSO concentration of 10 %. The optimal cell density is one million cells per ml.
3. Transfer the cell suspensions into cryogenic vials (1 ml per vial) and incubate on ice for 15 min.
4. Store cryogenic vials at −80 °C for 24 h. Afterwards transfer cryogenic cultures to liquid nitrogen (−196 °C) for long-term storage.

In order to thaw cryopreserved cells, follow the instructions below:

1. Quickly place cryo-cultures in a water bath pre-warmed to 37 °C.
2. Transfer cell suspensions to BATi growth medium (approximately ten times the volume of the cryo-culture).
3. Centrifuge for 5 min at $200 \times g$.
4. Resuspend cell pellets in BATi growth medium and seed on well plates.

3.1.5 Brown Adipogenic Differentiation of BATi

Differentiate immortalized brown preadipocytes according to the protocol depicted in Fig. 4 and described below.

1. Seed 1.8×10^5 cells per well of a 6-well plate in BATi growth medium (day −4).
2. After 48 h (day −2) remove the medium and add BATi differentiation medium.
3. Induce adipogenic differentiation after additional 48 h (day 0) by exchanging BATi differentiation medium with BATi induction medium. Prepare the medium freshly before use. Cultures need to be confluent when inducing adipogenesis (*see* **Notes 3** and **4**).

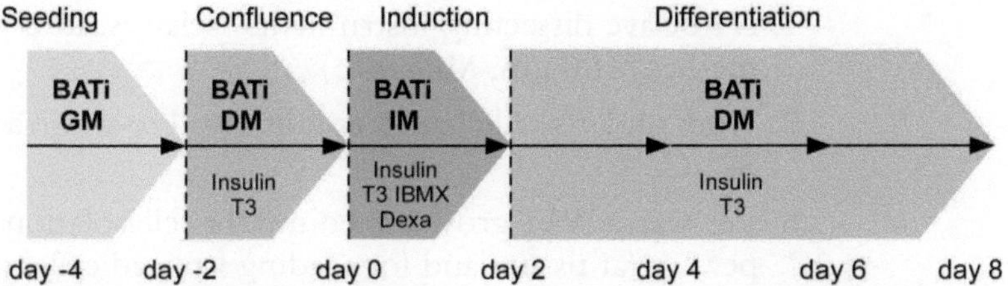

Fig. 4 Protocol of *brown* adipogenic differentiation of immortalized brown preadipocytes. Different media (*BATi GM* BATi growth medium, *BATi DM* BATi differentiation medium, *BATi IM* BATi induction medium, *Dexa* dexamethasone, *IBMX* 3-isobutyl-1-methylxanthine, *T3* triiodothyronine) are used at the indicated stages of adipogenic differentiation (day 4 to day 8)

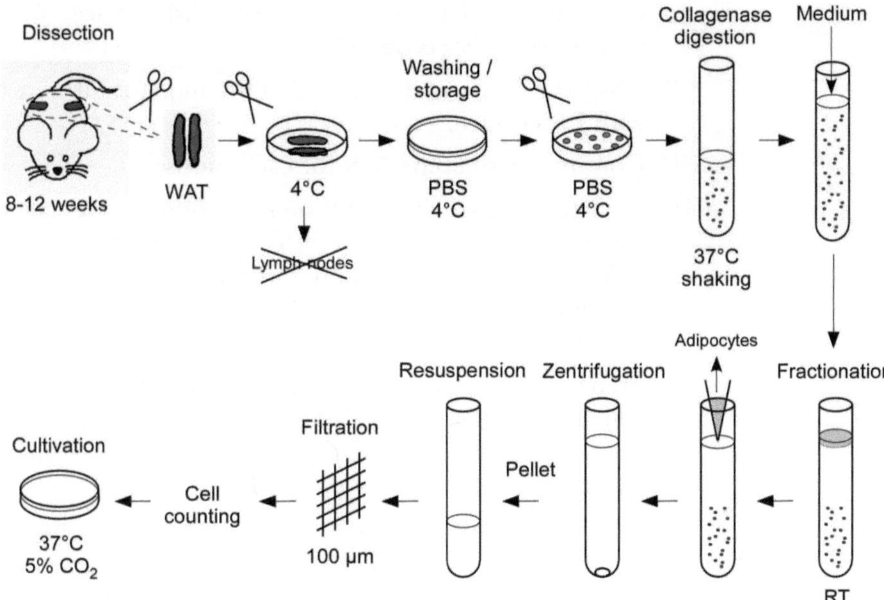

Fig. 5 Isolation of WAT MSCs/stromal cells from white fat pads of 8–12-week-old mice. WAT pads of mice are dissected and collagenase digested. WAT stromal cells are separated from mature *white* adipocytes and seeded on 6-well plates

4. After a 48 h induction phase (day 2), maintain cells in BATi differentiation medium. Replenish medium every second day until day 8, until the cells are differentiated into mature brown adipocytes (*see* **Note 5**).

3.2 Model of White Adipogenesis

3.2.1 Isolation of WAT Stromal Cells

Isolate WAT MSCs/stromal cells from inguinal and gonadal white fat of 8–12-week-old mice (Figs. 1 and 5) following a protocol modified from Ostertag et al. and Rodeheffer et al. [13–15] (*see* also **Note 6**). Approximately four mice are required for 1 g of adipose tissue.

1. Autoclave dissecting instruments (scissors, forceps) and nylon meshes (100 μm; Millipore).
2. Work under sterile bench conditions. Pre-warm a water bath to 37 °C.
3. Pre-warm WAT growth medium for cell isolation (about 10 ml per 1 g fat tissue) and for seeding isolated cells to 37 °C.
4. Provide a beaker with 70 % ethanol to sterilize the dissecting instruments during the isolation of WAT stromal cells.
5. Provide two dishes with PBS and two dishes without PBS on ice, sterile 15 ml falcon tubes, 5 ml syringes, and cannulas.
6. Sacrifice a mouse.
7. Remove the two inguinal white fat pads and the gonadal WAT per mouse and transfer them to a dish on ice. Remove the lymph nodes (*see* **Note** 7).
8. Wash the tissue in a dish with PBS and store it in a new dish with PBS on ice.
9. Continue with **steps 6–8** for each mouse.
10. Transfer the tissue to a dish on ice. Chop the fat pads with scissors and transfer the tissue into 15 ml falcon tubes containing WAT collagenase digestion buffer. Each falcon tube should contain 1 g fat tissue per 7 ml WAT collagenase digestion buffer.
11. Transfer the falcon tubes into the pre-warmed water bath for collagenase digestion. Shake them thoroughly every 5 min. Incubate the suspensions for about 30–45 min until all bigger tissue remnants have disappeared.
12. Add 7 ml of pre-warmed WAT growth medium to each falcon tube.
13. Incubate at room temperature for 10 min.
14. Remove the floating adipocyte fraction by taking off the upper phase (about 1 ml) using a 1,000 μl pipette tip.
15. Centrifuge at 200 × *g* for 10 min at room temperature.
16. Discard supernatants.
17. Resuspend the pellets (corresponding to 1 g of adipose tissue) with 2 ml WAT growth medium per pellet.
18. Suck the suspensions into 5 ml syringes using cannulas (0.9 × 70 mm; 20 G × 2 ¾).
19. Remove the cannulas, fix one 100 μm nylon mesh at each syringe using a cutoff pipette tip, and filtrate the suspension through the nylon mesh into a new 15 ml falcon tube.
20. Count the cells and seed 2×10^5 cells per well of a 6-well plate in WAT growth medium.

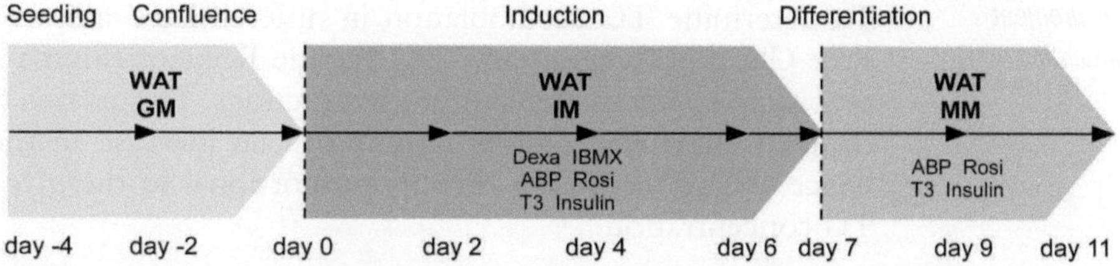

Fig. 6 Protocol of white adipogenic differentiation of primary WAT stromal cells. Different media (*WAT GM* WAT growth medium, *WAT IM* WAT induction medium, *WAT MM* WAT maintenance medium, *ABP* ascorbate, biotin, pantothenate mixture, *Dexa* dexamethasone, *IBMX* 3-isobutyl-1-methylxanthine, *Rosi* rosiglitazone, *T3* tri-iodothyronine) are used at the indicated stages of adipogenic differentiation (day 4 to day 11)

21. Incubate primary WAT stromal cells at 37 °C and 5 % CO_2 for 24 h.
22. Wash cells two to three times with WAT growth medium to remove dead cells, tissue remnants, and mature fat cells. Maintain cells in WAT growth medium at 37 °C and 5 % CO_2.

3.2.2 White Adipogenic Differentiation of WAT Stromal Cells

Differentiate primary WAT stromal cells according to the protocol depicted in Fig. 6 and described below.

1. Seed 2×10^5 cells per well of a 6-well plate in WAT growth medium (*see* **Note 8**).
2. Culture cells in WAT growth medium until they are confluent (day 2).
3. Provide cells with new growth medium (day 2) and incubate for 48 h.
4. After 48 h (day 0) induce adipogenic differentiation by exchanging WAT growth medium with WAT induction medium for 7 days (*see* **Note 3**). Change the medium every other day and prepare it always freshly before use.
5. After 7 days (day 7) maintain cells in WAT maintenance medium for additional 4 days. Replenish medium every other day until day 11, when the cells are differentiated into mature white adipocytes. Prepare medium always freshly before use.

3.3 Analysis of Adipogenic Differentiation

Oil Red O is a dye used for staining of lipids in tissues and cells. Lipid droplets stained by accumulated Oil Red O appear in a red color (*see* **Note 9**).

3.3.1 Oil Red O Staining

1. Wash differentiated adipocytes with PBS.
2. Fix cells with PFA fixative for 15 min at room temperature.
3. Wash cells three times with PBS.
4. Incubate cells with Oil Red O working solution (1 ml per well of a 6-well plate) for 4 h at room temperature.
5. Wash cells three times with PBS.

3.3.2 Measurement of TG Accumulation

To determine TG accumulation in differentiated adipocytes, use Free Glycerol Reagent and Triglyceride Reagent (Sigma-Aldrich) according to the manufacturer's instructions. The assay is based on coupled enzyme reactions resulting in an increase in the absorbance at 540 nm that is directly proportional to the glycerol and TG concentration.

1. Wash cells with PBS.
2. Add 200 μl TG-assay buffer per well of a 6-well plate and freeze at −80 °C.
3. Prepare TG-assay reagent. Prepare 1 ml per sample, for the standard and the negative control. Allow to warm to room temperature.
4. Thaw cells on ice, scrape, resuspend, and transfer them into a 1.5 ml reaction tube.
5. Centrifuge at 15,000 × *g* and 4 °C for 10 min.
6. Remove 10 μl of the supernatant for protein content determination and store at 4 °C.
7. Resuspend the remaining part of the samples, which is about 170 μl.
8. Prepare 170 μl of TG-assay buffer as negative control.
9. Prepare the standard by diluting 10 μl of the glycerol standard solution with 160 μl TG-assay buffer.
10. Add 1 ml TG-assay reagent to each sample, to the standard and negative control.
11. Incubate at 37 °C for 5 min.
12. Centrifuge at 15,000 × *g* and room temperature for 2 min.
13. Measure absorbance at 540 nm using a photometer.
14. Determine the protein concentration of the samples (e.g., using the Bradford protein assay).
15. Calculate glycerol/TG concentration per mg protein according to the manufacturer's instructions.

3.3.3 Analysis of Adipogenic and Thermogenic Marker Expression

Prepare total protein lysates from differentiated fat cells as follows:

1. Wash cells with 4 °C PBS.
2. Add 200 μl of 4 °C lysis buffer per well of a 6-well plate.
3. Scrape, resuspend, and transfer the cells into a 1.5 ml reaction tube.
4. Centrifuge lysates at 15,000 × *g* and 4 °C for 10 min.
5. Discard the upper phase composed of floating lipids.
6. Use the remaining supernatant for determination of protein concentration (e.g., using the Bradford method).

7. Prepare protein samples for Western blotting containing 50 μg protein by adding 3× Laemmli buffer, incubating at 97 °C for 5 min, and cooling down on ice.
8. Store samples at −20 °C for up to 2 weeks or directly subject them to SDS-PAGE.

Perform a discontinuous gel electrophoresis to separate proteins according to their molecular size under denaturating conditions. Perform SDS-PAGE and Western blot analysis. Detect adipogenic as well as thermogenic markers (*see* Table 2) as follows:

1. Block membranes in blocking buffer for 2 h and wash with TBST for 5 min at room temperature.
2. Incubate with the primary antibody (*see* Table 2) at 4 °C overnight in primary antibody buffer.
3. Wash three times with TBST for 5 min.
4. Incubate with a horseradish peroxidase (HRP)-coupled secondary antibody for 1 h at room temperature in secondary antibody buffer.
5. Wash three times with TBST for 5 min.
6. Subject the membranes to chemiluminescence-based detection using ECL reagent and chemiluminescence films (Amersham Biosciences).

3.4 Analysis of cGMP Signaling in Preadipocytes and Mature Fat Cells

3.4.1 Analysis of PKG Activity

PKG activity in undifferentiated and differentiated brown as well as white adipocytes can be determined by Western blot analysis of RhoA and VASP phosphorylation [7, 16]. Both the small Rho-GTPase RhoA and the focal adhesion protein VASP (vasodilator-stimulated phosphoprotein) are substrates of PKG and are phosphorylated upon cGMP stimulation. In order to examine PKG activity, an antibody specific for the phosphorylation of RhoA at serine 188, which is a phosphorylation site of PKA and PKG, or an antibody specific for the phosphorylation of VASP at serine 239, which is exclusively phosphorylated by PKG, can be used (*see* Table 3).

1. Seed BAT–MSCs, BATi, or WAT stromal cells as described above.
2. In order to analyze undifferentiated cells, culture cells for 2 days to reach a confluence of at least 70–80 %.
3. For the analysis of differentiated cells, differentiate the cells according to the protocols described above until day 8 (mature brown adipocytes) or day 11 (mature white adipocytes), respectively.
4. Prepare total protein lysates as described above (Subheading 3.3.3).

5. Perform SDS-PAGE and Western blot analysis as described above (Subheading 3.3.3) to determine phospho-RhoA serine 188 and phospho-VASP serine 239 levels (*see* Table 3).
6. Use total RhoA, total VASP, and tubulin levels as loading controls (*see* Table 3).

3.4.2 Measurement of Cellular cGMP Levels

Cellular cGMP levels can be measured at different stages of adipogenic differentiation and serve as an indicator for the activity of cGMP-producing guanylyl cyclases (GCs) and cGMP-degrading phosphodiesterases (PDEs). In order to measure the cellular cGMP content, perform ELISAs as follows:

1. Seed BAT–MSCs, BATi, or WAT stromal cells for determination of cGMP and protein content as described above.
2. In order to analyze undifferentiated cells, culture cells for about 2–3 days until they reach a confluence of 100 %.
3. For the analysis of differentiated cells, differentiate the cells according to the protocols described above.
4. Measure cellular cGMP content using a Cyclic GMP EIA Kit (Cayman). Follow the manufacturer's instructions.
5. Determine protein content from reference wells by preparing total protein lysates as described above (Subheading 3.3.3).
6. Calculate cellular cGMP content per mg protein following the instructions of the cGMP ELISA Kit manufacturer.

4 Notes

1. Always isolate BAT–MSCs from newborn mice as soon as possible after birth. Mice are altricial animals that are characterized by minor amounts of differentiated brown adipocytes at birth, but the tissue is successively recruited during the first few days after birth [6]. Therefore, the amount of BAT–MSCs decreases after birth.
2. Pre-warm only the required amount of BAT–MSC culture medium and BATi differentiation medium before use (and not the whole medium bottle) as insulin and triiodothyronine are not stable at 37 °C.
3. It is a crucial requirement that BATi cells are confluent for induction of adipogenesis to ensure efficient adipogenic differentiation. In contrast, WAT stromal cells need to be 2 days post-confluence for induction.
4. After the 48 h induction phase during differentiation of BATi cells, parts of the monolayer at the edge of the well may detach and the pH of the medium might be slightly acid.

5. Change media very carefully to avoid detachment of the cell layer or lipid droplet containing cells, especially if the cells have already started to differentiate. Position the pipette close to the cell layer in the middle of a well and carefully release the medium drop by drop.
6. WAT stromal cells can be isolated either from pooled inguinal and gonadal white fat pads of 8–12-week-old mice or from inguinal or gonadal WAT separately depending on the scientific questions to be addressed. For example, isolation of WAT stromal cells from inguinal WAT is recommended for the analysis of "browning" (formation of multilocular, UCP-1 positive adipocytes) in white fat because inguinal WAT is particularly prone to "browning" [17].
7. It is important to remove the lymph nodes from the adipose tissue to avoid cocultivation of lymphocytes together with the isolated WAT stromal cells. The "yellowish"-appearing nodules (usually 1–2 per white fat pad) have to be removed using forceps and scissors.
8. We use uncoated 6-well plates from Sarstedt for BAT–MSCs and BATi cells. According to our experience, uncoated 6-well plates from Techno Plastic Products AG (TPP) are most suitable to culture WAT stromal cells in order to ensure optimal growth conditions.
9. For quantification of the staining, isopropanol is added to the wells to extract the Oil Red O. Optical density (OD) is then measured at 520 nm wavelength.

Acknowledgment

This work was supported by the DFG.

References

1. Finucane MM, Stevens GA, Cowan MJ et al (2011) National, regional, and global trends in body-mass index since 1980: systematic analysis of health examination surveys and epidemiological studies with 960 country-years and 9.1 million participants. Lancet 377:557–567
2. Ahima RS (2011) Digging deeper into obesity. J Clin Invest 121:2076–2079
3. Karastergiou K, Mohamed-Ali V (2010) The autocrine and paracrine roles of adipokines. Mol Cell Endocrinol 318:69–78
4. Wymann MP, Schneiter R (2008) Lipid signalling in disease. Nat Rev Mol Cell Biol 9:162–176
5. Deng Y, Scherer PE (2010) Adipokines as novel biomarkers and regulators of the metabolic syndrome. Ann N Y Acad Sci 1212: E1–E19
6. Cannon B, Nedergaard J (2004) Brown adipose tissue: function and physiological significance. Physiol Rev 84:277–359
7. Haas B, Mayer P, Jennissen K et al (2009) Protein kinase G controls brown fat cell differentiation and mitochondrial biogenesis. Sci Signal 2:ra78
8. Nikolic DM, Li Y, Liu S et al (2011) Overexpression of constitutively active PKG-I protects female, but not male mice from

diet-induced obesity. Obesity (Silver Spring) 19:784–791

9. Zhang X, Ji J, Yan G et al (2010) Sildenafil promotes adipogenesis through a PKG pathway. Biochem Biophys Res Commun 396: 1054–1059
10. Sengenes C, Bouloumie A, Hauner H et al (2003) Involvement of a cGMP-dependent pathway in the natriuretic peptide-mediated hormone-sensitive lipase phosphorylation in human adipocytes. J Biol Chem 278: 48617–48626
11. Rosen ED, MacDougald OA (2006) Adipocyte differentiation from the inside out. Nat Rev Mol Cell Biol 7:885–896
12. Nechad M (1983) Development of brown fat cells in monolayer culture. II. Ultrastructural characterization of precursors, differentiating adipocytes and their mitochondria. Exp Cell Res 149:119–127
13. Ostertag A, Jones A, Rose AJ et al (2010) Control of adipose tissue inflammation through TRB1. Diabetes 59:1991–2000
14. Rodeheffer MS, Birsoy K, Friedman JM (2008) Identification of white adipocyte progenitor cells in vivo. Cell 135:240–249
15. Mitschke et al (2013) Increased cGMP promotes healthy expansion and browning of white adipose tissue. FASEB J 27(4): 1621–1630
16. Butt E, Abel K, Krieger M et al (1994) cAMP- and cGMP-dependent protein kinase phosphorylation sites of the focal adhesion vasodilator-stimulated phosphoprotein (VASP) in vitro and in intact human platelets. J Biol Chem 269:14509–14517
17. Seale P, Conroe HM, Estall J et al (2011) Prdm16 determines the thermogenic program of subcutaneous white adipose tissue in mice. J Clin Invest 121:96–105

Chapter 12

A Genetic Strategy for the Analysis of Individual Axon Morphologies in cGMP Signalling Mutant Mice

Hannes Schmidt, Gohar Ter-Avetisyan, and Fritz G. Rathjen

Abstract

One of the many physiological functions of cyclic guanosine 3′,5′ monophosphate (cGMP) signalling is the regulation of a specific mode of axonal branching. The bifurcation of axons from dorsal root ganglion (DRG) neurons at the dorsal root entry zone of the embryonic spinal cord is triggered by a cGMP signalling pathway comprising the ligand C-type natriuretic peptide (CNP), the cGMP-producing natriuretic peptide receptor 2 (Npr2), and the cGMP-dependent protein kinase Iα (cGKIα). Absence of any of these components causes a loss of bifurcation and sensory axons instead only turn in either a rostral or a caudal direction. In this chapter we describe a genetic strategy to study the impact of cGMP signalling on the arborization of individual DRG neurons in mice. Expression of an alkaline phosphatase (AP) reporter is selectively induced in Npr2-positive DRG neurons by tamoxifen-dependent activation of a Cre recombinase under the control of the Npr2 promoter. This approach might also be employed for the analysis of axonal branching in neuronal subsets expressing Npr2 elsewhere in the nervous system.

Key words Axonal branching, DRG, Sparse neuronal labelling, CNP, Npr2, cGKI, cGMP, Tamoxifen, Cre recombinase

1 Introduction

Axonal branching is a key principle for the establishment of neural connectivity and contributes to the enormous complexity of the mature nervous system. An individual neuron could thereby connect with several targets which provides the physical framework for parallel transmission of information. As impairments in axonal branching might lead to neurodevelopmental disorders the elucidation of underlying molecular mechanisms has become the subject of intense studies during recent years [1–4].

The complicated branching patterns of some types of neurons however can pose a challenge for the analysis at molecular levels. Arborization of outgrowing axons might occur at intermediate target regions of their trajectories as well as in the termination zones.

Thomas Krieg and Robert Lukowski (eds.), *Guanylate Cyclase and Cyclic GMP: Methods and Protocols*, Methods in Molecular Biology, vol. 1020, DOI 10.1007/978-1-62703-459-3_12,

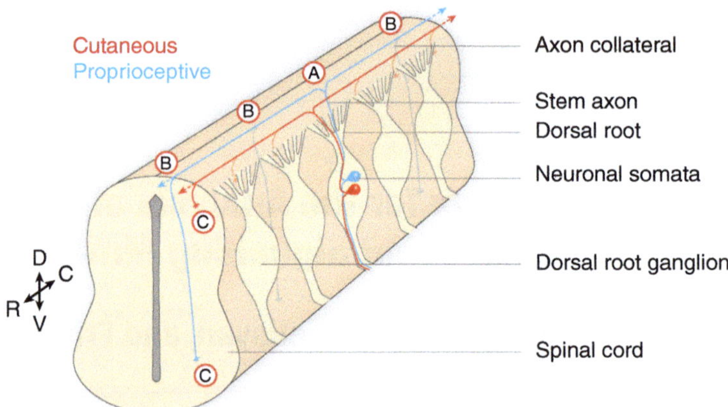

Fig. 1 Schematic overview of the central projection patterns of DRG neurons. The axonal trajectories of sensory neurons into the embryonic spinal cord display a sequence of at least three spatiotemporally distinct modes of branch formation. First, axons of DRG neurons bifurcate (**a**) upon reaching the DREZ of the spinal cord in a rostral and a caudal arm. Only after a waiting period collateral branches (**b**) from these daughter axons penetrate the spinal cord. Finally, terminal arborization (**c**) of the cutaneous and proprioceptive neuron projections occurs in the respective dorsal and ventral target areas. (*D* dorsal, *V* ventral, *R* rostral, *C* caudal)

Depending on the cellular structure where branches form, two modes of axonal branching might be distinguished: (1) branching of the extending growth cone as exemplified by axonal bifurcation and (2) collateral branching, also termed interstitial branching, from the axon shaft. The projections of sensory dorsal root ganglion (DRG) neurons into the embryonic spinal cord display a simple arborization pattern that integrates these two different modes of axonal branching in a spatiotemporally separated manner (*see* Fig. 1). Therefore it provides an accessible system for the analysis of the molecular mechanisms underlying axonal branching. In mice, the central projecting axons of DRG neurons reach the dorsal root entry zone (DREZ) of the spinal cord between embryonic days (E) 10 and E13 where their growth cones bifurcate into a rostral and a caudal arm. The two daughter axons then further extend at the lateral margin of the cord in opposite directions [5]. Only after a waiting period are collaterals generated from these stem axons that then penetrate the gray matter. Finally, terminal arborization of nociceptive and mechanoreceptive collaterals occurs in specific layers of the dorsal horn while proprioceptive collaterals terminate on motor neurons in the ventral spinal cord. A detailed analysis of axonal branching of DRG neurons unravelled a cyclic guanosine 3′,5′ monophosphate (cGMP) signalling cascade composed of the ligand C-type natriuretic peptide (CNP), the receptor guanylyl cyclase natriuretic peptide receptor 2 (Npr2), and the protein kinase cGMP-dependent protein kinase Iα (cGKIα) that is essential for the bifurcation of sensory axons at the DREZ of the spinal cord [6–10]. In the absence of one of these components, sensory axons no longer bifurcate and display instead only right-angled turns in either a rostral or a caudal direction, whereas the formation of collaterals is not affected.

A prerequisite to study axonal branching is the ability to discern the single neuronal arbor from the dense forest of surrounding axons. Advancing the scope of established methods for the analysis of neuronal morphologies such as traditional Golgi staining [11], horseradish peroxidase and neurobiotin injection [12, 13], the use of carbocyanine tracers [14], or the random expression of fluorescent proteins under the control of the Thy1-promoter [15], a genetic technology based on the coexpression of site-specific recombinases [16] with transgenic reporters [17, 18] combines the labelling of individual neurons within genetically defined cellular subsets with the manipulation of selected genes. This protocol focuses on the comparison of branching patterns of DRG neurons from Npr2 heterozygous mice which resemble the wild-type phenotype with those of homozygous Npr2-deficient animals. The visualization of single DRG neuron morphologies is achieved by a tamoxifen-dependent CreER^{T2} recombinase [19] under the control of the Npr2 promoter which secures expression in all DRG neurons. This mutant is crossed with the transgenic reporter line Z/AP [20]. Activation of CreER^{T2} by tamoxifen application then induces AP-expression by removal of a transcriptional terminator flanked by two loxP sites within the Z/AP reporter transgene (*see* Fig. 2). The low recombination efficiency of the Z/AP reporter [21, 22] triggers expression of AP in only a small number of DRG neurons (*see* **Note 1**) which is then visualized by a histochemical staining procedure in a whole-mount preparation of E12.5 spinal cord (*see* Fig. 3). This protocol might serve as a general guideline in future studies that should reveal whether the cGMP signalling pathway that underlies sensory axon bifurcation may extend to other neuronal populations.

2 Materials

2.1 Mouse Lines

Three mouse lines are required for crossbreeding (*see* Fig. 2): the Npr2-CreER^{T2} mouse which we generated by knocking in a CreER^{T2} expression cassette after the start codon of the Npr2 gene (manuscript in preparation), the spontaneous Npr2-cn mutant [23], and the Z/AP reporter mouse [20] (the latter two are available from the Jackson Laboratory; Jax stock numbers 003913 and 003919, respectively).

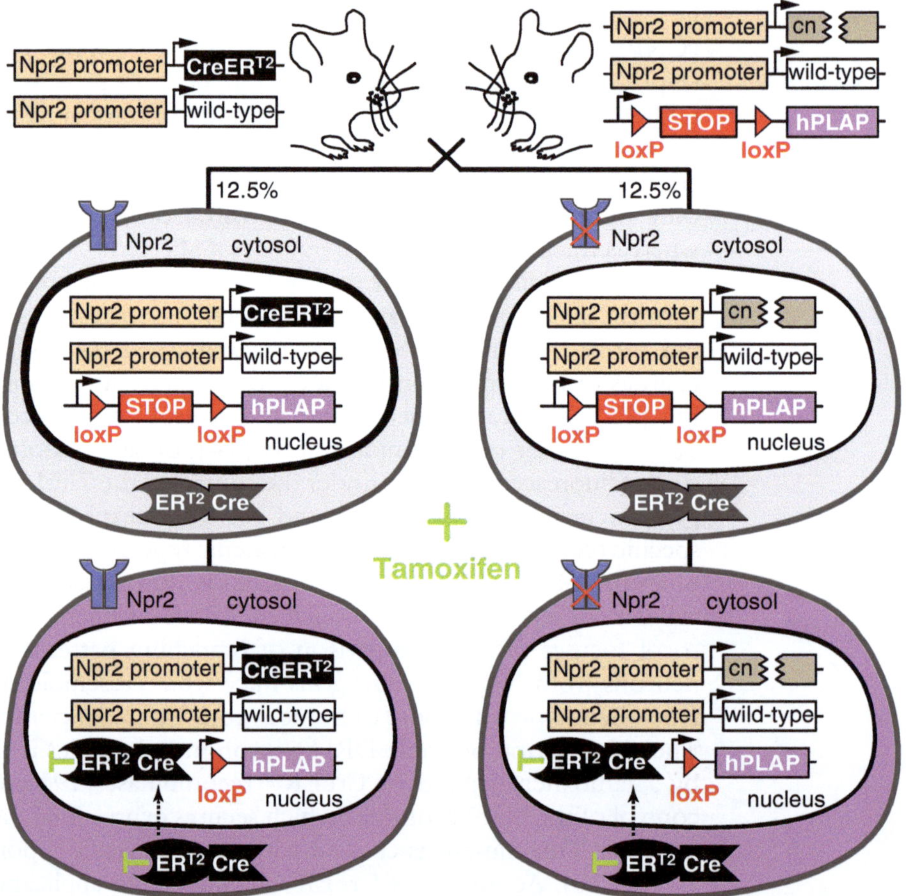

Fig. 2 An approach for transgenic labelling of GC-B-expressing cells. Schematic illustration of the genetic strategy to induce AP reporter gene expression in Npr2-positive cells by tamoxifen-dependent activation of CreER^{T2}. Crossbreeding of Npr2-CreER^{T2} heterozygous mice with mice heterozygous for both Npr2-cn and the Z/AP reporter (hPLAP, human placental alkaline phosphatase) produces experimental animals that are either *Npr2*$^{CreERT2/+}$; *Z/AP*$^{+}$ or *Npr2*$^{CreERT2/cn}$; *Z/AP*$^{+}$ with an expected yield of 12.5 % of each in the offspring. Administration of tamoxifen leads to nuclear translocation of CreER^{T2}. Cre-mediated excision of a loxP-flanked transcriptional termination signal (STOP) thereby activates expression of the AP reporter gene

2.2 Induction of AP Reporter Gene Expression by Administration of Tamoxifen

1. Tamoxifen.
2. Ethanol.
3. Corn oil.
4. 2-ml plastic tube.
5. Ultrasonic bath with heating.
6. 1 ml disposable syringe with 0.1 ml gradation.
7. Reusable feeding needle.

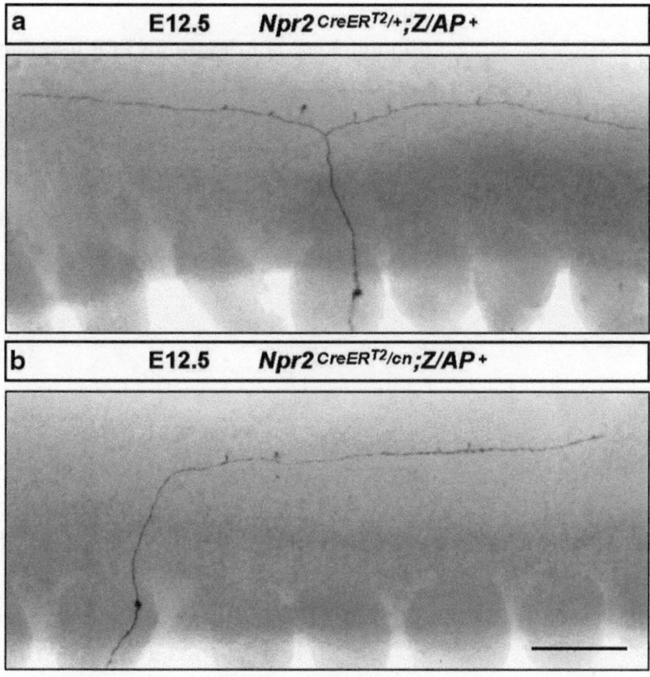

Fig. 3 Visualization of single DRG neurons to study the impact of cGMP signalling on axon bifurcation. The comparison of axonal projections of DRG neurons from (**a**) Npr2 heterozygous mice and (**b**) Npr2 homozygous mutant animals demonstrates a loss of axon bifurcation at the dorsal root entry zone of the spinal cord due to impaired cGMP signalling. Histochemical staining for AP activity reveals the morphology of individual DRG neurons in a whole-mount preparation of the mouse embryonic spinal cord with attached DRG at E12.5 (dorsolateral views depicted; the rostral side is to the *left*, caudal to the *right*, dorsal side *up*, ventral side *down*, DRG at the *bottom*). (Scale bar, 200 μm)

2.3 Whole-Mount Preparation of E12 Mouse Spinal Cord with Attached DRG

1. Surgical instruments for dissection (scissors, large forceps, and Dumont No. 5 dissecting forceps with inside-polished tips).
2. Dissecting stereomicroscope.
3. 100-mm Sylgard-coated Petri dish.
4. Sheet of filter paper.
5. 12-well plates and 100-mm petri dishes.
6. Ice-cold PBS: 137 mM NaCl, 2.7 mM KCl, 8 mM Na_2HPO_4, 1.47 mM KH_2PO_4.
7. Fixation buffer for subsequent AP staining: 2 % (v/v) PFA, 0.2 % (v/v) glutaraldehyde, 100 mM $MgCl_2$ in 1× PBS.

Table 1
Oligonucleotide primers for PCR genotyping

Name	Sequence
Npr2-WT-F	CTCAGATTCCTCCCTTCTCG
Npr2-WT-R	GGCATAGCTCAGGTTGTGTT
Npr2-CreER^{T2}-R	TTGGACATGGTGGAATTCAT
Npr2-cn-F	TTCACAGCGCTGTCAGCTGAG
Npr2-cn-R	ACTTAGGGAGCGCTGACTGTGG
ZAP-CMV-F	TCCGCGTTACATAACTTACGG
ZAP-CMV-R	ATGGGGAGAGTGAAGCAGAAC

2.4 Genotyping

1. High Pure PCR Template Preparation Kit (Roche) for purification of gDNA.
2. 10× PCR buffer (including dNTPs): 200 mM Tris–HCl, pH 8.4, 500 mM KCl, 15 mM $MgCl_2$, 2 mM of each dNTP (dATP, cCTP, dGTP, dTTP).
3. Taq polymerase (5 U/μl).
4. Oligonucleotide primers, each dissolved at 50 μM in H_2O: Npr2-WT-F, Npr2-WT-R, Npr2-CreER^{T2}-R, Npr2-cn-F, Npr2-cn-R, ZAP-CMV-F, and ZAP-CMV-R (*see* Table 1).
5. Restriction enzyme: *AflIII* (5 U/μl).

2.5 Detection of AP Reporter Gene Expression by Histochemical Staining

1. Stock solutions stored at room temperature: 1 M Tris–HCl, pH 9.5, 2.5 M NaCl, 1 M $MgCl_2$, 20 % (v/v) NP-40, and 10 % (w/v) sodium deoxycholate.
2. Stock solutions kept at −20 °C:
   (a) 25 mg/ml 5-bromo-4-chloro-3-indolyl phosphate (BCIP): Weigh 0.25 g BCIP in a 20 ml glass vial, add 10 ml of 100 % dimethylformamide, and vortex shortly. BCIP dissolves easily giving a yellow-colored solution. Store the dissolved BCIP protected from light at −20 °C.
   (b) 25 mg/ml nitroblue tetrazolium (NBT): Weigh 0.25 g NBT in a 20 ml glass vial, add 10 ml of 70 % (v/v in H_2O) dimethylformamide, and vortex shortly. NBT dissolves easily giving a colorless clear solution. Store the dissolved NBT protected from light at −20 °C.
3. 12-well plates.
4. PBS.
5. Water bath at 65 °C.
6. 100 mM $MgCl_2$ in PBS.

7. AP washing buffer: 10 mM Tris–HCl, pH 9.5, 100 mM NaCl, 10 mM $MgCl_2$.
8. AP staining solution: 100 mM Tris–HCl, pH 9.5, 100 mM NaCl, 50 mM $MgCl_2$. Fill up with water to 2/3 of the final volume. Add 0.02 % (v/v) NP-40 and 0.01 % sodium deoxycholate and vortex shortly. Add 337 μg/ml NBT and 175 μg/ml BCIP, fill up with water to the final volume.
9. PBS with 0.1 % Tween 20.
10. Stereomicroscope.
11. Mouse surgical instruments.
12. Vannas-style spring scissors.
13. ImmuMount (Thermo Scientific).

3 Methods

3.1 Crossbreeding of Mouse Lines

1. In a first step, breed (*see* **Note 2**) the Z/AP reporter line (*see* **Note 3**) with heterozygous Npr2-cn mice (*see* **Note 4**). Wean the offspring at the age of three weeks and genotype from ear punches or tail biopsies using PCR analysis (*see* Subheading 3.4).
2. Select $Npr2^{cn/+}$; Z/AP^{+} bigenic mice for breeding with Npr2-CreER^{T2} heterozygous animals in order to generate the experimental mice. $Npr2^{CreERT2/+}$; Z/AP^{+} and $Npr2^{CreERT2/cn}$; Z/AP^{+} (each with an expected yield of 12.5 % of the offspring) will be used for the comparison of DRG neuron morphologies in Npr2 heterozygous with those in Npr2 homozygous mutant animals.
3. For the breeding of timed pregnant mice set up one male with two female mice for mating and check for vaginal plugs every 12 h. The time point of plug detection is determined as E 0.5.
4. Weigh the plugged mouse to enable the determination of pregnancy at E9.5 (*see* **Note 5**).

3.2 Induction of AP Reporter Gene Expression by Administration of Tamoxifen

1. Set up a 20 mg/ml stock solution of tamoxifen (*see* **Notes 6** and 7) by suspending 40 mg of tamoxifen in 0.1 ml ethanol in a 2 ml plastic tube. Add 3.9 ml of corn oil, vortex the suspension vigorously, and sonicate for 5 min at 37 °C. Divide the stock solution into 0.15 ml aliquots and store protected from light up to 4 weeks at −20 °C.
2. Reweigh the plugged mice at E9.5. If pregnancy is likely (*see* **Note 5**) the animal weight is used to calculate the amount of tamoxifen to be administered (0.1 mg tamoxifen per g body weight) (*see* **Note 8**).

3. Sonicate the tamoxifen stock solution after thawing for 5 min at 37 °C. Measure out the required volume of stock solution (*see* **step 2** of Subheading 3.2) and bring to a final volume of 0.15 ml by addition of corn oil containing 5 % ethanol.
4. For oral gavage delivery (*see* **Note 9**) fill the tamoxifen suspension into a 1 ml syringe and attach the feeding needle.
5. Restrain the pregnant mouse by scruffing the skin behind its ears with your thumb and index finger. The animal should not be able to move its head. Additionally immobilize the mouse in your hand by pinning the tail between little finger and palm.
6. Gently without force insert the feeding needle into the mouse stomach by slightly rotating the syringe. Empty the syringe by slowly pressing the plunger. After removal of the feeding needle the animal should be under surveillance for 5 min. If a gag reflex is observed during this time, the gavaging procedure has to be repeated (*see* **Note 10**).

3.3 Whole-Mount Preparation of E12 Mouse Spinal Cord with Attached DRG

1. After administration of tamoxifen allow at least 3 days for reporter gene expression.
2. Euthanize the pregnant mouse at E12.5 according to the regulations approved by the local authorities for animal welfare and dissect the embryos from the uterine horns (*see* **Note 11**).
3. Decapitate each embryo and transfer the torso into a specified well of a 12-well plate with ice-cold PBS. Save a part of the corresponding head for the isolation of genomic DNA to allow subsequent PCR genotyping.
4. Place one embryonic torso at a time with its dorsal side up on a piece of filter paper in a sylgard-coated dish. Wet the tissue with a few drops of PBS to prevent it from drying.
5. Beginning in the middle of the torso, carefully pinch the skin above the spinal cord with two fine-tipped forceps and gently tear it apart. Proceed first towards the tail and then resuming from the middle towards the anterior side.
6. To facilitate the removal of the whole spinal cord with connected DRG insert the polished tip of a dissecting forceps in the middle of the embryo's right side just beneath the cord. Carefully detach the tissue with horizontal sliding movements first towards the tail and then to the anterior end. Repeat the procedure on the other side of the embryo.
7. Finally, pull out the loosened spinal cord with the attached DRG from the cervical part to the caudal end of the torso.
8. Submerge the isolated spinal cord in an appropriately labelled well of a 12-well plate filled with 2.5 ml of AP fixative and incubate for 30 min on ice with gentle shaking. Proceed

with the dissection of spinal cords from the remaining embryos.

3.4 Detection of AP Reporter Gene Expression by Histochemical Staining

1. Following 30 min of fixation transfer the whole-mount preparations of embryonic spinal cord into a 12-well plate with PBS and incubate for 1 h at 65 °C in a water bath to inactivate endogenous alkaline phosphatase (*see* **Note 12**).
2. Thereafter rinse the tissues in PBS with 100 mM $MgCl_2$ and incubate overnight at 4 °C.
3. The following day equilibrate the cords in AP washing buffer for 30 min at room temperature.
4. Incubate the cords in AP staining solution for up to 4 days in the dark with gentle agitation at room temperature. After AP-stained axons become visible, wash the whole-mount preparations in PBS containing 0.1 % Tween 20 and postfix overnight in 4 % PFA at 4 °C.
5. Wash twice with PBS at room temperature.
6. Using a dissection microscope position a stained cord with its ventral side up in a drop of PBS. Using spring-scissors cut the floor plate on the whole length of the spinal cord to allow flat-mounting of the cord in an open book mode.
7. Aspirate excess fluid and embed the flattened sample in an aqueous mounting medium (ImmuMount).
8. Image the samples using indirect illumination before a white background.

3.5 Genotyping

1. Purify genomic DNA from ear punches, tail biopsies, or embryonic tissue using the High Pure PCR Template Preparation Kit (Roche) according to the instructions of the manufacturer.
2. Components for PCR genotyping reactions (25 μl total volume per reaction):

	Npr2-CreER^{T2}	Npr2-cn	Z/AP
H_2O	16.3 μl	16.7 μl	17.1 μl
10× PCR buffer	2.5 μl	2.5 μl	2.5 μl
dNTP (2,5 mM each)	2 μl	2 μl	2 μl
$MgCl_2$ (50 mM)	0.75 μl	0.75 μl	0.75 μl
Oligonucleotide primers (50 μM)	0.4 μl Npr2-WT-F[a]	0.4 μl Npr2-cn-F[a]	0.2 μl ZAP-CMV-F[a]
	0.4 μl Npr2-WT-R[a]	0.4 μl Npr2-cn-R[a]	0.2 μl ZAP-CMV-R[a]
	0.4 μl Npr2-CreER^{T2}[a]	-	-
gDNA template (0,2 μg/μl)	2 μl	2 μl	2 μl
Taq polymerase (5 U/μl)	0.25 μl	-	0.25 μl

[a]Please *see* Table 1 for the sequences of PCR primers

3. PCR genotyping of Npr2-CreER^{T2}:
   (a) Standard PCR program: Initial denaturation step for 3 min at 95 °C; then 35 cycles of denaturation (30 s at 94 °C), annealing (30 s at 57 °C), and elongation (30 s at 72 °C); a final polymerization step for 5 min at 72 °C; store at 4 °C.
   (b) Expected product size: 456 bp for the wild type and 356 bp for the transgene.
4. PCR genotyping of Npr2-cn:
   (a) Hotstart PCR program: Initial denaturation step for 3 min at 95 °C; add 0.25 μl Taq polymerase (5 U/μl) at 80 °C; then 35 cycles of denaturation (35 s at 94 °C), annealing (25 s at 62 °C), and elongation (35 s at 72 °C); a final polymerization step for 5 min at 72 °C; store at 4 °C.
   (b) Subsequently, digest the PCR product with 0.25 μl *AflIII* (5 U/μl) in the appropriate buffer for 2 h at 37 °C (*see* **Note 13**).
   (c) Expected fragment size: 348 bp for the wild type, and 160 and 188 bp for the mutant.
5. PCR genotyping of Z/AP:
   (a) Standard PCR program: Initial denaturation step for 4 min at 95 °C; then 35 cycles of denaturation (45 s at 95 °C), annealing (30 s at 58 °C), and elongation (40 s at 72 °C); a final polymerization step for 5 min at 72 °C; store at 4 °C.
   (b) Expected fragment size: 347 bp for the transgene.
6. Analyze PCR products and restriction fragments by agarose gel electrophoresis.

4 Notes

1. Staining of 30 spinal cords with attached DRG from E12.5 mouse embryos with an appropriate genotype detected on average 5 AP-expressing DRG neurons per whole-mount preparation after a single dose of tamoxifen. No AP-positive cells have been observed in the absence of tamoxifen stimulation. This low recombination efficiency, that is optimal for the unambiguous analysis of axonal projection patterns, corresponds well with previously reported numbers of AP-stained retina cells in R26CreER; Z/AP mice [22]. Labelling has also been detected in other Npr2-expressing cells, e.g., neurons of the cranial sensory ganglia, demonstrating the general utility of this method for the analysis of axonal morphologies. Additionally, this approach might be adopted to a range of

other applications by the use of Npr2-CreER^{T2} mice in conjunction with high-efficiency fluorescent reporter lines.

2. Experimental use of mice should follow officially approved guidelines for the care and use of laboratory animals.
3. The Z/AP reporter mouse line encodes a double-reporter system: Constitutive expression of lacZ under control of the CMV enhancer/chicken actin promoter is switched to AP expression by Cre recombinase activity [20]. The Z/AP reporter line requires heterozygous maintenance because of homozygous lethality. In general, other reporter mouse lines with a loxP-flanked transcriptional termination cassette before the reporter transgene might be used for the analysis of axonal branching. However, the recombination efficiencies should be expected to vary and therefore the amount of tamoxifen applied for activation of Cre recombinase must be determined individually.
4. Npr2-cn heterozygous animals are used for breeding because *Npr2*$^{cn/cn}$ mice display a dwarfed phenotype as the result of disrupted guanylyl cyclase activity [23].
5. Pregnancy might be expected if the plugged mice gain in weight more than 2.5 g between plug detection and E9.5.
6. Tamoxifen is a known human carcinogen, teratogen, and mutagen. Therefore, its use must follow the officially approved security guidelines.
7. Alternatively, 4-hydroxytamoxifen-which is presumably the active ligand-might be used to induce Cre recombinase activity. However, note that tamoxifen is cheaper than 4-hydroxytamoxifen and somewhat easier to dissolve.
8. Administration of tamoxifen is usually regarded an animal experiment that requires a veterinary license from the local authorities.
9. Administration of tamoxifen by oral delivery has been reported to reduce litter loss associated with intraperitoneal inflammation after i.p. injection [18].
10. The tamoxifen-treated mice should be kept in separate cages and the animal bedding has to be collected for secure disposal.
11. A detailed procedure of the whole-mount spinal cord preparation with an accompanying video-protocol has been published elsewhere [24].
12. After fixation and heat inactivation it is possible to store the spinal cord preparations for longer time periods before continuing with the staining procedure.
13. The Npr2-cn mutation is caused by a T to G transversion [23] that creates a recognition site for *AflIII*.

Acknowledgements

We thank Dr. Alistair Garratt (Max Delbrück Center, Berlin) for a critical reading of the manuscript. Work in the authors' laboratory was supported by a grant from the Deutsche Forschungsgemeinschaft (SFB665).

References

1. Acebes A, Ferrus A (2000) Cellular and molecular features of axon collaterals and dendrites. Trends Neurosci 23:557–565
2. Schmidt H, Rathjen FG (2010) Signalling mechanisms regulating axonal branching in vivo. Bioessays 32:977–985
3. Gibson DA, Ma L (2011) Developmental regulation of axon branching in the vertebrate nervous system. Development 138:183–195
4. Gallo G (2011) The cytoskeletal and signaling mechanisms of axon collateral branching. Dev Neurobiol 71:201–220
5. Ozaki S, Snider WD (1997) Initial trajectories of sensory axons toward laminar targets in the developing mouse spinal cord. J Comp Neurol 380:215–229
6. Schmidt H, Werner M, Heppenstall PA et al (2002) cGMP-mediated signaling via cGKIalpha is required for the guidance and connectivity of sensory axons. J Cell Biol 159:489–498
7. Schmidt H, Stonkute A, Juttner R et al (2007) The receptor guanylyl cyclase Npr2 is essential for sensory axon bifurcation within the spinal cord. J Cell Biol 179:331–340
8. Schmidt H, Stonkute A, Juttner R et al (2009) C-type natriuretic peptide (CNP) is a bifurcation factor for sensory neurons. Proc Natl Acad Sci USA 106:16847–16852
9. Zhao Z, Wang Z, Gu Y et al (2009) Regulate axon branching by the cyclic GMP pathway via inhibition of glycogen synthase kinase 3 in dorsal root ganglion sensory neurons. J Neurosci 29:1350–1360
10. Zhao Z, Ma L (2009) Regulation of axonal development by natriuretic peptide hormones. Proc Natl Acad Sci USA 106:18016–18021
11. Golgi C (1873) Sulla struttura della sostanza grigia del cervello. Gazzetta Medica Italiana-Lombardia 33:244–246
12. Kristensson K, Olsson Y (1971) Retrograde axonal transport of protein. Brain Res 29:363–365
13. Lapper SR, Bolam JP (1991) The anterograde and retrograde transport of neurobiotin in the central nervous system of the rat: comparison with biocytin. J Neurosci Methods 39:163–174
14. Honig MG, Hume RI (1989) Dil and diO: versatile fluorescent dyes for neuronal labelling and pathway tracing. Trends Neurosci 12:333–1
15. Feng G, Mellor RH, Bernstein M et al (2000) Imaging neuronal subsets in transgenic mice expressing multiple spectral variants of GFP. Neuron 28:41–51
16. Birling MC, Gofflot F, Warot X (2009) Site-specific recombinases for manipulation of the mouse genome. Methods Mol Biol 561:245–263
17. Badea TC, Wang Y, Nathans J (2003) A noninvasive genetic/pharmacologic strategy for visualizing cell morphology and clonal relationships in the mouse. J Neurosci 23:2314–2322
18. Joyner AL, Zervas M (2006) Genetic inducible fate mapping in mouse: establishing genetic lineages and defining genetic neuroanatomy in the nervous system. Dev Dyn 235:2376–2385
19. Feil R, Wagner J, Metzger D et al (1997) Regulation of Cre recombinase activity by mutated estrogen receptor ligand-binding domains. Biochem Biophys Res Commun 237:752–757
20. Lobe CG, Koop KE, Kreppner W et al (1999) Z/AP, a double reporter for cre-mediated recombination. Dev Biol 208:281–292
21. Badea TC, Nathans J (2004) Quantitative analysis of neuronal morphologies in the mouse retina visualized by using a genetically directed reporter. J Comp Neurol 480:331–351
22. Badea TC, Hua ZL, Smallwood PM et al (2009) New mouse lines for the analysis of neuronal morphology using CreER(T)/loxP-directed sparse labeling. PLoS One 4:e7859
23. Tsuji T, Kunieda T (2005) A loss-of-function mutation in natriuretic peptide receptor 2 (Npr2) gene is responsible for disproportionate dwarfism in cn/cn mouse. J Biol Chem 280:14288–14292
24. Schmidt H, Rathjen FG (2011) DiI-labeling of DRG neurons to study axonal branching in a whole mount preparation of mouse embryonic spinal cord. J Vis Exp 58:e3667

Chapter 13

Receptor Binding Assay for NO-Independent Activators of Soluble Guanylate Cyclase

Peter M. Schmidt and Johannes-Peter Stasch

Abstract

The characterization of the interaction between a ligand and its receptor is crucial for a broad variety of applications in academia as well as in the pharmaceutical industry. Although various sophisticated high-throughput technologies have been established to investigate the binding of ligands to their receptors, classical filtration-based receptor binding assays still have some advantages when smaller number of samples need to be tested. Here we describe a technically easy, cheap, and reliable receptor binding assay that was successfully applied to determine the binding constant of the NO-independent activator of soluble guanylate cyclase, cinaciguat, and the impact of other small molecules on its interaction with the enzyme.

Key words Receptor binding assay, Filter plates, Radioligand, Soluble guanylate cyclase, cinaciguat, riociguat, BAY 58-2667, sGC, Binding constant

1 Introduction

Characterizing the interaction of a ligand with its receptor is of crucial importance for the understanding of countless biological processes including intra- and extracellular signaling cascades, cell–cell and cell–matrix interactions, or intercellular communication to name a few. A detailed knowledge of the binding mechanism of a ligand to its receptor opens the opportunity to modulate this interaction in a positive or a negative way. This understanding is the basis for various applications including the development of new drugs.

Especially to cater to the needs of the pharmaceutical industry to screen vast amounts of potential drug candidates, various highly developed approaches have been established, each with its own strength and weaknesses. Assays such as Alpha screen [1, 2], time-resolved fluorescence [3, 4], complementation assays [5, 6], surface

Conflict of Interest: Peter M. Schmidt was employed from 2000 to 2003 by Bayer Pharma AG and is currently a full-time employee of CSL. Johannes-Peter Stasch is currently a full-time employee of Bayer Pharma AG.

Thomas Krieg and Robert Lukowski (eds.), *Guanylate Cyclase and Cyclic GMP: Methods and Protocols*, Methods in Molecular Biology, vol. 1020, DOI 10.1007/978-1-62703-459-3_13, © Springer Science+Business Media, LLC 2013

plasmon resonance [7, 8], scintillation proximity assays [9], isothermal calorimetry [10], or micro-thermophoresis [11] are undoubtedly powerful approaches. However, they also have several drawbacks including the need of chemically modified ligands, which can alter the interaction with the receptor; the amount of receptor and/or ligand needed to get a robust signal; or the need to buy very specialized hardware or commercial kits [12].

The present protocol is not suitable for high-throughput campaigns due to the different manual steps needed. However, this classical receptor binding assay developed for the NO-independent sGC activator cinaciguat (BAY 58-2667) uses only microgram amounts of its receptor sGC and does not rely on chemically modified ligands or specialized expensive hardware. It is suitable for projects aiming to address how small-molecule sGC activators such as cinaciguat bind to the enzyme or, alternatively, whether they interfere with the binding of cinaciguat. By using this protocol, we determined the binding affinity of cinaciguat [13] and chemical derivatives to its receptor sGC [14] and were able to show that the NO-independent sGC-stimulator riociguat did not interfere with the binding of cinaciguat suggesting a different binding site for both compounds [13]. Furthermore, we found that removal of the enzyme's prosthetic heme group [15] or the addition of compounds interacting with the heme pocket [16] had direct impact on the binding of cinaciguat suggesting a new mechanism of sGC activation by cinaciguat [17]. In the same fashion the present protocol could be useful to investigate other small-molecule sGC activators [17, 18] or to facilitate the search for potential endogenous cinaciguat-like sGC activators [19].

2 Materials

All chemicals used are of highest analytical grade and prepared in ultrapure water. Cinaciguat can be stored at room temperature. 3H-cinaciguat and cinaciguat were prepared as described [20, 21]. BAY 41–2272 and purified sGC can be obtained from Enzo Life Sciences (www.enzolifesciences.com). Cinaciguat (BAY 58-2667) is available via Axxora (www.axxora.com). It is important to keep in mind that some NO-independent sGC activators such as cinaciguat are affected by the oxidation state of the prosthetic heme group of sGC. NO-sensitive (reduced) sGC shows a characteristic absorption peak at 431 nm (Soret peak), which is shifted to 392 nm once the heme group is oxidized. In order to maintain reducing conditions the incubation buffer in the present protocol contains dithiothreitol. However, in general it is advised to confirm the heme oxidation state of sGC by spectroscopy before using an enzyme preparation (e.g. the data sheet for sGC from Enzo Life Sciences shows the respective absorption spectrum).

2.1 Dilution Series of cinaciguat

1. Cinaciguat stock solution: Weigh out required amount (e.g., 1 mg) into a 1.5 ml Eppendorf tube. Dissolve compound in a 60 % (v/v) mixture of acetonitrile and water (ACN:H_2O) to obtain a concentration of 2 mM (*see* **Note 1**). Stock solutions can be stored at room temperature.
2. Tritium-labelled cinaciguat (3H-cinaciguat): 10 mM stock solution in ACN:H_2O with a specific activity of about 5.4 Ci/mmol (*see* **Note 2**).
3. 3H-cinaciguat pre-dilution: Create a 100 μM pre-dilution of 3H-cinaciguat by mixing 10 μl of the stock solution (10 mM) with 990 μl ACN:H_2O.
4. Concentration series of 3H-cinaciguat: Create a 10 and 3 μM dilution of 3H-cinaciguat by mixing 100 μl/30 μl of the pre-dilution (100 μM) with 900 μl/970 μl ACN:H_2O. Sub-dilute 100 μl of each concentration with 900 μl of ACN:H_2O to obtain 3H-cinaciguat concentrations of 1 μM and 300 nM, respectively. Repeat steps to make up 3H-cinaciguat concentrations of 100, 30, 10, and 3 nM (*see* **Note 3**). Store dilutions at room temperature.
5. Concentration series of 3H-cinaciguat with a fixed concentration of non-labelled cinaciguat: Create a 20 μM/6 μM dilution of 3H-cinaciguat by mixing 200 μl/60 μl of the pre-dilution (100 μM) with 800 μl/940 μl ACN:H_2O. Sub-dilute 100 μl of each concentration with 900 μl of ACN:H_2O to obtain 3H-cinaciguat concentrations of 2 μM and 600 nM, respectively. Repeat steps to make up 3H-cinaciguat concentrations of 200, 60, 20, and 6 nM. Finally, mix each 3H-cinaciguat concentration with an equal volume of nonradioactive cinaciguat (2 mM) in ACN:H_2O (e.g., 500 μl + 500 μl). Store dilutions at room temperature.

2.2 Filtration Components

1. Standard vacuum manifold for 96-well plates.
2. 96 Well glass fiber filter membrane plates (Multiscreen FC 1 μm, Millipore).
3. Coating solution: 0.5 % (w/v) polyvinylpyrolidone (PVP), 0.1 % (v/v) Tween-20 in water (*see* **Note 4**).

2.3 Assay Components

1. 1× wash buffer: 100 mM NaCl, 10 mM Tris–HCl, pH 7.2. Store at room temperature. Can be used for multiple assays.
2. 5× incubation buffer: 0.5 mM EDTA, 5 mM dithiothreitol, 15 mM $MgCl_2$, 250 mM triethanolamine (TEA)-HCl, pH 7.5. Store at room temperature. Make up fresh for every assay as dithiothreitol is not stable.
3. IgG-buffer: 1 % (w/v) IgG in 1× incubation buffer (*see* **Note 5**).
4. PEG-buffer: 35 % (w/v) PEG 8000 in 1× incubation buffer (*see* **Note 6**).

5. Cut pipette tips: 200 μl Eppendorf tips with a cut tip (*see* **Note 7**).

2.4 Components for the Detection of 3H-cinaciguat

1. Thermomixer for 1.5 ml reaction tubes.
2. Liquid scintillation cocktail.
3. Scintillation vials.
4. Liquid scintillation counter.

3 Methods

Carry out all procedures at room temperature unless otherwise specified. Make sure that your laboratory is certified to allow handling of radioactive substances. Discuss all procedures with the respective radiation safety officer and make sure that all radioactive waste is collected and disposed in an adequate way.

3.1 Coating Glass Fiber Filter Plates

1. Prepare 20 ml of coating solution for one 96-well filter plate.
2. Transfer 200 μl of coating solution into the wells of the filter plate.
3. Incubate plate for at least 3 h (*see* **Note 8**).
4. Aspirate the coating solution by using a vacuum manifold.
5. Wash coated filter plates 6 times with 200 μl ice-cold 1× wash buffer (*see* **Note 9**).
6. Plates can now be used for the receptor binding assay.

3.2 Receptor Binding Assay

1. The assay mixture for a single well without the addition of cinaciguat consists of 20 μl 5× incubation buffer, 68 μl water, and 2 μl purified sGC (=90 μl). For an entire concentration response curve (8 concentrations of 3H-cinaciguat) in duplicate and in the absence and presence of nonradioactive cinaciguat prepare a 32× mastermix: 640 μl 5× incubation buffer, 64 μl purified sGC, and 2,176 μl water resulting in a total volume of 2,880 μl (*see* **Note 10**). Transfer 90 μl of the mastermix into 32× 1.5 ml reaction tubes labelled according to the respective 3H-cinaciguat concentrations.
2. Add 10 μl of the prepared 10× concentrated 3H-cinaciguat ligand mixtures to the tubes containing the sGC-assay mixture. Make up all samples in duplicate (8 concentrations 3H-cinaciguat in duplicate = 16 tubes).
3. Repeat procedure with the prepared 3H-cinaciguat concentrations premixed with an at least 100-fold excess of nonradioactive cinaciguat (*see* **Note 11**).
4. Incubate all 32 samples for 10 min at 37 °C.

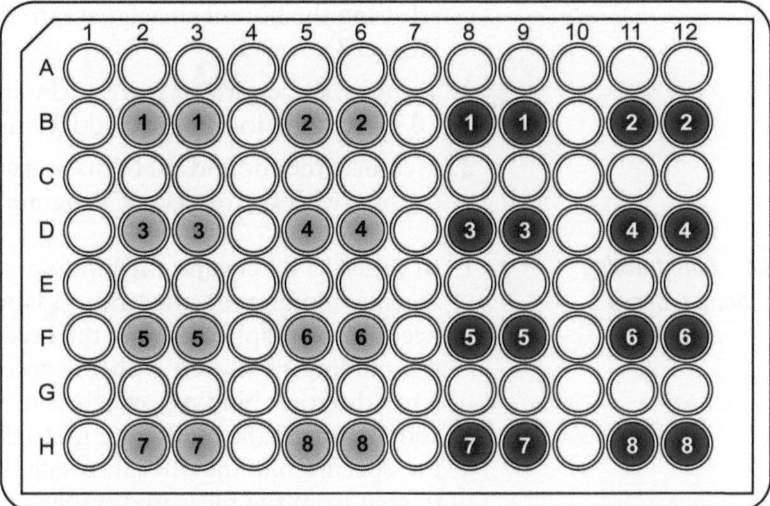

Fig. 1 Layout of the 96-well filtration plate. After mixing the sGC assay mixture (90 μl per tube; *see* Subheading 3.2, **step 1**) with 10 μl of the tenfold concentrated cinaciguat solution (*see* Subheading 3.2, **step 2**) and subsequent incubation and precipitation (*see* Subheading 3.2, **steps 4** and **7**) the assay mixtures (200 μl) are transferred into the pre-coated 96-well filter plate. A possible layout of this plate is shown in this scheme. Wells 1–8 represent the decreasing concentration of 3H-cinaciguat (1 μM, 300, 100, 30, 10, 3, 1, and 0.3 nM) in duplicate in the absence (*light grey*) or the presence (*dark grey*) of an at least 100-fold excess of nonradioactive cinaciguat (*see* Subheading 2.1)

5. Place samples on ice for 10 min.
6. Add 15 μl of ice-cold IgG-buffer to each sample resulting in a final IgG concentration per sample of 0.75 μg/μl (*see* **Note 5**).
7. Add 85 μl of ice-cold PEG-buffer to each sample to precipitate the proteins. Final PEG concentration in the assay mixture is 15 % (w/v) (*see* **Note 6**).
8. Transfer the entire assay mixtures (200 μl) containing the precipitated protein into the wells of the pre-coated filter plate. Separate unbound cinaciguat from the precipitated sGC by filtration (*see* Fig. 1; **Note 12**).
9. Wash wells twice with 200 μl ice-cold 1× incubation buffer.
10. Remove the filter membranes of all wells by using a 200 μl pipette tip and transfer the membranes into 1.5 ml reaction tubes (*see* Fig. 1; **Note 13**).

3.3 Measuring Bound cinaciguat

1. Add 500 μl of ACN:H_2O to each of the 32 tubes and shake tubes for 3 h at 40 °C and 1,000 rpm in a thermo mixer. The vigorous shaking disintegrates the glass fiber filters entirely

releasing the bound cinaciguat into the ACN:H_2O mixture (*see* **Note 13**).

2. Transfer the entire mixture including the filter debris into a 20 ml scintillation vial and add 10 ml of scintillation cocktail.
3. Measure the bound 3H-cinaciguat by using a scintillation counter with the respective program for tritium.

3.4 Constructing a Binding Curve

1. In order to determine the binding constant for cinaciguat the specific binding of the tritium-labelled compound to sGC needs to be graphed against the used concentration. However, the binding measured with the receptor binding assay represents the total binding, which consists of the specific binding and the unspecific binding (e.g., to the filter matrix). To obtain the specific binding the unspecific binding needs to be subtracted from the measured results.
2. The unspecific binding of a radioligand is measured in the presence of a vast excess of nonradioactive ligand. The concentration of the nonradioactive ligand should be at least 100-fold (better 1,000-fold) higher than the highest concentration of radioligand. Due to its limited solubility a 100-fold excess of cinaciguat was used for this assay.
3. Subtract for all concentrations of 3H-cinaciguat the measured signal in the presence of nonradioactive cinaciguat from the respective values obtained in its absence.
4. Graph the calculated values against the 3H-cinaciguat concentrations to get the sigmoid binding curve shown in Fig. 2a.
5. The 3H-cinaciguat concentration that results in 50 % of the maximum specific (saturable) binding to the target enzyme sGC represents the KD value for cinaciguat. In the present case the KD value was determined to be 13.4 nM.
6. For a protein without specific binding sites, the total binding equals the unspecific binding resulting in a specific binding curve that is flat. This case is shown in Fig. 2b for a heme-containing model enzyme: glucose-oxidase (GOD).

4 Notes

1. Due to its amphipathic character cinaciguat has limited solubility in aqueous solutions. However, the compounds can be readily dissolved in DMSO or, as in this study, in 60 % (v/v) ACN in water. For the present study ACN was chosen as 3H-cinaciguat was obtained in this solvent. Stock solutions can be divided into aliquots and frozen at −20 °C without loss of activity.

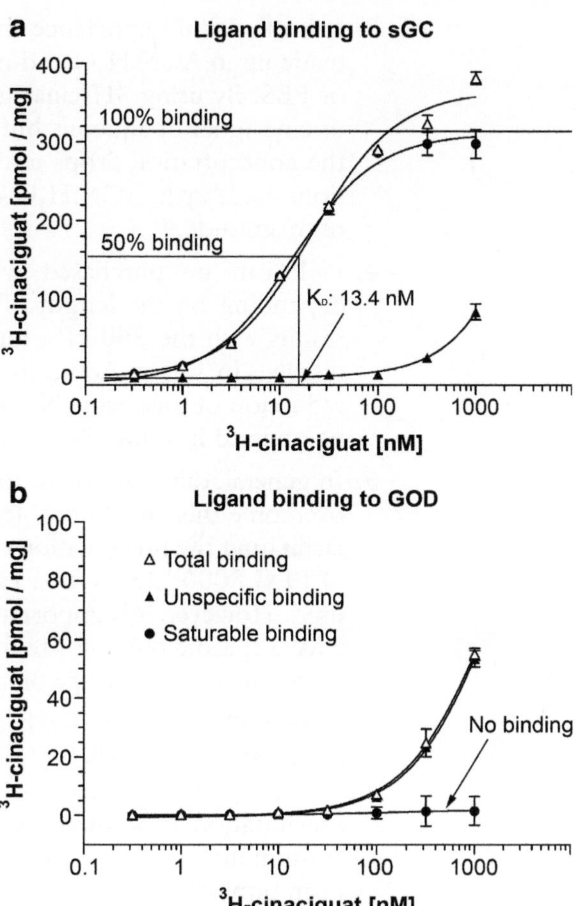

Fig. 2 Determining the binding constant for cinaciguat. In the absence of nonradioactive cinaciguat the measured radioactivity increases with the concentration of 3H-cinaciguat. This so-called total binding cannot be saturated as shown in (**a**) (*white triangles*) as it includes the unspecific binding of 3H-cinacguat to assay components (*black triangles*), which will increase with higher concentrations of radioligand. To obtain correct values for the specific binding of 3H-cinaciguat, the unspecific binding needs to be subtracted from the total binding. The unspecific binding is determined by performing the assay with a vast excess of nonradioactive cinaciguat. The resulting specific (saturable) binding curve of 3H-cinaciguat to sGC allows determining the binding constant K_D. For a protein without specific binding site for cinaciguat such as glucose oxidase (GOD) the total binding equals the unspecific binding resulting in no specific binding at all (**b**). Reprinted from Analytical Biochemistry, 314 (1), Schmidt P, Schramm M, Schroder H, Stasch JP, "*Receptor binding assay for nitric oxide- and heme-independent activators of soluble guanylate cyclase*," 162–165, 2003, with permission from Elsevier

2. In general carbon-14-labelled cinaciguat could be used as well but due to its higher specific activity 3H-cinaciguat is easier to detect at low concentrations.

3. It is of critical importance that all cinaciguat sub-dilutions are made up in ACN:H_2O and not in aqueous buffer such as TBS or PBS. By using 3H-cinaciguat we found that serial dilutions of cinaciguat in aqueous buffers become very inaccurate once the concentration drops under 1 μM. In contrast, serial dilutions made up in ACN:H_2O show a high accuracy over 6 orders of magnitude.
4. PVP can be purchased with different molecular weights depending on the length of the polymer. We obtained good results with the 360 kDa form. PVP takes some time to fully dissolve in water. Addition of Tween-20 resulted in further reduction of unspecific binding to the filter [22], but it might be omitted if it interferes with some compounds.
5. In general, diluted protein solutions are hard to precipitate. To overcome this known problem, IgG is added as carrier protein facilitating the precipitation of sGC by the subsequent addition of PEG 8000 [23]. Other carrier proteins than IgG might be used. However, it is important that the carrier protein does not have a specific binding site for cinaciguat as this would impact on the measured specific binding.
6. PEG 8000 showed the best precipitation results at a final concentration of 15 % (w/v). In general, PEG has been chosen as precipitant in contrast to other established protein precipitation procedures (e.g., ethanol, trichloroacetic acid, ammonium sulfate) mainly for two reasons: Firstly, PEG has been published to be a gentle non-denaturing precipitant that is hardly affected by buffer conditions and, in addition, shows minimal interaction with the protein to be precipitated [24, 25]. Secondly, the binding of cinaciguat to sGC strongly depends on the ionic interaction of its negatively charged carboxylic groups with the enzyme [26]. In order to minimize potential interferences with this interaction a nonionic precipitant such as PEG was favored. However, the drawback of PEG is its high viscosity which makes it harder to handle.
7. The 35 % (w/v) PEG 8000 solution is very viscous and hard to mix with the binding assay. To facilitate the handling and to avoid the introduction of bubbles, PEG 8000 is added to the assay mixture with cut 200 μl pipette tips. A box with cut tips was prepared before the assay.
8. We found that an incubation time of 3 h resulted in a low unspecific binding of 3H-cinaciguat to the filter plates. Longer incubation times did not reduce the background signal any further.
9. Tween-20 has been commonly used to remove the non-covalently bound prosthetic heme moiety of sGC [27, 28]. To avoid inadvertent loss of the enzyme's prosthetic group by the Tween-20 in the coating buffer, plates were washed extensively.

10. A 32× mastermix should be enough to set-up 32 tubes. However, it is more likely that a 34× mastermix will be needed to ensure to have enough assay buffer for all 32 tubes.
11. In order to measure a reliable binding constant, cinaciguat's association to and dissociation from its receptor sGC need to be at equilibrium. As this state is reached faster if cinaciguat and 3H-cinaciguat are added to the assay mixture simultaneously, both compounds are premixed.
12. As the assay mixture is very viscous due to the addition of PEG 8000 it can be problematic to aspirate unbound 3H-cinaciguat through the filter by using the vacuum manifold. In case a well gets clogged use a 1 ml pipette with a cut 1,000 μl tips to apply pressure onto a well while the filter plate is still hooked up on the vacuum manifold.
13. In theory liquid scintillation cocktail can be added directly to the wells to measure bound radioligand. However, we found that only a fraction of the signal can be detected this way. Therefore the glass fiber filters are removed from the filtration plate by using a 200 μl pipette tip and subsequently incubated under rigorous shaking with ACN:H_2O. This procedure releases bound radioligand from the filter membranes into the ACN:H_2O solution.

References

1. Eglen RM, Reisine T, Roby P, Rouleau N, Illy C, Bosse R, Bielefeld M (2008) The use of AlphaScreen technology in HTS: current status. Curr Chem Genomics 1:2–10
2. Ullman EF, Kirakossian H, Singh S, Wu ZP, Irvin BR, Pease JS, Switchenko AC, Irvine JD, Dafforn A, Skold CN et al (1994) Luminescent oxygen channeling immunoassay: measurement of particle binding kinetics by chemiluminescence. Proc Natl Acad Sci USA 91(12):5426–5430
3. Bazin H, Trinquet E, Mathis G (2002) Time resolved amplification of cryptate emission: a versatile technology to trace biomolecular interactions. J Biotechnol 82(3):233–250
4. Degorce F, Card A, Soh S, Trinquet E, Knapik GP, Xie B (2009) HTRF: a technology tailored for drug discovery—a review of theoretical aspects and recent applications. Curr Chem Genomics 3:22–32
5. Olson KR, Eglen RM (2007) Beta galactosidase complementation: a cell-based luminescent assay platform for drug discovery. Assay Drug Dev Technol 5(1):137–144
6. Rothkegel C, Schmidt PM, Atkins DJ, Hoffmann LS, Schmidt HH, Schroder H, Stasch JP (2007) Dimerization region of soluble guanylate cyclase characterized by bimolecular fluorescence complementation in vivo. Mol Pharmacol 72(5):1181–1190
7. Jason-Moller L, Murphy M, Bruno J (2006) Overview of Biacore systems and their applications. Current protocols in protein science Chapter 19:Unit 19.14. doi: 10.1002/0471142301.ps1914s45
8. Navratilova I, Hopkins AL (2011) Emerging role of surface plasmon resonance in fragment-based drug discovery. Future Med Chem 3(14):1809–1820
9. Glickman JF, Schmid A, Ferrand S (2008) Scintillation proximity assays in high-throughput screening. Assay Drug Dev Technol 6(3):433–455
10. O'Neill MA, Gaisford S (2011) Application and use of isothermal calorimetry in pharmaceutical development. Int J Pharm 417(1–2):83–93
11. Jerabek-Willemsen M, Wienken CJ, Braun D, Baaske P, Duhr S (2011) Molecular interaction studies using microscale thermophoresis. Assay Drug Dev Technol 9(4):342–353
12. Schmidt PM (2009) Biochemical detection of cGMP from past to present: an overview. Handb Exp Pharmacol 191:195–228
13. Schmidt P, Schramm M, Schroder H, Stasch JP (2003) Receptor binding assay for nitric oxide- and

heme-independent activators of soluble guanylate cyclase. Anal Biochem 314(1):162–165
14. Stasch JP, Schmidt P, Alonso-Alija C, Apeler H, Dembowsky K, Haerter M, Heil M, Minuth T, Perzborn E, Pleiss U, Schramm M, Schroeder W, Schroder H, Stahl E, Steinke W, Wunder F (2002) NO- and haem-independent activation of soluble guanylyl cyclase: molecular basis and cardiovascular implications of a new pharmacological principle. Br J Pharmacol 136(5):773–783
15. Stasch JP, Schmidt PM, Nedvetsky PI, Nedvetskaya TY, Kumar HSA, Meurer S, Deile M, Taye A, Knorr A, Lapp H, Muller H, Turgay Y, Rothkegel C, Tersteegen A, Kemp-Harper B, Muller-Esterl W, Schmidt HH (2006) Targeting the heme-oxidized nitric oxide receptor for selective vasodilatation of diseased blood vessels. J Clin Invest 116(9):2552–2561
16. Hoffmann LS, Schmidt PM, Keim Y, Schaefer S, Schmidt HH, Stasch JP (2009) Distinct molecular requirements for activation or stabilization of soluble guanylyl cyclase upon haem oxidation-induced degradation. Br J Pharmacol 157(5):781–795
17. Evgenov OV, Pacher P, Schmidt PM, Hasko G, Schmidt HH, Stasch JP (2006) NO-independent stimulators and activators of soluble guanylate cyclase: discovery and therapeutic potential. Nat Rev 5(9):755–768
18. Schmidt HH, Schmidt PM, Stasch JP (2009) NO- and haem-independent soluble guanylate cyclase activators. Handb Exp Pharmacol 191:309–339
19. Wang Y, Liu H, McKenzie G, Witting PK, Stasch JP, Hahn M, Changsirivathanathamrong D, Wu BJ, Ball HJ, Thomas SR, Kapoor V, Celermajer DS, Mellor AL, Keaney JF Jr, Hunt NH, Stocker R (2010) Kynurenine is an endothelium-derived relaxing factor produced during inflammation. Nat Med 16(3): 279–285
20. Hahn MG, Alonso-Alija C, Stoll F, Heil M, Mittendorf M, Schlemmer KH, Wunder F, Stasch JP (2007) Design and synthesis of the first NO- and haem-independent sGC activator BAY 58-2667 for the treatment of acute decompensated heart failure. BMC Pharmacol 7:P25
21. Seidel D, Pleiss U (2010) Labelling of the guanylate cyclase activator cinaciguat (BAY 58-2667) with carbon-14, tritium and stable isotopes. J Label Compd Radiopharm 53:130–139
22. Scott CW, Gomes BC, Hubbs SJ, Koenigbauer HC (1995) A filtration-based assay to quantitate granulocyte-macrophage colony-stimulating factor binding. Anal Biochem 228(1):150–154
23. Demoliou-Mason CD, Barnard EA (1984) Solubilization in high yield of opioid receptors retaining high-affinity delta, mu and kappa binding sites. FEBS Lett 170(2):378–382
24. Atha DH, Ingham KC (1981) Mechanism of precipitation of proteins by polyethylene glycols. Analysis in terms of excluded volume. J Biol Chem 256(23):12108–12117
25. Bhat R, Timasheff SN (1992) Steric exclusion is the principal source of the preferential hydration of proteins in the presence of polyethylene glycols. Protein Sci 1(9):1133–1143
26. Schmidt PM, Schramm M, Schroder H, Wunder F, Stasch JP (2004) Identification of residues crucially involved in the binding of the heme moiety of soluble guanylate cyclase. J Biol Chem 279(4):3025–3032
27. Foerster J, Harteneck C, Malkewitz J, Schultz G, Koesling D (1996) A functional heme-binding site of soluble guanylyl cyclase requires intact N-termini of alpha 1 and beta 1 subunits. Eur J Biochem 240(2):380–386
28. Schmidt P, Schramm M, Schroder H, Stasch JP (2003) Preparation of heme-free soluble guanylate cyclase. Protein Expr Purif 31(1):42–46

Chapter 14

Direct Intrathecal Drug Delivery in Mice for Detecting In Vivo Effects of cGMP on Pain Processing

Ruirui Lu and Achim Schmidtko

Abstract

Intrathecal delivery of drugs is an important method in pain research in order to investigate pain-relevant effects in the spinal cord in vivo. Here, we describe a method of intrathecal drug delivery by direct lumbar puncture in mice. The procedure does not require surgery, is rapidly performed, and does not produce neurological deficits. If cGMP analogs are injected, a state of transient hindpaw hypersensitivity can be induced which is quantifiable by measurement of hindpaw withdrawal latency in response to mechanical stimulation.

Key words Direct lumbar puncture, Subarachnoid space, Spinal delivery, Pain, Mice

1 Introduction

A large body of evidence indicates that cGMP essentially contributes to the processing of persistent pain in the spinal cord. One pain-relevant cGMP-dependent signaling pathway is initiated by release of NO and subsequent cGMP production by NO-sensitive guanylyl cyclase (also called soluble guanylyl cyclase) in spinal cord neurons [1]. Recent studies indicate that cGMP is also produced by particulate guanylyl cyclase types A and B upon activation by natriuretic peptides in primary afferent neurons that terminate in the spinal cord [1–3]. cGMP exerts its pain-relevant effects by activation of cGMP-dependent protein kinase Iα [4–7], cyclic nucleotide-gated channels [8], and probably additional, so far unidentified targets. Hence, different cGMP sources and cGMP targets in the spinal cord contribute to pain processing during persistent pain.

In order to study the effects of cGMP-modulating drugs and other compounds on pain processing in the spinal cord, it is useful to apply the substances directly to the site of interest, and to observe the induced behavioral effects [9]. Anatomically, the spinal cord, which extends from the foramen magnum to the level of the first or second lumbar vertebrae, is embedded in the vertebral column

Thomas Krieg and Robert Lukowski (eds.), *Guanylate Cyclase and Cyclic GMP: Methods and Protocols*, Methods in Molecular Biology, vol. 1020, DOI 10.1007/978-1-62703-459-3_14, © Springer Science+Business Media, LLC 2013

and surrounded by connective meninges (dura mater, arachnoid mater, and pia mater). The space between the arachnoid mater and the pia mater is named the subarachnoid space and filled with cerebrospinal fluid. Lumbar delivery of drugs into the subarachnoid space (i.e., intrathecal delivery) allows for pharmacological investigations at the lumbar spinal cord level. The method was first described by Hylden and Wilcox [9] and later upgraded by Fairbanks [10]. With this technique, drugs can be administered efficiently, rapidly, and without need of surgical manipulations. Here we describe in detail a modified version of the direct lumbar puncture method in mice, and we demonstrate how this method can be used for detecting in vivo effects of cGMP in pain processing.

2 Materials

2.1 Equipment

1. Vapor-tight glass tank for isoflurane anesthesia.
2. Luer tip syringe (10 μl).
3. Sterile disposable 30 gauge ½ in. needles.
4. Dynamic Plantar Aesthesiometer (Ugo Basile, Comerio, Italy) for the determination of mechanical paw sensitivity (*see* **Note 1**).

2.2 Animals

We used male C57BL/6N mice (6–12 weeks old) for our study. All experiments have to be approved by the local Ethics Committee for Animal Research (*see* **Note 2**).

2.3 Chemicals

1. Isoflurane.
2. Skin antiseptic.
3. cGMP analog, for example 8-(4-chlorophenylthio) guanosine-3′,5′-cyclic monophosphate (8-pCPT-cGMP; dissolved in sterile 0.9 % saline) (*see* **Note 3**).

3 Methods

3.1 Determination of Baseline Mechanical Thresholds

1. Place the mice into the enclosures of a Dynamic Plantar Aesthesiometer at least 2 h before the experiment to habituate them to the measurement conditions.
2. Measure the mechanical sensitivity of the plantar hindpaw prior to drug injection (i.e., at baseline). The Dynamic Plantar Aesthesiometer pushes a thin rod (diameter 0.5 mm) with increasing force through a wire-grated floor against the plantar surface of a hindpaw from beneath. It stops automatically and records the latency time after which the animal withdraws the paw (Fig. 1). The force increases from 0 to 5 g within 10 s

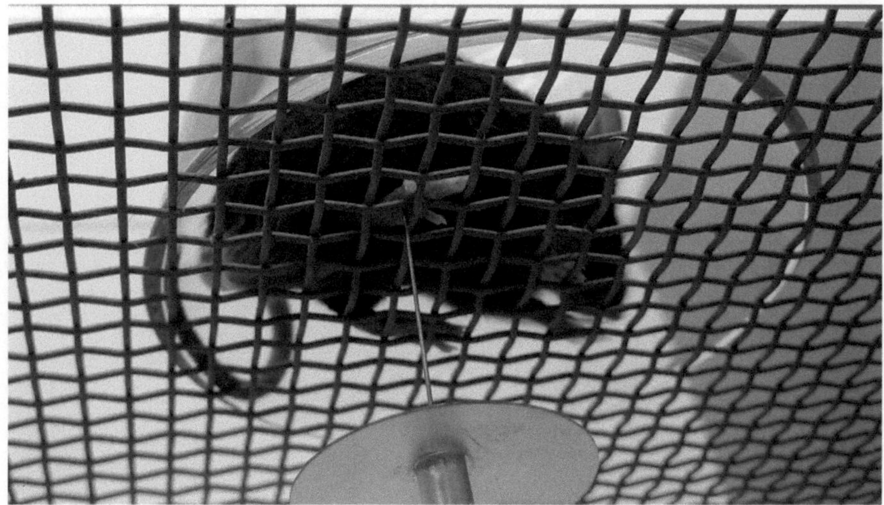

Fig. 1 Measurement of mechanical paw withdrawal latency times using a Dynamic Plantar Aesthesiometer. The mouse is placed in an enclosure with a perforated bottom. The stimulator of the device is positioned under a hindpaw, and after starting the measurement a thin rod rises with increasing force to touch the plantar surface of a paw. The measurement stops automatically and the latency time after which the animal withdraws the paw is recorded

(ramp 0.5 g/s) and is then held at 5 g for an additional 10 s. Baseline paw withdrawal latency time is calculated as the mean of 4–6 consecutive stimulations of the left and right hindpaws at intervals of at least 20 s. In 6- to 12-week-old C57BL/6N mice, the mean baseline paw withdrawal latency time is normally 7–9 s.

3.2 Intrathecal Drug Delivery

The intrathecal drug delivery is performed using a technique similar to that described previously [9, 10] (*see* **Note 4**):

1. Use short inhalation anesthesia to facilitate the direct intrathecal puncture. This is accomplished by placing the mouse in a small glass tank, which contains an isoflurane-soaked paper towel. To prevent exposure of the experimenter to isoflurane, the procedure is performed in a ventilated fume hood. The procedure requires vigilant monitoring, since the depth of the isoflurane anesthesia may change rapidly.
2. Once the mouse is anesthetized, remove it quickly from the glass tank and shave it on the back (*see* **Note 5**).
3. Place the mouse on a flat operating surface in the prone position, disinfect the back with a skin antiseptic, and place a nitrile glove (cut in half) over the head and the upper body of the anesthetized mouse (Fig. 2). During the succeeding steps the mouse gradually wakes up, but the darkness of the glove keeps the mouse calm and prevents it biting.

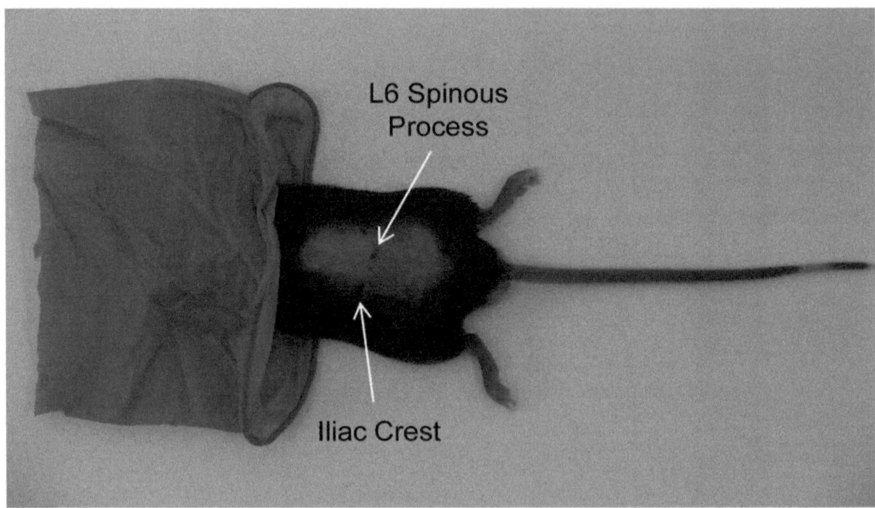

Fig. 2 Skin surface injection site. The mouse was anesthetized with isoflurane, its back shaved, and a nitrile glove (cut in half) placed over the head and the upper body. A skin marker was used to mark the L6 spinous process and the iliac crests. The area between the L5 and L6 spinous process is used as the skin puncture site

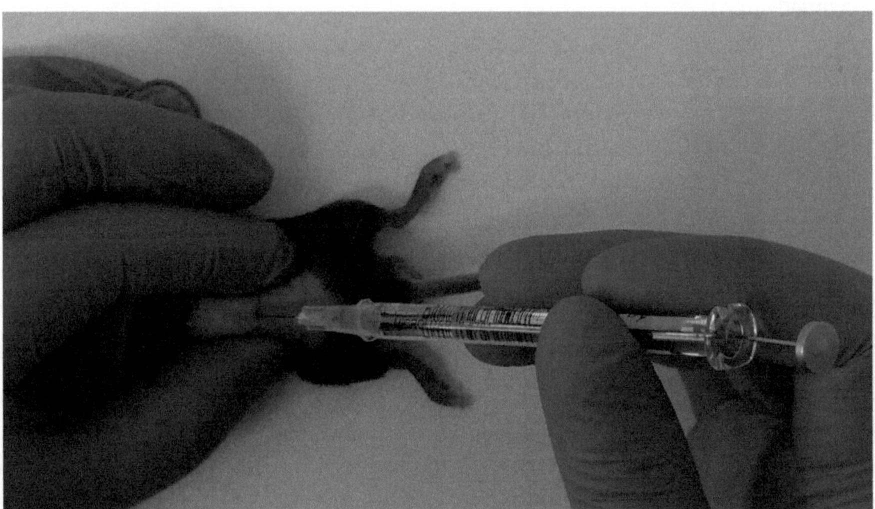

Fig. 3 Intrathecal drug delivery. The mouse is grasped at the iliac crest, using the thumb and the index finger of one hand (*left hand* in the figure), so that both hind legs move outward and downward. Then the skin surface injection site is localized with the other hand and the syringe/needle is inserted between the L5 and L6 spinous process as described in the main text

4. Grip the animal tightly by the iliac crest, using the thumb and the index finger of one hand (Fig. 3, left hand). The grip should fix the mouse by moving both hind legs outward and downward.
5. Tense the skin above the lumbar spine to present the horizontal plane (supracristal plane), which appears at the highest point

of the iliac crests and crosses over the L5 vertebrae. With the other hand, palpate the surface landmarks on the dorsum to identify the lowest lumbar spinous process (L6).

6. Gently trace the L6 spinous process and curve the column to open up the intervertebral space. The area between the L5 and L6 spinous process is used as the skin puncture site. The site of injection between the L5 and L6 spinous processes corresponds to the cauda equina (the bundle of nerves distal to the end of the spinal cord) and hence minimizes the possibility of spinal cord damage. Normally the L6 spinous process remains visible on the skin for a few seconds after tracing, which facilitates orientation. If the L6 spinous process is difficult to detect, drawing an imaginary line between the highest points of both iliac crests, the cross point between this line and the midline of the spinal column can also be used to locate the puncture site (Fig. 2) (*see* **Note 6**).
7. After the puncture site has been located, insert a 10 μl syringe, connected to a sterile disposable 30 gauge ½ in. needle, vertically into the skin between the L5 and L6 spinous process. Once contact with the bone of the spinal column is sensed, reduce the syringe angle to ~30° and push the syringe carefully forward into the intervertebral space. Puncture of the dura mater is reliably indicated by a reflexive flick of the tail or formation of an "S" shape by the tail [10]. The mouse should be allowed to wake up again to ensure proper reactions in response to the dura puncture.
8. Once an indication of dura puncture is observed, inject the drug solution slowly (within ~2 s) in a volume of 5 μl (Fig. 3). After waiting for ~5 s, rotate the needle slightly on withdrawal (*see* **Notes 7–9**).
9. Finally, observe the mouse for 2 min in order to ensure that there are no injuries. In control experiments we did not observe any motor impairments, weight loss, or any sign of discomfort in mice during a 7-day period after intrathecal injection of 0.9 % saline.

3.3 Effects of Intrathecally Injected cGMP Analogs

1. Immediately after intrathecal injection, using the method described above, place the mouse into the enclosure of the Dynamic Plantar Aesthesiometer.
2. Measure the mechanical thresholds at 10 min intervals up to 80 min after injection. Typical time courses of mechanical thresholds after intrathecal injection of 25 mmol 8-pCPT-cGMP and 0.9 % saline are shown in Fig. 4. The reduced paw withdrawal latency times, as compared to baseline values, indicate that a transient mechanical hypersensitivity was induced by 8-pCPT-cGMP, suggesting pronociceptive effects of cGMP in vivo (*see* **Note 10**).

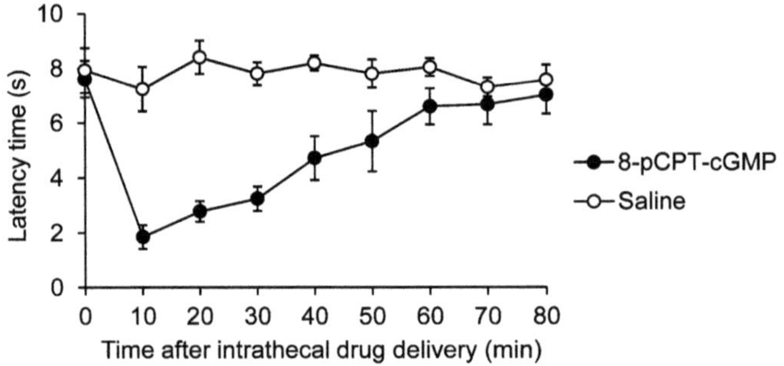

Fig. 4 Pronociceptive effects of 8-pCPT-cGMP. The time course of mechanical hindpaw hypersensitivity induced by intrathecal injection of the cGMP analog 8-pCPT-cGMP (25 nmol in 5 μl) or vehicle (0.9 % saline) is shown. The reduced paw withdrawal latency times indicate that intrathecal 8-pCPT-cGMP evoked a transient mechanical hindpaw hypersensitivity. Data are presented as means ± SEM. $n = 5–6$ animals per group

4 Notes

1. Instead of the Dynamic Plantar Aesthesiometer, other devices such as von Frey hairs can be used to measure the mechanical thresholds. In addition, other sensory thresholds can be determined to quantify the drug-induced nociceptive response.
2. If other mouse strains are used, differences in the response to drug injection may occur due to a genetic variability in nociception [11, 12].
3. Water-based vehicles (0.9 % saline or artificial cerebrospinal fluid, ACSF) are recommended, because other vehicles, such as ethanol or DMSO, can affect the nociceptive behavior.
4. As an alternative to direct lumbar puncture, a chronic indwelling catheter can be implanted into the spinal subarachnoidal space [13, 14]. This technique, however, requires surgery and is more time-consuming than direct lumbar puncture.
5. Experienced injectors can also perform the lumbar puncture without shaving the back of the mouse.
6. Efficient positioning of the mouse for further manipulation is the key to successful injection. The mouse should always be given enough space to breathe.
7. An injection volume of 5 μl has proven to be the optimal volume in mice.
8. Due to the dead volume of the needle, at least 100 μl drug solution is needed to fill the syringe.
9. The injection itself can sometimes result in transient stress-induced changes in the nociceptive sensitivity of the mice. Therefore, it is critical to include vehicle controls on every experimental day on which this technique is applied.

10. In addition to effects of cGMP analogs, we observed pronociceptive effects after intrathecal injection of the NO donor NOC-5 and C-type natriuretic peptide [1, 8]. However, depending on the doses and pharmacological properties of injected compounds, antinociceptive effects can also occur after intrathecal drug delivery of cGMP analogs [13, 15], probably due to the presence of both pronociceptive and antinociceptive cGMP effectors in the nociceptive system [8, 13, 15, 16].

Acknowledgment

Supported by the Deutsche Forschungsgemeinschaft (SFB 815-A14).

References

1. Schmidtko A, Gao W, Konig P et al (2008) cGMP produced by NO-sensitive guanylyl cyclase essentially contributes to inflammatory and neuropathic pain by using targets different from cGMP-dependent protein kinase I. J Neurosci 28:8568–8576
2. Schmidt H, Stonkute A, Juttner R et al (2009) C-type natriuretic peptide (CNP) is a bifurcation factor for sensory neurons. Proc Natl Acad Sci USA 106:16847–16852
3. Zhang FX, Liu XJ, Gong LQ et al (2010) Inhibition of inflammatory pain by activating B-type natriuretic peptide signal pathway in nociceptive sensory neurons. J Neurosci 30:10927–10938
4. Tao YX, Hassan A, Haddad E et al (2000) Expression and action of cyclic GMP-dependent protein kinase Ialpha in inflammatory hyperalgesia in rat spinal cord. Neuroscience 95:525–533
5. Schmidtko A, Ruth P, Geisslinger G et al (2003) Inhibition of cyclic guanosine 5′-monophosphate-dependent protein kinase I (PKG-I) in lumbar spinal cord reduces formalin-induced hyperalgesia and PKG upregulation. Nitric Oxide 8:89–94
6. Tegeder I, Del Turco D, Schmidtko A et al (2004) Reduced inflammatory hyperalgesia with preservation of acute thermal nociception in mice lacking cGMP-dependent protein kinase I. Proc Natl Acad Sci USA 101: 3253–3257
7. Luo C, Gangadharan V, Bali KK et al (2012) Presynaptically localized cyclic GMP-dependent protein kinase 1 is a key determinant of spinal synaptic potentiation and pain hypersensitivity. PLoS Biol 10:e1001283
8. Heine S, Michalakis S, Kallenborn-Gerhardt W et al (2011) CNGA3: a target of spinal nitric oxide/cGMP signaling and modulator of inflammatory pain hypersensitivity. J Neurosci 31:11184–11192
9. Hylden JL, Wilcox GL (1980) Intrathecal morphine in mice: a new technique. Eur J Pharmacol 67:313–316
10. Fairbanks CA (2003) Spinal delivery of analgesics in experimental models of pain and analgesia. Adv Drug Deliv Rev 55:1007–1041
11. Mogil JS, Wilson SG, Bon K et al (1999) Heritability of nociception II. "Types" of nociception revealed by genetic correlation analysis. Pain 80:83–93
12. Mogil JS, Wilson SG, Bon K et al (1999) Heritability of nociception I: responses of 11 inbred mouse strains on 12 measures of nociception. Pain 80:67–82
13. Schmidtko A, Gao W, Sausbier M et al (2008) Cysteine-rich protein 2, a novel downstream effector of cGMP/cGMP-dependent protein kinase I-mediated persistent inflammatory pain. J Neurosci 28:1320–1330
14. Wu WP, Xu XJ, Hao JX (2004) Chronic lumbar catheterization of the spinal subarachnoid space in mice. J Neurosci Methods 133:65–69
15. Tegeder I, Schmidtko A, Niederberger E et al (2002) Dual effects of spinally delivered 8-bromo-cyclic guanosine mono-phosphate (8-bromo-cGMP) in formalin-induced nociception in rats. Neurosci Lett 332:146–150
16. Schmidtko A, Tegeder I, Geisslinger G (2009) No NO, no pain? The role of nitric oxide and cGMP in spinal pain processing. Trends Neurosci 32:339–346

Chapter 15

The Geisler Method: Tracing Activity-Dependent cGMP Plasticity Changes upon Double Detection of mRNA and Protein on Brain Slices

Wibke Singer, Hyun-Soon Geisler, and Marlies Knipper

Abstract

We recently demonstrated that an increase of guanosine 3′,5′-cyclic monophosphate (cGMP) signaling could protect the inner ear from noise-induced hair cell damage. Noise exposure not only damages hair cells but also alters the central responsiveness to sound leading to plasticity changes. cGMP signaling has long been known to play a crucial role for plasticity changes and long-term potentiation (LTP). To get a first insight into the role of cGMP for noise-induced plasticity changes we aimed to co-trace the mRNA and protein of plasticity-related genes as, e.g., the immediate early gene Arc (activity-regulated cytoskeletal protein) with markers for the cGMP pathway. We developed a method that permits the simultaneous monitoring of mRNA and protein through light microscopy to visualize gene expression in neurons and synapses of its processes. Accordingly, different from previous fluorescence-based assays that detect, e.g., fluorochrome-labeled Arc antibodies and Arc mRNA, we describe here a methodology that allows the detection of mRNA and protein of synaptic genes using nonfluorescent stable tracers for high-resolution observation of activity-dependent plasticity changes using light microscopy even after weeks or months.

Key words mRNA, Protein, Riboprobes, Antibody, Immunohistochemistry, In situ hybridization

1 Introduction

In the heart, lung, and other organs, cGMP has been described to facilitate protective processes in response to traumatic events [1]. In sensory and neuronal systems, the role of cGMP is understood to a lesser extent. In the ear, we showed that an increase of cGMP protects the sensory outer hair cells and inner hair cell synapses from damage upon acoustic overstimulation [2]. Noise damage can lead to hyperacusis or tinnitus [3]. Noise damage can also alter the central responsiveness to sound [4, 5] and trigger long-lasting aberrant plasticity changes in the auditory cortex [3, 6]. It has moreover been described that hearing disorders as, e.g., tinnitus can develop on a background of emotional trauma [7–9] that

Thomas Krieg and Robert Lukowski (eds.), *Guanylate Cyclase and Cyclic GMP: Methods and Protocols*,
Methods in Molecular Biology, vol. 1020, DOI 10.1007/978-1-62703-459-3_15,

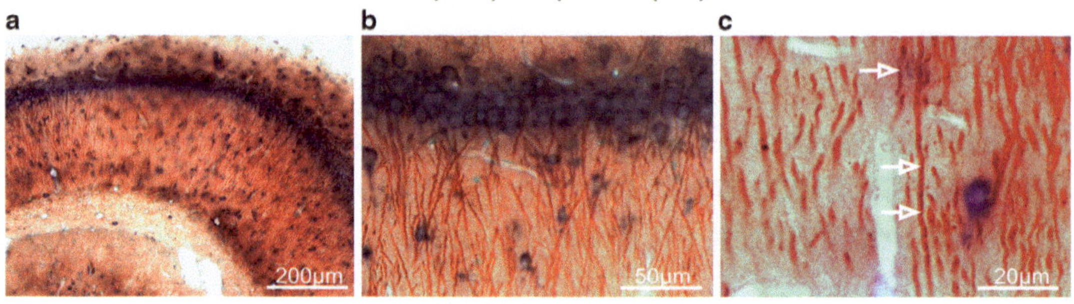

Fig. 1 Visualization of Arc mRNA and Arc protein in pyramidal neurons of the CA1 region in the rat hippocampus. (**a**) Shows a 10× magnification of the CA1 region of the rat hippocampus stained for Arc mRNA (*blue*) and Arc protein (*red*). (**b**) Shows a higher magnification of (**a**). The Arc mRNA (*blue*) can be observed in the cell soma and the Arc protein (*red*) more in the dendrites. (**c**) Shows a 120× magnification of the CA1 region. *Arrows* indicate Arc mRNA (*blue*) in the dendrites

involve cognitive changes and altered activity patterns in the hippocampus [10, 11] and the amygdala [12, 13]. Regarding this aspect, we recently demonstrated that noise-induced plasticity changes encompass not only the auditory cortex but also the hippocampus and the basolateral amygdala and vary with changes in auditory sound responsiveness [3, 14]. The different plasticity changes could be traced by co-detection of mRNA and protein of the immediate early gene Arc (Figs. 1, 2, and 3). Arc has been documented to be mobilized in glutamatergic neurons within minutes following LTP-like activity [15]. Following LTP-like activity, Arc mRNA is directly transported to distal dendrites [16], where through sustained translation for 2–4 h, Arc protein scales down surface AMPA receptors in dendritic spines, a process essential for LTP consolidation [15, 17, 18]. Accordingly the detection and monitoring of Arc mRNA and protein in brain tissue following, e.g., trauma-induced plasticity changes will provide a most useful tool to get a first insight into LTP/LTD-like changes and presumptive differences in homeostatic plasticity. Plasticity-related protein expression as the one of Arc is controlled by soluble guanylyl cyclase or the cGMP effector kinase PKG [19]. Indeed, cGMP signaling has long been known to play a crucial role for LTP [1, 20, 21]. We aim to validate a correlation and differential relation of cGMP pathways and Arc mobilization during plasticity changes linked with hearing disorders. We therefore describe here our developed methodology for simultaneous monitoring of expression changes in mRNA and protein in brain tissue (Figs. 1, 2, and 3). Different from previous assays [22] that use a fluorescence-based double detection of mRNA and protein the methodology described here allows the detection of mRNA and protein using nonfluorescent tracers. We provide a detailed protocol of the methodology and

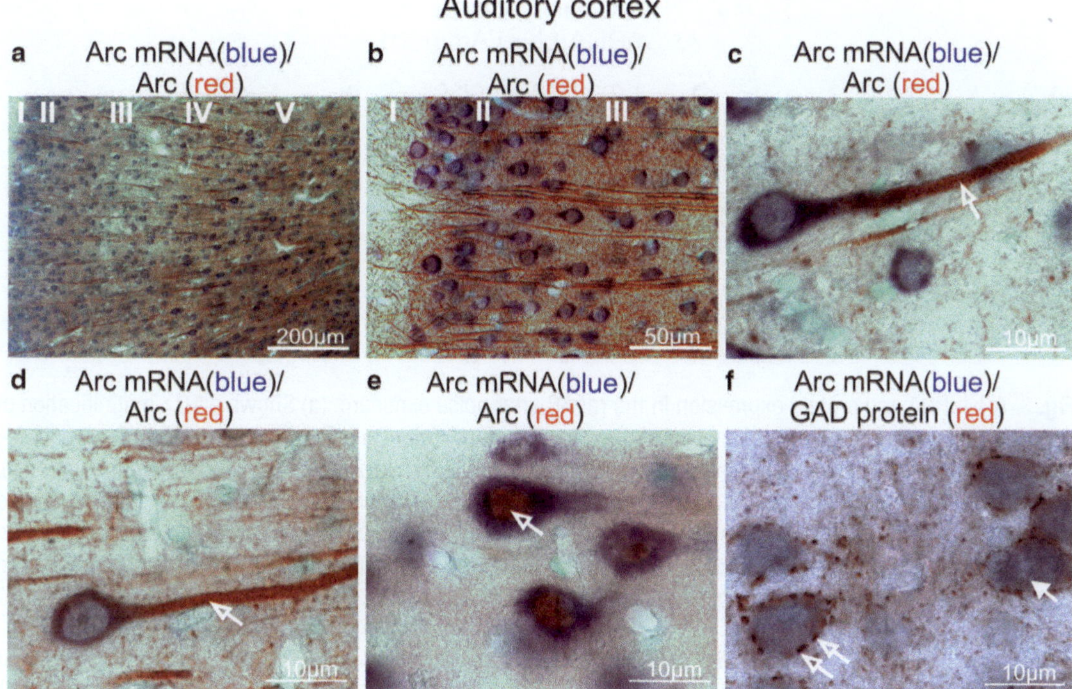

Fig. 2 Visualization of Arc mRNA and Arc protein in pyramidal neurons in the rat auditory cortex (AC). (**a**) Shows a 10× magnification of layers I–V of the AC stained for Arc mRNA (*blue*) and Arc protein (*red*). (**b**) Shows a higher magnification of (**a**). The Arc mRNA (*blue*) can be observed in the cell soma and the Arc protein (*red*) more in the dendrites. (**c**)–(**f**) Shows a 120× magnification of pyramidal neurons in layer III of the AC. (**c**) *Arrows* indicate Arc mRNA (*blue*) in the dendrites. (**d** and **e**) Staining for Arc mRNA (*blue*) and Arc protein (*red*) was performed. (**d**) The antibody used stains Arc protein (*red, open arrow*) in the dendrites, whereas the antibody used in (**e**) shows Arc protein (*red, open arrow*) expression in the cell soma. (**f**) Staining for Arc mRNA (*blue*) and glutamic acid decarboxylase (GAD) protein (*red*) was performed. The Arc mRNA (*blue, arrow*) is located in the cell soma whereas the GAD protein (*red, open arrows*) is expressed dot-like around the soma, which could indicate synaptic contacts

provide few examples that illustrate the detection of Arc mRNA and associated proteins of choice in the hippocampus (Fig. 1) and the auditory cortex (Figs. 2 and 3). The readership may feel free to either replace the riboprobe or the antibody by molecules of the cGMP pathway or any other genes of interest. Examples of different antibodies are shown in Fig. 2d–f. In Fig. 2d, e different antibodies for Arc are used that either label the dendrites (Fig. 2d) or the cell soma (Fig. 2e). In Fig. 2f Arc mRNA, typically expressed in glutamatergic neurons [15], is colabeled with an antibody against glutamic acid decarboxylase (GAD) to identify the GABAergic interneurons [23]. The differential expression of Arc mRNA and protein after exposure to different noise intensities for 2 h could be observed 2 weeks after exposure. Figure 3 shows an increase in Arc expression after moderate noise exposure whereas traumatic noise leads to a reduced Arc expression. The described

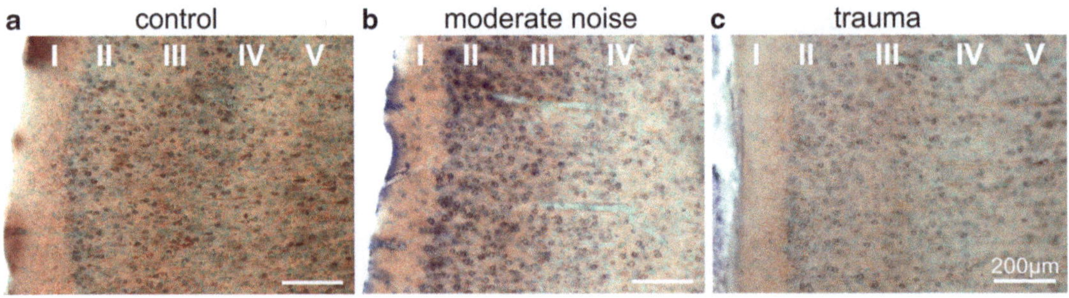

Fig. 3 Arc mRNA and protein expression in the rat AC after noise exposure. (**a**) Shows a 10× magnification of the AC of a control rat (no noise exposure) stained for Arc mRNA (*blue*) and Arc protein (*red*). (**b**) Shows a 10× magnification of the AC of a rat exposed to moderate noise (80 dB 10 kHz for 2 h) stained for Arc mRNA (*blue*). (**c**) Shows a 10× magnification of the AC of a rat exposed to traumatic noise (120 dB, 10 kHz, for 2 h) stained for Arc mRNA (*blue*) and Arc protein (*red*). An increase in Arc expression can be observed after moderate noise exposure in comparison to the control animal. In the traumatized animal no increase or even a decrease in Arc expression is observed in comparison to control animals [14]

technique permits high-resolution observation (Figs. 1 and 2) of activity-dependent changes (Fig. 3) in, e.g., Arc mRNA and cGMP signaling proteins in neurons or dendrites which enables us to elucidate the involvement of the cGMP pathway in noise-induced plasticity changes.

2 Materials

Prepare all solutions using ultrapure water (prepared by purifying deionized water to attain a sensitivity of 18 MΩ cm at 25 °C). Prepare and store all reagents at room temperature (RT) (unless indicated otherwise). Diligently follow all waste disposal regulations when disposing waste materials. All solutions should be autoclaved unless otherwise indicated.

Make sure that all experiments involving animals are in accordance with relevant guidelines and regulations.

2.1 Tissue Fixation/Embedding

1. 2 %/4 % Paraformaldehyde (PFA): 20 g/40 g PFA, 900 ml 1× phosphate-buffered saline (PBS). Leave at 54 °C in a water bath until the powder is completely dissolved (*see* **Note 1**). Adjust the volume to 1 L with 1× PBS, prepare 50 ml aliquots, and store at −20 °C. Do not autoclave.
2. 10× PBS: 88.9 g $Na_2HPO_4 \times 2H_2O$ (0.5 M), 87.6 g NaCl (1.5 M), 900 ml ultrapure water (prepared by purifying deionized water to attain a sensitivity of 18 MΩ cm at 25 °C), pH 7.0. Adjust the pH with HCl. Adjust the volume to 1 L with water. For 1× PBS dilute 10× PBS in ddH_2O.

3. 4 % Agarose: 4 g Agarose, 100 ml 1× PBS + 0.4 % PFA (*see* **Note 2**). Heat the agarose in the microwave until it melts (temperature around 55 °C). Prepare fresh every time. Do not autoclave.

2.2 In Situ Hybridization

1. Washing buffer: 12.1 g Tris (100 mM), 8.8 g NaCl (150 mM) pH 7.5. Add water to a volume of 900 ml. Mix and adjust the pH using 25 % HCl. Adjust the volume to 1 L with water.
2. Tris–HCl + 0.05 % Tween 20 (Applichem, Darmstadt, Germany): Add 500 μl Tween 20 to 1 L Tris–HCl.
3. 1× PBS prepared using DMPC water.
4. 2× SSC:17.5 g NaCl (0.3 M), 8.8 g sodium citrate (0.03 M), pH 7.0. Prepare the solution as in previous step. For 1× and 0.5× SSC dilute 2× SCC with DMPC water.
5. DMPC water: 1 ml DMPC, 1 L H_2O. Incubate at 37 °C overnight and autoclave.
6. Blocking solution: 0.25 % acetic anhydrate (Sigma, Darmstadt, Germany), 25 ml Tris–HCl. Dilute 62.5 μl acetic anhydrate in 25 ml Tris–HCl. Do not store it; prepare fresh every time (*see* **Note 3**).
7. Blocking reagent (Roche, Mannheim, Germany): 1.16 g maleic acid (0.12 M), 877 mg NaCl (1.6 M), pH 7.0. Add water to 80 ml and dissolve the powder in the microwave. Adjust the pH. Adjust the volume to 100 ml with water. After autoclaving prepare 1 ml aliquots and store at −20 °C.
8. Blocking buffer: 1 ml blocking reagent, 9 ml Tris–HCl, 30 μl Triton (Sigma, Munich, Germany). Keep for 30 min at 37 °C in a water bath.
9. De-/rehydration: Use 100, 95, and 70 % ethanol. For 95 and 70 % ethanol dilute ethanol absolute (abs.) with DMPC water.
10. Chloroform to remove lipophilic substances (*see* **Note 3**).
11. 2× Prehybridization buffer (Sigma, Munich, Germany). Dilute 2× prehybridization buffer to 1× using 2× SSC. Store at −20 °C.
12. Riboprobes: Riboprobes have to be digoxigenin labeled, either self-made or purchased (*see* **Note 4**).
13. Hybridization buffer: Microarray Hybridization buffer (GE Healthcare, Freiburg, Germany), ddH_2O, formamide in a ratio of 1:1:2. Prepare 1 ml aliquots and store at −20 °C.
14. Anti-digoxigenin antibody: Directly before use, dilute Anti-Digoxigenin-AP, Fab fragments (Roche, Mannheim, Germany) 1:750 in blocking buffer. Store on ice.
15. AP buffer: 12.1 g Tris (100 mM), 5.8 g NaCl (100 mM), pH 9.5. Prepare 1 L solution as described above. Before use add 1.02 g MgCl.

16. Substrate: 50 mg/ml BCIP (5-Bromo-4-chloro-3-indolyl phosphate, p-toluidine salt, Sigma-Aldrich, Deutschland) in 100 % *N,N*-dimethylformamide, 50 mg/ml NBT (Nitro Blue Tetrazolium, Biomol, Hamburg Germany) in 70 % *N,N*-dimethylformamide (*see* **Note 3**), 10 ml AP buffer. Add 35 μl of BCIP and 45 μl of NBT to 10 ml AP buffer; incubate in the dark.
17. Fixation: To fix the staining use 95 % ethanol.

2.3 Immunohistochemistry

1. Washing buffers: 1× PBS + 0.05 % Tween 20 (Applichem, Darmstadt, Germany). 1× PBS.
2. Blocking buffer: 30 % H_2O_2, 1× PBS. Dissolve 30 % H_2O_2 in 1× PBS 1:10.
3. Blocking solution: 0.5 g bovine serum albumin (BSA), 10 ml 1× PBS. Prepare 5 % BSA (*see* **Note 5**). Store at 4 °C.
4. Streptavidin/biotin Blocking kit (Vector Laboratories, Burlingame, CA, USA). Store at 4 °C.
5. Antibody dilution: 5 % BSA, 1× PBS (*see* **Note 6**). Dilute 5 % 1:5 with 1× PBS to obtain 1 % BSA.
6. Primary antibody (*see* **Note 7**).
7. Secondary antibody: Biotinylated antibody against the species the primary antibody was generated in diluted 1:500 in 1 % BSA. Here biotinylated anti-rabbit antibody was used (Vector Laboratories, Burlingame, CA, USA). Store at 4 °C.
8. Streptavidin–Horseradish peroxidase (Vector Laboratories, Burlingame, CA, USA) diluted 1:300 in 1 % BSA (*see* **Note 8**). Store at 4 °C.
9. Chromogen: Peroxidase substrate kit AEC (Vector Laboratories, Burlingame, CA, USA). Store at 4 °C.

2.4 Nuclear Staining (Optional)

1. Methyl green (Applichem, Darmstadt, Germany): For stock solution dilute 2 g methyl green in 20 ml ethanol (abs.), 80 ml ddH_2O (*see* **Note 9**).

2.5 Mounting Medium

1. Kaisers Gelatin (Merck, Darmstadt, Germany) soluble at 50 °C.

3 Methods

Carry out all steps at RT unless otherwise indicated.

3.1 Tissue Preparation (See Note 10)

1. After decapitation and opening of the skull, dissect the complete brain in one piece and transfer immediately to 50 ml 2 % or 4 % PFA (for mice 2 %, for rats 4 % PFA).

2. Fix the tissue for 48 h at 4 °C on a shaker. After 24 h change the PFA.
3. Take the tissue out of the PFA and rinse with ddH_2O.
4. Embed the tissue in 4 % agarose (*see* **Note 11**). Store the embedded tissue in 1× PBS + 0.4 % PFA at 4 °C.
5. Brains are sectioned at 40–60 μm on a vibratome VT1000S (Leica, Wetzlar, Germany).
6. Slices are stored in a 24-well plate in 1× PBS at 4 °C (*see* **Note 12**).

3.2 In Situ Hybridization (See Note 13)

1. Transfer the slices with a paintbrush to a glass dish (*see* **Note 14**).
2. Wash twice for 5 min with 1× PBS.
3. To block positive charges incubate the slices for 5 min in 0.25 % acetic anhydrate.
4. Wash twice for 5 min with 1× PBS.
5. To dehydrate incubate the tissue first in 70 %, then in 95 %, and finally twice in 100 % ethanol, each for 1 min.
6. For removing lipophilic substances, incubate slices for 5 min in chloroform (*see* **Note 3**).
7. To rehydrate incubate the tissue twice in 100 %, then in 95 %, and finally in 70 % ethanol, each for 1 min (*see* **Note 15**).
8. Transfer the slices to a 24-well plate (*see* **Note 16**).
9. Wash twice for 5 min in 2× SSC.
10. Incubate the slices for 1 h in prehybridization buffer at 37 °C (*see* **Note 17**).
11. Dilute your riboprobes in hybridization buffer (*see* **Note 17**) and incubate them for 10 min at 65 °C (*see* **Note 18**). Store them on ice until ready to use.
12. Apply the riboprobes to the slices. Put the well plate to a moist formamide chamber and incubate them at 55–58 °C (*see* **Note 19**) overnight.
13. Wash 20 min in 2× SSC, 20 min in 1× SSC, and 20 min in 0.5× SSC, respectively, at hybridization temperature.
14. Wash 5 min in Tris–HCl + 0.05 % Tween 20 (*see* **Note 20**).
15. Incubate the slices for 30 min in blocking reagent.
16. Apply the anti-Dig antibody to the slices and incubate 30 min at 37 °C.
17. Wash twice for 15 min in Tris–HCl.
18. Rinse the slices in AP buffer (*see* **Note 21**).
19. Incubate each slice with 500 μl of substrate (20 min to 5 h, *see* **Note 22**).

20. Stop the reaction by applying ddH_2O. Wash additionally twice for 15 min with ddH_2O.
21. Fix the staining using 95 % ethanol for 15 min, followed by two washing steps with ddH_2O for 10 min (*see* **Note 23**).

3.3 Immunohistochemistry

1. Wash the slices twice for 5 min in 1× PBS.
2. Block the endogenous peroxidases with 3 % H_2O_2 for 10 min.
3. Wash twice for 5 min in 1× PBS.
4. Incubate slices for 30 min in blocking solution (*see* **Note 24**).
5. Apply 6 drops of the streptavidin blocking solution to each slice and incubate for 15 min (*see* **Note 25**).
6. Wash twice for 5 min in 1× PBS.
7. Apply 6 drops of the biotin blocking solution to each slice and incubate for 15 min (*see* **Note 25**).
8. Wash twice for 5 min in 1× PBS.
9. Apply your primary antibody (*see* **Note 7**) diluted in 1 % BSA to the slices. Put the well plate to a moist chamber and incubate them at 4 °C overnight.
10. Wash three times 15 min in 1× PBS + 0.05 % Tween 20.
11. Apply the secondary antibody diluted 1:500 in 1% BSA to the slices (*see* **Note 17**) and incubate for 1 h.
12. Wash three times 15 min in 1× PBS + 0.05 % Tween 20.
13. Apply streptavidin–horseradish peroxidase diluted 1:300 in 1 % BSA to the slices (*see* **Note 8**) and incubate for 30 min.
14. Wash three times 15 min in 1× PBS + 0.05 % Tween 20.
15. Apply the chromogen (*see* **Note 26**) to the slices (500 μl per slice) and incubate 10–30 min (*see* **Note 27**).
16. Stop the reaction in ddH_2O. Wash additionally twice for 10 min with ddH_2O.

3.4 Nuclear Staining (Optional)

1. Apply 200 μl methyl green to each slice (*see* **Note 9**).
2. Stop the reaction using ddH_2O. Exchange the ddH_2O until no color is left in the liquid, and then wash additionally twice for 10 min in ddH_2O.

3.5 Mounting

1. Prepare a glass cuvette with ddH_2O and heat to 60 °C in a water bath.
2. Dissolve the Kaisers Gelatin at 60 °C.
3. To apply the tissue place them one at a time in the glass cuvette filled with warm water and use a paintbrush to apply them to microscope slides.
4. Mount them using the warm Kaisers Gelatin.
5. Mounted slices can be stored at 4 °C for several months.

4 Notes

1. Shake the bottles in between to help the powder to dissolve.
2. The amount of agarose depends on the number of samples that have to be embedded.
3. Work under a hood as the substance used is harmful and toxic (acetic anhydrate, chloroform, *N,N*-dimethylformamide).
4. Here Arc (accession number NM-019361) was amplified using the primers (for: 5′-CAATGTGATCCTGCAGATTG-3′; rev: 5′-TGATGGCATAGGGGCTAACA-3′) giving a probe length of 449 bp. Always use a sense probe as a negative control for the experiment.
5. Use a magnetic stir bar.
6. Store the rest of the 1 % BSA at 4 °C for diluting the secondary antibody and the streptavidin–horseradish peroxidase the next day.
7. The primary antibody depends on the gene of interest. Follow the manufacturer's instructions for storage and dilution. Here we used two different polyclonal rabbit antibodies against Arc (Synaptic Systems, Göttingen, Germany and for Fig. 2e Santa Cruz, Heidelberg, Germany) diluted 1:400 and a polyclonal rabbit antibody against GAD 65/67 (Fig. 2f: Alpha Diagnostics International, San Antonio, TX, USA) diluted 1:100.
8. Streptavidin–horseradish peroxidase binds the biotin of the biotinylated secondary antibody. Streptavidin–horseradish peroxidase will react with the peroxidase AEC substrate leading to an oxidation resulting in a red color.
9. After our experience the staining with methyl green can be very strong; we recommend diluting methyl green at least to 1:50 in ddH_2O and maximally incubate slices for 1 min. The final concentration and incubation time should be determined by the investigator.
10. Animal care, procedures, and treatments were performed in accordance with institutional and national guidelines following approval by the University of Tübingen, Veterinary Care Unit, and the Animal Care and Ethics Committee of the regional board of the Federal State Government of Baden-Württemberg, Germany.
11. For embedding use a 12-well plate. Apply 4 % agarose at approximately 55 °C to the well and place the whole brain into the agarose leaving a little space between the bottom of the well and the tissue. When the agarose is hard remove the agarose including the brain from the well using a spatula and place it in a 50 ml tube containing 1× PBS + 0.4 % PFA. Store at 4 °C.
12. The slices should not be stored longer than 1 month.

13. As RNA will be detected use only autoclaved/sterile instruments and change the pipette tip every time.
14. Use a glass dish as chloroform will destroy plastic.
15. Apply 100 % ethanol very carefully after the chloroform step as the slices start to move in a circle due to the different conditions of the solutions.
16. From now onwards do all the steps on a shaker set in a way that the slices move in the liquid unless otherwise indicated.
17. The volumes should be chosen such that the slices are covered and the liquid does not evaporate; we suggest 300 μl per well.
18. Incubation on 65 °C helps to prevent nonspecific binding in between the riboprobes.
19. The hybridization temperature depends on the riboprobe used and has to be determined by the investigator. Normally a good result is achieved when using 55–58 °C; if necessary you can go up until 60 °C. For our Arc riboprobe 56.5 °C is used.
20. This will permeabilize the tissue for the application of the anti-Dig antibody.
21. AP buffer is used to change the pH from neutral to alkaline so that the substrate can work at its pH optimum of 9.5.
22. The incubation time has to be determined by the investigator; if necessary it can be extended until the next day. The reaction has to be stopped when the sense probe gives a signal.
23. Slices can be stored for a few days in 1× PBS at 4 °C or you can continue immediately on the same slices with the immunohistochemistry.
24. Do not wash after removing the blocking solution.
25. Follow the manufacturer's instructions for use.
26. Prepare the AEC chromogen according to the manufacturer's instructions:

    5 ml ddH$_2$O plus 2 drops of buffer stock, 3 drops of AEC stock solution, and 2 drops of hydrogen peroxidase solution.
27. The final incubation time has to be determined by the investigator, generally 10–30 min gives a good result.

References

1. Domek-Lopacinska K, Strosznajder JB (2005) Cyclic GMP metabolism and its role in brain physiology. J Physiol Pharmacol 56(Suppl 2): 15–34
2. Jaumann M, Dettling J, Gubelt M, Zimmermann U, Gerling A, Paquet-Durand F, Feil S, Wolpert S, Franz C, Varakina K, Xiong H, Brandt N, Kuhn S, Geisler HS, Rohbock K, Ruth P, Schlossmann J, Hutter J, Sandner P, Feil R, Engel J, Knipper M, Ruttiger L (2012) cGMP-Prkg1 signaling and Pde5 inhibition shelter cochlear hair cells and hearing function. Nat Med 18(2):252–259. doi:10.1038/nm.2634nm.2634 [pii]

3. Knipper M, Müller M, Zimmermann U (2012) Molecular mechanism of tinnitus. J.J. Eggermont et al. (eds.), Tinnitus, Springer Handbook of Auditory Research 47, Springer Science+Business Media New York 2012. doi: 10.1007/978-1-4614-3728-4_3
4. Zuccotti A, Kuhn S, Johnson SL, Franz C, Singer W, Hecker D, Geisler HS, Köpschall I, Rohbock K, Gutsche K, Dlugaiczyk J, Schick B, Marcotti W, Rüttiger L, Schimmang T, Knipper M (2012) Lack of brain-derived neurotrophic factor hampers inner hair cell synapse physiology, but protects against noise induced hearing loss. J Neurosci 32(25):8545–8553
5. Rüttiger L, Singer W, Panford-Walsh R, Matsumoto M, Lee SC, Zuccotti A, Zimmermann U, Jaumann M, Rohbock K, Xiong H, Knipper M (2013) The reduced cochlear output and the failure to adapt the central auditory response causes tinnitus in noise exposed rats. PLoS One 8(3):e57247. doi:10.1371/journal.pone.0057247
6. Knipper M, Zimmermann U, Müller M (2010) Molecular aspects of tinnitus. Hear Res 266 (1–2):60–69. doi:S0378-5955(09)00193-2 [pii] 10.1016/j.heares.2009.07.013
7. Hinton DE, Chhean D, Pich V, Hofmann SG, Barlow DH (2006) Tinnitus among Cambodian refugees: relationship to PTSD severity. J Trauma Stress 19(4):541–546. doi:10.1002/jts.20138
8. Neigh GN, Gillespie CF, Nemeroff CB (2009) The neurobiological toll of child abuse and neglect. Trauma Violence Abuse 10(4): 389–410. doi:1524838009339758 [pii] 10.1177/1524838009339758
9. Jastreboff PJ (1990) Phantom auditory perception (tinnitus): mechanisms of generation and perception. Neurosci Res 8(4):221–254
10. Landgrebe M, Langguth B, Rosengarth K, Braun S, Koch A, Kleinjung T, May A, de Ridder D, Hajak G (2009) Structural brain changes in tinnitus: grey matter decrease in auditory and non-auditory brain areas. Neuroimage 46(1):213–218. doi:S1053-8119(09)00141-4 [pii] 10.1016/j.neuroimage.2009.01.069
11. Lockwood AH, Salvi RJ, Coad ML, Towsley ML, Wack DS, Murphy BW (1998) The functional neuroanatomy of tinnitus: evidence for limbic system links and neural plasticity. Neurology 50(1):114–120
12. Shulman A, Strashun AM, Afriyie M, Aronson F, Abel W, Goldstein B (1995) SPECT imaging of brain and tinnitus-neurotologic/neurologic implications. Int Tinnitus J 1(1):13–29
13. Mirz F, Gjedde A, Sodkilde-Jrgensen H, Pedersen CB (2000) Functional brain imaging of tinnitus-like perception induced by aversive auditory stimuli. Neuroreport 11(3): 633–637
14. Singer W, Zuccotti A, Jaumann M, Lee SC, Panford-Walsh R, Xiong H, Zimmermann U, Franz C, Geisler HS, Köpschall I, Rohbock K, Varakina K, Verpoorten S, Reinbothe T, Schimmang T, Rüttiger L, Knipper M (2013) Noise-induced inner hair cell ribbon loss disturbs central arc mobilization: a novel molecular paradigm for understanding tinnitus. Mol Neurobiol 47(1):261–279. doi:10.1007/s12035-012-8372-8
15. Bramham CR, Alme MN, Bittins M, Kuipers SD, Nair RR, Pai B, Panja D, Schubert M, Soule J, Tiron A, Wibrand K (2010) The Arc of synaptic memory. Exp Brain Res 200(2): 125–140. doi:10.1007/s00221-009-1959-2
16. Link W, Konietzko U, Kauselmann G, Krug M, Schwanke B, Frey U, Kuhl D (1995) Somatodendritic expression of an immediate early gene is regulated by synaptic activity. Proc Natl Acad Sci USA 92(12):5734–5738
17. Bramham CR, Worley PF, Moore MJ, Guzowski JF (2008) The immediate early gene arc/arg3.1: regulation, mechanisms, and function. J Neurosci 28(46):11760–11767
18. Tzingounis AV, Nicoll RA (2006) Arc/Arg3.1: linking gene expression to synaptic plasticity and memory. Neuron 52(3):403–407
19. Gallo EF, Iadecola C (2011) Neuronal nitric oxide contributes to neuroplasticity-associated protein expression through cGMP, protein kinase G, and extracellular signal-regulated kinase. J Neurosci 31(19):6947–6955. doi:10.1523/JNEUROSCI.0374-11.2011
20. Brackmann M, Schuchmann S, Anand R, Braunewell KH (2005) Neuronal Ca2+ sensor protein VILIP-1 affects cGMP signalling of guanylyl cyclase B by regulating clathrin-dependent receptor recycling in hippocampal neurons. J Cell Sci 118(Pt 11):2495–2505. doi:10.1242/jcs.02376
21. Hindley S, Juurlink BH, Gysbers JW, Middlemiss PJ, Herman MA, Rathbone MP (1997) Nitric oxide donors enhance neurotrophin-induced neurite outgrowth through a cGMP-dependent mechanism. J Neurosci Res 47(4):427–439
22. Ramirez-Amaya V, Vazdarjanova A, Mikhael D, Rosi S, Worley PF, Barnes CA (2005) Spatial exploration-induced Arc mRNA and protein expression: evidence for selective, network-specific reactivation. J Neurosci 25(7):1761–1768
23. Margeta-Mitrovic M, Mitrovic I, Riley RC, Jan LY, Basbaum AI (1999) Immunohistochemical localization of GABA(B) receptors in the rat central nervous system. J Comp Neurol 405(3):299–321

Chapter 16

Detection of cGMP in the Degenerating Retina

Stylianos Michalakis, Jianhua Xu, Martin Biel, and Xi-Qin Ding

Abstract

Cyclic guanosine 3′-5′-monophosphate (cGMP) plays a key role in the physiological process of light detection in photoreceptor cells of the retina. However, there is also growing evidence that cGMP may be critically involved in some pathophysiological processes of the retina since degenerating photoreceptors in mouse models of retinitis pigmentosa and achromatopsia accumulate high levels of cGMP. Here, we describe methods that allow the detection, subcellular localization, and quantification of cGMP in the retina and propose that cGMP accumulation can be used as a biomarker for photoreceptor degeneration.

Key words Cyclic guanosine 3′-5′-monophosphate, cGMP, Cyclic nucleotide-gated channel, CNG channel, Retinal degeneration, Retinitis pigmentosa, Achromatopsia

1 Introduction

Vision begins with the detection of light by photopigments in the photoreceptor outer segments of the retina [1, 2]. This in turn triggers changes of intracellular cGMP levels that are translated into an electrical signal by cyclic nucleotide-gated (CNG) ion channels [3]. CNG channels control the membrane potential and the calcium concentration of photoreceptors. In the dark the channels are maintained in the open state by a high concentration of cGMP generated by the retina-specific membrane guanylyl cyclases E and F (GUCY2E and GUCY2F) [4, 5]. The resulting influx of Na^+ and Ca^{2+} ("dark current") depolarizes the photoreceptor and promotes synaptic transmission. Light-induced hydrolysis of cGMP by retinal phosphodiesterases (PDE6A in rods and PDE6C in cones) leads to closure of the CNG channel [6]. As a result, the photoreceptor hyperpolarizes and shuts off synaptic glutamate release. Closure of the CNG channels also decreases the photoreceptor Ca^{2+} concentration because it shuts down Ca^{2+} entry while the Na^+/Ca^{2+}-K^+ exchanger continues to clear Ca^{2+} from the cytosol [7]. This low Ca^{2+} concentration contributes to the recovery from light

Thomas Krieg and Robert Lukowski (eds.), *Guanylate Cyclase and Cyclic GMP: Methods and Protocols*, Methods in Molecular Biology, vol. 1020, DOI 10.1007/978-1-62703-459-3_16,

response through activation of the guanylyl cyclase by Ca^{2+}-sensing guanylyl cyclase-activating proteins (GCAPs) [8].

The importance of a precise regulation of cGMP signaling is highlighted by the fact that mutations in the genes encoding the CNG channels, the GCAPs, the guanylyl cyclases, or the phosphodiesterases all cause severe eye diseases [4, 5, 8–10]. Indeed, loss-of-function mutations of retinal phosphodiesterases are known to cause elevation of the retinal cGMP levels such as that seen in *rd1* mice [11–13] or *cpfl1* mice [14]. Moreover, loss of CNG channel function in rods (e.g., in CNGB1 knockout mice [15]) or in cones (e.g., in CNGA3 knockout mice [16]) also leads to elevation of the photoreceptor cGMP levels [17, 18] (likely via the Ca^{2+}/GCAP/guanylyl cyclase pathway). Depending on the affected gene and cell type (rod and/or cone photoreceptor) the pathology varies from achromatopsia to cone dystrophy and retinitis pigmentosa. A common pathological feature of these diseases is a progressive degeneration of photoreceptors, which can be rather slow, e.g., in achromatopsia [19, 20], or very fast like in some forms of retinitis pigmentosa [21, 22].

cGMP signaling was implicated in the regulation of cell death and survival in various cell types [23] including photoreceptors [13, 24, 25]. Interestingly, others and we have observed accumulation of cGMP in degenerating photoreceptors of various mouse models of retinal diseases [14, 18, 26–29]. Although it is well established that high levels of cGMP accelerate photoreceptor cell death via activation of CNG channels [30], it is less clear how other cGMP targets, like cGMP-dependent kinases, which are also expressed in photoreceptors [31, 32], contribute to cell death and/or survival [13].

Here, we describe methods that allow the detection, subcellular localization, and quantification of cGMP in the retina and facilitate the investigation of the role of cGMP in retinal degeneration.

2 Materials

2.1 Cyclic GMP ELISA

We use the cyclic GMP Complete Kit (Assay Designs, Farmingdale, NY, USA) to measure cGMP levels in the mouse retina.

1. 0.1 M HCl: 0.1 M hydrochloric acid in water.
2. Cyclic GMP standard (5,000 pmol/mL cGMP).
3. Triethylamine.
4. Acetic anhydride.
5. Goat anti-rabbit IgG microtiter plate.
6. Neutralizing reagent.
7. Rabbit polyclonal antibody to cGMP.
8. cGMP conjugate (a blue solution of cGMP conjugated to alkaline phosphatase).

9. Wash buffer concentrate: Tris-buffered saline containing detergents.
10. pNpp substrate (a solution of p-nitrophenyl phosphate).
11. Stop solution (a solution of trisodium phosphate in water).
12. Plate sealer.
13. Disposable pellet mixer: cordless motor (VWR), 1.5 ml pestle, and 1.5 ml microtube.
14. Microplate reader: SPECTRAmax® 190 Microplate Spectrophotometer (Molecular Devices Corporation, Sunnyvale, CA, USA) (other microplate reader also works as long as it has wavelength of 405 nm).

2.2 Cyclic GMP Immunofluorescence

1. Super PAP-Pen Mini liquid blocker (Science Services).
   (a) M Phosphate buffer (PB): dissolve 14.24 g Na_2HPO_4 $2H_2O$ and 2.76 g NaH_2PO_4 H_2O in 1,000 ml water; adjust pH with NaOH 7.4.
2. 4 % PFA fixation buffer: 4 % paraformaldehyde/PB, heat to 60 °C, and stir to dissolve.
3. Blocking buffer: use either 5 % BSA/0.5 % Triton X-100/PB or 5 % ChemiBLOCKER™ (Merck Millipore)/0.5 % Triton X-100/PB.
4. Incubation buffer: use either 1 % BSA/0.3 % Triton X-100/PB or 2 % ChemiBLOCKER™/0.3 % Triton X-100/PB.
5. Primary antibody: sheep polyclonal anti-cGMP [33] (can be obtained from Prof. Dr. Harry W. M. Steinbusch, Maastricht University, Netherlands, working dilution: 1:3,000 in incubation buffer).
6. Secondary antibody: either Alexa-Fluor® 488 donkey anti-sheep IgG (Invitrogen, working dilution: 1:1,000) or Cy3 donkey anti-sheep IgG (Jackson ImmunoResearch, working dilution: 1:300 in incubation buffer).
7. Nuclear dye: either 4 ,6-diamidino-2-phenylindole (DAPI, Sigma, working dilution: 1:10,000 in PB) or Hoechst 33342 (Invitrogen, working dilution: 1:2,000 in PB).
8. Mounting medium: either Vectashield (Vector Laboratories, Inc) or Cell Lab Fluoromount-G™ (Beckman Coulter).

3 Methods

3.1 Cyclic GMP ELISA

The method described below is modified from the cyclic GMP Complete Kit Manual (Assay Designs). This colorimetric competitive immunoassay quantitatively determines the extracellular or intracellular cyclic GMP in cells and tissue treated with 0.1 M

HCl. The acid treatment is to stop endogenous phosphodiesterase activity and stabilize the released cyclic GMP. Following acid treatment, the samples are analyzed directly in a microtiter plate without extraction, drying, and reconstitution.

1. Bring all reagents to room temperature for at least 30 min prior to use.
2. Weigh mouse retinas in a microtube (*see* **Note 1**).
3. Add 10 volumes of cold 0.1 M HCl.
4. Homogenize the tissues using a disposable pellet mixer for 1 min (on ice).
5. Centrifuge at 600 × *g* for 10 min at room temperature. Transfer the supernatant into a clean tube. The samples can be diluted with 0.1 M HCl if needed.
6. Prepare cGMP standard: Allow the 5,000 pmol/mL cGMP standard solution to warm up to room temperature. Label five 1.5 ml tubes #1 through #5. Add 450 μL 0.1 M HCl into tube #1 and 400 μL of 0.1 M HCl into tubes #2–5. Add 50 μL of the 5,000 pmol/mL standard to tube #1. Vortex thoroughly (this is critical!). Add 100 μL of tube #1 to tube #2 and vortex thoroughly. Continue this for tubes #3 through #5. The concentration of cGMP in tubes #1 through #5 will be 500, 100, 20, 4, and 0.8 pmol/mL, respectively (*see* **Note 2**).
7. Add 50 μL of the neutralizing reagent into each well, except the total activity (TA) and Blank wells.
8. Add 100 μL of 0.1 M HCl into the nonspecific binding (NSB) and the "0 pmol/mL standard" (Bo) wells.
9. Add 100 μL of standards into the appropriate wells (*see* **Note 3**).
10. Add 100 μL of the samples into the appropriate wells (*see* **Notes 3** and **4**).
11. Add 50 μL of 0.1 M HCl into the NSB wells (*see* **Note 5**).
12. Add 50 μL of blue conjugate into each well except the TA and Blank wells (*see* **Note 5**).
13. Add 50 μL of yellow antibody into each well, except the Blank, TA and NSB wells. Note: Every well used should be green in color except the NSB wells which should be blue. The Blank and TA wells are empty at this point and have no color (*see* **Note 5**).
14. Incubate the plate at room temperature for 2 h on a plate shaker at ~500 rpm. The plate may be covered with the plate sealer.
15. Empty the contents of the wells and wash twice with wash buffer.
16. After the final wash, empty or aspirate the wells, and firmly tap the plate on a lint-free paper towel to remove any remaining wash buffer.
17. Add 5 μL of the blue conjugate to the TA wells.

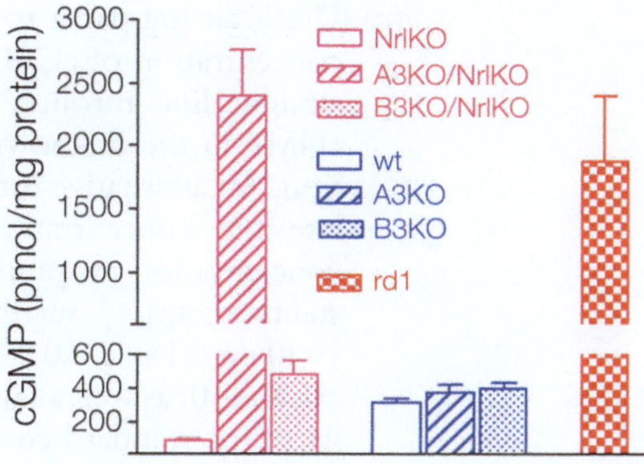

Fig. 1 Detection of cGMP in retinal degeneration using ELISA. The assay was performed using retinal homogenates prepared from mice at postnatal 30 days. Compared to the levels in Nrl knockout (KO) mice (a cone-dominant mouse line with deficiency of the neural retina leucine zipper transcription factor), CNGA3/Nrl double KO (A3KO/NrlKO) mice exhibited a remarkable elevation in retinal cGMP levels while CNGB3/Nrl double KO (B3KO/NrlKO) mice showed a moderate increase. The cGMP level was about 25-fold higher in A3KO/NrlKO retina than that in NrlKO retina. Only a small increase was detected in CNGA3 KO (A3KO) and CNGB3 KO (B3KO) mice, compared to the wild type (wt). This is likely because of the small proportion of cones in the rod-dominant retina. In addition, cGMP level in the WT retina was significantly higher than the level in NrlKO retinas. Retinas of *rd1* mice (lacking rod phosphodiesterase activity) were included as an assay control. Data are represented as means ± SEM of measurements from four independent experiments using retinas from 8 to 10 mice. This research was originally published in Journal of Biological Chemistry (Thapa et al., Journal of Biological Chemistry, 2012; 287: 18018–18029. © the American Society for Biochemistry and Molecular Biology)

18. Add 200 μL of the pNpp substrate solution to every well. Incubate at room temperature for 1 h without shaking (*see* **Note 6**).
19. Add 50 μL of stop solution to every well. This stops the reaction and the plate should be read immediately.
20. Blank the plate reader against the Blank wells and read the optical density at 405 nm.
21. Calculation of results (*see* **Note 7**) (Fig. 1):
    (a) Calculate the average net optical density (OD) bound for each standard and sample by subtracting the average NSB OD from the average OD bound:

$$\text{Average net OD} = \text{average bound OD} - \text{average NSB OD}$$

    (b) Calculate the binding of each pair of standard wells as a percentage of the maximum binding wells (Bo), using the following formula:

$$\text{Percent Bound} = \text{Net OD} / \text{Net Bo OD} \times 100$$

(c) Use logit-log paper to plot percent bound (B/Bo) versus concentration of cGMP for the standards. Approximate a straight line through the points. The concentration of cGMP in the unknowns can be determined by interpolation. An alternative is to use Sigma Plot program to calculate the concentration of cGMP. Input the standard concentration (x, pmol/mL) and average net OD of the standard (y). Use the 4-parameter logistic function, $y = y0 + a/(1 + (x/x0)^{\wedge} b)$, to get the four parameters, a, b, $x0$, and $y0$, as well as an R^2 value (which indicates the quality of the standard curve), and input the y values (average net OD of unknown samples) into the formula $x = x0 \times$ power $((a/(y - y0) - 1, 1/b)$ to determine cGMP concentrations of unknown samples.

3.2 Cyclic GMP Immunofluorescence

We use 10–14 μm retinal cryosections on adhesion glass slides (e.g., Superfrost Plus slides, Thermo Scientific) for this detection (*see* **Notes 8–10**).

1. Warm slides at room temperature for 30 min.
2. Encircle every retinal cryosection using Super PAP-Pen Mini liquid blocker and allow 5 min to dry.
3. Wash in PB for 5 min (*see* **Note 11**).
4. Fix in 4 % PFA for 10 min.
5. Wash in PB for 5 min, three times.
6. Block retinal sections using blocking buffer for 60 min at room temperature in a humidified chamber.
7. Make primary antibody solution: sheep anti-cGMP in incubation buffer.
8. Remove excess fluid from slide.
9. Incubate sections with primary antibody solution overnight at 4 °C in a humidified chamber (*see* **Notes 12–13**).
10. Wash and rinse slide with PB for three times for 5–10 min, each.
11. Make secondary antibody solution: goat anti-sheep in incubation solution.
12. Remove excess fluid from slide. Don't let sections dry out!
13. Incubate for 60 min at room temperature in a lightproof humidified chamber.
14. Wash with PB containing nuclear dye for 5 min.
15. Wash with PB for 5 min, two times.
16. Remove excess fluid from slide (*see* **Note 14**).

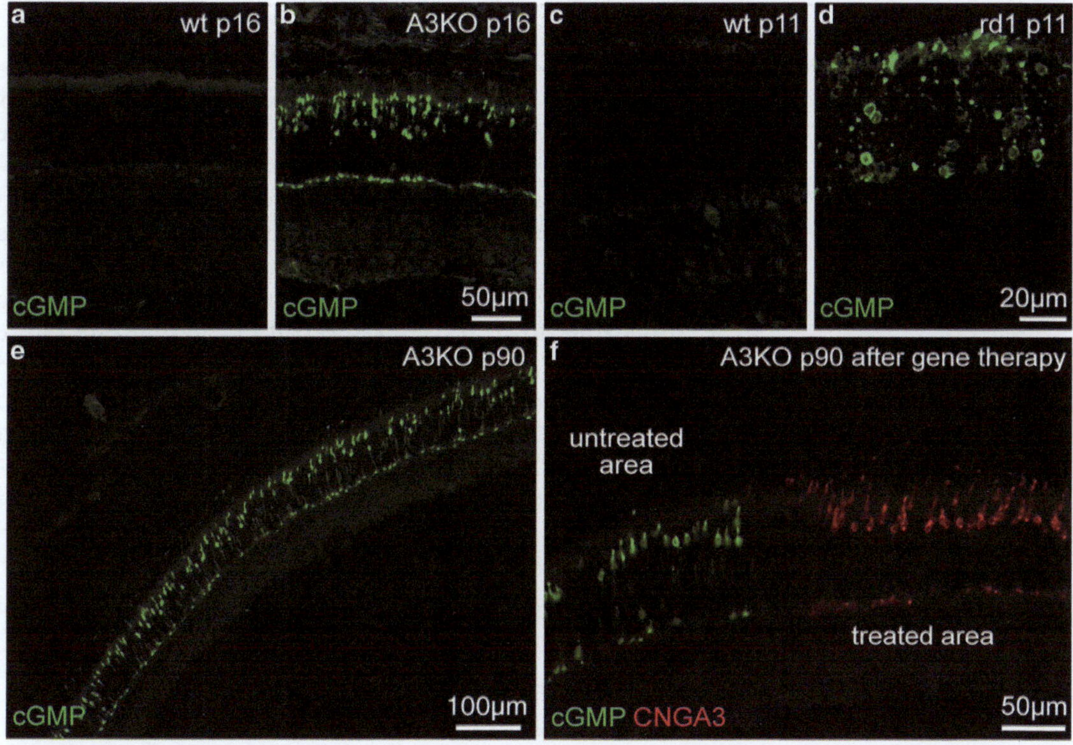

Fig. 2 Detection of cGMP in retinal degeneration using immunohistochemistry. (**a**–**b**) Fluorescence images from 16-day-old (p16) (**a**) wild-type (wt) and (**b**) CNGA3 knockout (A3KO) retina stained for cGMP (*green*). Cone photoreceptors in the A3KO mouse, which is a mouse model for achromatopsia [16, 34], are strongly labeled for cGMP. (**c**–**d**) Confocal scans from p11 (**c**) wt and (**d**) *rd1* retinal slices stained for cGMP. Rod photoreceptors in the rd1 mouse, which is a mouse model of retinitis pigmentosa caused by mutations in PDE6B [11], also accumulate high levels of cGMP. (**e**) Overview image of an A3KO retina at p90 shows that all cone photoreceptors still contain high levels of cGMP. (**f**) High-magnification image of a retinal slice from an A3KO mouse after adeno-associated virus (AAV)-mediated gene replacement therapy [18] stained for cGMP (*green*) and CNGA3 (*red*). The picture taken from the treatment-border area clearly demonstrates that AAV-mediated expression of CNGA3 (*red*) lowered cGMP to wild-type levels by activating the Ca^{2+}-triggered feedback inhibition of the guanylyl cyclase (*see* Subheading 1) in A3KO cones after treatment

17. Mount with mounting medium (~3 μL on each section).
18. Cover slip with cover glass and fingernail polish around edges.
19. Clean slides using Kimwipes and store in lightproof box at 4 °C protected from bleaching until use.
20. Collect images with an appropriate fluorescence or confocal microscope (Fig. 2).

4 Notes

1. Retinal sample isolation and handling for cGMP ELISA: After the mouse is euthanized, pull back the skin around the eye until the eyeball pops out from the socket. Gently squeeze forceps behind the eye until it comes out completely. Use the tip of a scalpel blade to make a quick cut down the middle of the cornea, and then push with side of blade until the lens pops out of the cut. Pull up with the forceps from the back of the eye, squeezing down the entire eyeball but letting out a little bit to keep from pulling the entire eye out. The retina should come out easily onto the forceps. Tap it out into liquid nitrogen to freeze; collect in tubes for storage at −80 °C until use. When processing, weigh a labeled empty tube and then transfer frozen retinas into the tube. Weigh the tube again to get the net weight. Use at least 4 retinas in each tube for the assay. Weigh samples as quickly as possible to avoid thawing.
2. Diluted cGMP standards should be used within 60 min of preparation.
3. All standards and samples should be run in duplicate.
4. Samples may need to be diluted for an assay if a high cGMP level is anticipated. For measurement of cGMP levels in retinas of CNG channel-deficient mice (CNGA3 knockout mice) and *rd1* mice, we dilute tissue homogenate with 0.1 M HCl at 1:5–10. Make sure to multiply sample concentrations by the dilution factor used during sample preparation.
5. Pipet the reagents to the sides of the wells to avoid possible contamination.
6. Prior to the addition of substrate, ensure there is no residual wash buffer in the wells (remaining wash buffer causes variation in assay results). Use a Kimwipe rolled up to gently touch the well bottom to absorb residual wash buffer.
7. To normalize for protein content, divide the resulting picomole per mL determinations (pmol/mL) by the total protein concentration (mg/mL) in each sample. This is expressed as pmol cGMP per mg of total protein.
8. Retinal sample isolation and handling for cGMP immunofluorescence: The method works with eyecups isolated from perfusion-fixed (4 % PFA) mice or with immersion-fixed eyecups (recommended fixation time: 45–90 min in ice-cold 4 % PFA). In the letter case a quick dissection procedure (1–2 min until the first fixation step) is recommended: After the mouse is euthanized, enucleate the eyeballs and transfer them to ice-cold PB. Make a small hole at the ora serrata using a 24G needle and place the eyes in ice-cold 4 % PFA for 5 min.

Subsequently, transfer the eyeballs to a wet Whatman paper and remove the cornea and lens by cutting along the ora serrata using micro Vanna scissors. Immediately transfer the eye to ice-cold 4 % PFA for the remaining fixation time (40–85 min). Wash the eyecups in PB, place them overnight in 30 % sucrose/PB and embed in optical cutting temperature (OCT) embedding medium for cryosectioning.

9. Due to the relatively low sensitivity of the anti-cGMP antibody, only weak staining of photoreceptor outer segments can be observed in the wild-type retina under normal light conditions. Make sure to stain wild-type retinal slices side by side with retinal degeneration sections as a negative control.
10. Make sure to include a "secondary antibody-only" control in your experiments where the same procedure was performed in the absence of anti-cGMP antibody. This staining procedure should result in no or very low background staining. If the background staining is too high adjust the secondary antibody dilution or change the supplier of the secondary antibody.
11. All washing steps can be performed in a glass beaker filled with enough buffer to cover the slices.
12. Primary antibody incubation should be done at 4 °C overnight. Incubation at room temperature for shorter period of time will lead to less binding.
13. Antibody incubation **steps 9** and **13** can be done with minimal volumes (we use 30–50 μl) of antibody incubation solution on the glass slides in a horizontal position. Make sure to use enough volume of solution to completely cover the retinal slice.
14. Let sections dry completely before adding mounting medium. Air-dry (under lightproof condition) for at least 20 min.

Acknowledgments

We thank Dr. Jan de Vente (University of Maastricht, Netherlands) for the gift of anti-cGMP antibody. This work was supported by the Deutsche Forschungsgemeinschaft (Bi484/4-1, Mi1238/1-2) and the National Eye Institute (P30EY12190 and R01EY019490).

References

1. Arshavsky VY, Burns ME (2012) Photoreceptor signaling: supporting vision across a wide range of light intensities. J Biol Chem 287(3): 1620–1626
2. Luo DG, Xue T, Yau KW (2008) How vision begins: an odyssey. Proc Natl Acad Sci USA 105(29):9855–9862
3. Biel M, Michalakis S (2009) Cyclic nucleotide-gated channels. Handb Exp Pharmacol 191:111–136
4. Karan S, Frederick JM, Baehr W (2010) Novel functions of photoreceptor guanylate cyclases revealed by targeted deletion. Mol Cell Biochem 334(1–2):141–155

5. Kuhn M (2009) Function and dysfunction of mammalian membrane guanylyl cyclase receptors: lessons from genetic mouse models and implications for human diseases. Handb Exp Pharmacol 191:47–69
6. Zhang X, Cote RH (2005) cGMP signaling in vertebrate retinal photoreceptor cells. Front Biosci 10:1191–1204
7. Yau KW, Nakatani K (1985) Light-induced reduction of cytoplasmic free calcium in retinal rod outer segment. Nature 313(6003): 579–582
8. Baehr W, Palczewski K (2009) Focus on molecules: guanylate cyclase-activating proteins (GCAPs). Exp Eye Res 89(1):2–3
9. Biel M, Michalakis S (2007) Function and dysfunction of CNG channels: insights from channelopathies and mouse models. Mol Neurobiol 35(3):266–277
10. Ionita MA, Pittler SJ (2007) Focus on molecules: rod cGMP phosphodiesterase type 6. Exp Eye Res 84(1):1–2
11. Bowes C, Li T, Danciger M, Baxter LC, Applebury ML, Farber DB (1990) Retinal degeneration in the rd mouse is caused by a defect in the beta subunit of rod cGMP-phosphodiesterase. Nature 347(6294):677–680
12. Farber DB, Lolley RN (1974) Cyclic guanosine monophosphate: elevation in degenerating photoreceptor cells of the C3H mouse retina. Science 186(4162):449–451
13. Paquet-Durand F, Hauck SM, van Veen T, Ueffing M, Ekstrom P (2009) PKG activity causes photoreceptor cell death in two retinitis pigmentosa models. J Neurochem 108(3):796–810
14. Trifunovic D, Dengler K, Michalakis S, Zrenner E, Wissinger B, Paquet-Durand F (2010) cGMP-dependent cone photoreceptor degeneration in the cpfl1 mouse retina. J Comp Neurol 518(17):3604–3617
15. Hüttl S, Michalakis S, Seeliger M, Luo DG, Acar N, Geiger H, Hudl K, Mader R, Haverkamp S, Moser M, Pfeifer A, Gerstner A, Yau KW, Biel M (2005) Impaired channel targeting and retinal degeneration in mice lacking the cyclic nucleotide-gated channel subunit CNGB1. J Neurosci 25(1):130–138
16. Biel M, Seeliger M, Pfeifer A, Kohler K, Gerstner A, Ludwig A, Jaissle G, Fauser S, Zrenner E, Hofmann F (1999) Selective loss of cone function in mice lacking the cyclic nucleotide-gated channel CNG3. Proc Natl Acad Sci USA 96(13):7553–7557
17. Koch S, Sothilingam V, Garcia Garrido M, Tanimoto N, Becirovic E, Koch F, Seide C, Beck SC, Seeliger MW, Biel M, Mühlfriedel R, Michalakis S (2012) Gene therapy restores vision and delays degeneration in the CNGB1−/− mouse model of retinitis pigmentosa. Hum Mol Genet 21(20):4486–4496
18. Michalakis S, Mühlfriedel R, Tanimoto N, Krishnamoorthy V, Koch S, Fischer MD, Becirovic E, Bai L, Huber G, Beck SC, Fahl E, Buning H, Paquet-Durand F, Zong X, Gollisch T, Biel M, Seeliger MW (2010) Restoration of cone vision in the CNGA3−/− mouse model of congenital complete lack of cone photoreceptor function. Mol Ther 18(12):2057–2063
19. Thomas MG, Kumar A, Kohl S, Proudlock FA, Gottlob I (2011) High-resolution in vivo imaging in achromatopsia. Ophthalmology 118(5):882–887. doi:10.1016/j.ophtha.2010.08.053.
20. Thiadens AA, Somervuo V, van den Born LI, Roosing S, van Schooneveld MJ, Kuijpers RW, van Moll-Ramirez N, Cremers FP, Hoyng CB, Klaver CC (2010) Progressive loss of cones in achromatopsia: an imaging study using spectral-domain optical coherence tomography. Invest Ophthalmol Vis Sci 51(11):5952–5957
21. Sahel J, Bonnel S, Mrejen S, Paques M (2010) Retinitis pigmentosa and other dystrophies. Dev Ophthalmol 47:160–167
22. Hartong DT, Berson EL, Dryja TP (2006) Retinitis pigmentosa. Lancet 368(9549): 1795–1809
23. Fiscus RR (2002) Involvement of cyclic GMP and protein kinase G in the regulation of apoptosis and survival in neural cells. Neuro-Signals 11(4):175–190
24. Sharma AK, Rohrer B (2007) Sustained elevation of intracellular cGMP causes oxidative stress triggering calpain-mediated apoptosis in photoreceptor degeneration. Curr Eye Res 32(3):259–269
25. Ulshafer RJ, Garcia CA, Hollyfield JG (1980) Sensitivity of photoreceptors to elevated levels of cGMP in the human retina. Invest Ophthalmol Vis Sci 19(10):1236–1241
26. Olshevskaya EV, Calvert PD, Woodruff ML, Peshenko IV, Savchenko AB, Makino CL, Ho YS, Fain GL, Dizhoor AM (2004) The Y99C mutation in guanylyl cyclase-activating protein 1 increases intracellular Ca2+ and causes photoreceptor degeneration in transgenic mice. J Neurosci 24(27):6078–6085
27. Ramamurthy V, Niemi GA, Reh TA, Hurley JB (2004) Leber congenital amaurosis linked to AIPL1: a mouse model reveals destabilization of cGMP phosphodiesterase. Proc Natl Acad Sci USA 101(38):13897–13902
28. Sancho-Pelluz J, Arango-Gonzalez B, Kustermann S, Romero FJ, van Veen T,

Zrenner E, Ekstrom P, Paquet-Durand F (2008) Photoreceptor cell death mechanisms in inherited retinal degeneration. Mol Neurobiol 38(3):253–269

29. Thapa A, Morris L, Xu J, Ma H, Michalakis S, Biel M, Ding XQ (2012) Endoplasmic reticulum stress-associated cone photoreceptor degeneration in cyclic nucleotide-gated channel deficiency. J Biol Chem 287(22):18018–18029
30. Paquet-Durand F, Beck S, Michalakis S, Goldmann T, Huber G, Muhlfriedel R, Trifunovic D, Fischer MD, Fahl E, Duetsch G, Becirovic E, Wolfrum U, van Veen T, Biel M, Tanimoto N, Seeliger MW (2011) A key role for cyclic nucleotide gated (CNG) channels in cGMP-related retinitis pigmentosa. Hum Mol Genet 20(5):941–947
31. Feil S, Zimmermann P, Knorn A, Brummer S, Schlossmann J, Hofmann F, Feil R (2005) Distribution of cGMP-dependent protein kinase type I and its isoforms in the mouse brain and retina. Neuroscience 135(3): 863–868
32. Gamm DM, Barthel LK, Raymond PA, Uhler MD (2000) Localization of cGMP-dependent protein kinase isoforms in mouse eye. Invest Ophthalmol Vis Sci 41(9):2766–2773
33. Tanaka J, Markerink-van Ittersum M, Steinbusch HW, De Vente J (1997) Nitric oxide-mediated cGMP synthesis in oligodendrocytes in the developing rat brain. Glia 19(4):286–297
34. Michalakis S, Geiger H, Haverkamp S, Hofmann F, Gerstner A, Biel M (2005) Impaired opsin targeting and cone photoreceptor migration in the retina of mice lacking the cyclic nucleotide-gated channel CNGA3. Invest Ophthalmol Vis Sci 46(4):1516–1524

INDEX

Thomas Krieg and Robert Lukowski (eds.), *Guanylate Cyclase and Cyclic GMP: Methods and Protocols*, Methods in Molecular Biology, vol. 1020, DOI 10.1007/978-1-62703-459-3,

MIX
Papier aus verantwortungsvollen Quellen
Paper from responsible sources
FSC® C105338

If you have any concerns about our products,
you can contact us on
ProductSafety@springernature.com

In case Publisher is established outside the EU,
the EU authorized representative is:
Springer Nature Customer Service Center GmbH
Europaplatz 3, 69115 Heidelberg, Germany

Printed by Libri Plureos GmbH
in Hamburg, Germany